Advanced Concurrency, Buffering, and Web-Sync in Sensor Platforms:

Zero-Loss Streams

Aquil Mirza Mohammed
https://orcid.org/0000-0001-7756-4363
The Hong Kong Polytechnic University, Hong Kong

IGI Global
Scientific Publishing
Publishing Tomorrow's Research Today

Vice President of Editorial	Melissa Wagner
Director of Acquisitions	Mikaela Felty
Director of Book Development	Jocelynn Hessler
Production Manager	Mike Brehm
Cover Design	Jose Rosado

Published in the United States of America by
IGI Global Scientific Publishing
701 East Chocolate Avenue
Hershey, PA, 17033, USA
Tel: 717-533-8845 | Fax: 717-533-7115
Website: https://www.igi-global.com E-mail: cust@igi-global.com

Library of Congress Cataloging-in-Publication Data

LCCN: 2026023807 (CIP Data Pending)
ISBN13: 979-8-2600-1101-0
Isbn13Softcover: 979-8-2600-1102-7
EISBN13: 979-8-2600-1103-4
British Cataloguing in Publication Data
A Cataloguing in Publication record for this book is available from the British Library.

Table of Contents

Detailed Table of Contents

This chapter presents an integrated framework for real-time sensor data aggregation and adaptive buffer management in concurrent systems. The architecture comprises a two-level buffer inspired by two-stage caching principles: a internal queue inside each sensor and a central buffer manager that receives data via unnamed pipes. The buffer manager employs dynamic capacity expansion to prevent overflow. The aggregation engine uses multiple worker threads that process readings concurrently; per-sensor windows are stored in a hash map protected by an RwLock. It supports three anomaly detection methods with explainable outputs. A data storage module provides thread-safe file persistence with automatic rotation, and a multi-threaded web server delivers a live dashboard, file statistics, and an API. Evaluation demonstrates zero data loss at 80k entries/s. The framework offers a blueprint for embedding cognitive capabilities into sensor-based research infrastructures.

This project initially aims to construct a consequent data processing pipeline. It should have the ability to collect readings from multiple simulated sensors, and perform a real-time statistical aggregation. It should also store the results durably, and serves them through a web interface. The system architecture is organized as a Rust workspace with four key components. The sensor_sim works for data generation, os_lib for buffer management, gateway for the main entry point, and dashboard for the web presentation. The core innovation in this project is the synchronization implementation. A lock-free, non-copy circular buffer is used, and enables the zero-blocking data transfer between the producer and consumer threads.

Chenyu Liu, The Hong Kong Polytechnic University, Hong Kong
Yaxin Luo, The Hong Kong Polytechnic University, Hong Kong
Jingyan Zeng, The Hong Kong Polytechnic University, Hong Kong
Liyuan Fan, The Hong Kong Polytechnic University, Hong Kong
Mingyuan Tang, The Hong Kong Polytechnic University, Hong Kong
Cong Wu, Liaoning Technical University, China

The ever-increasing volume of data and the need for real-time data processing make preventing data loss a major challenge. This chapter aims to solve this problem. A distributed sensor data aggregation platform based on Rust was designed and implemented. This aggregation platform enables the entire process of efficiently acquiring real-time data from multiple sensors, processing the data, storing it, and accessing it via the web. Its main supporting technology is a one-thread reading strategy for each sensor, a shared queue to absorb the problem of mismatched producer and consumer rates, and finally, multi-threaded aggregation for calculation within a fixed window. Through stress testing, data loss verification, and bench marking, this platform achieved zero data loss, correct and stable aggregation, and safe web access. Overall, this platform provides a scalable foundation for future deployment in industrial IoT environments, which require continuous and low-latency data access, reliable buffering, and strong fault tolerance.

This chapter, per the authors, presents a distributed sensor data aggregation platform in Rust demonstrating operating systems concepts including inter-process communication and concurrent data processing. The system employs a process-based architecture with ten child processes: five sensor readers connected via anonymous pipes, four aggregation workers sharing a named FIFO pipe, and one web server. A collector process groups sensor readings into fixed-duration time windows forwarded through POSIX shared memory with atomic counters. The aggregation engine applies Welford's online variance algorithm to compute streaming statistics; minimum, maximum, mean, and standard deviation; without a second data pass. Anomaly detection uses a configurable z-score threshold per sensor type. Five benchmark trials confirm zero data loss across 698 readings, a throughput of 71.0 requests per second, a p99 latency of 14.08 milliseconds, and peak buffer utilisation of 0.01%, validating the platform's reliability guarantees under the intended workload.

This chapter implements a Rust-based sensor data aggregation platform, achieving zero data loss in the case of multiple sensors running at the same time. The system faces three main challenges: preventing overflow in each sensor's 128-element internal buffer, supporting real-time statistical aggregation, and ensuring storage integrity during concurrent writes. In order to solve these problems, we adopt a multi-layer design: the front end is per-sensor reader threads, and the reading frequency is dynamically adjusted in combination with adaptive polling; a bounded mpsc::sync_channel is used as the intermediate buffer to provide backpressure when downstream processing becomes slow. A multi-worker aggregation engine processes sensor data in time-sliced windows. Additionally, we also implemented advanced features including real-time visualization, adaptive thread pooling and automated benchmarking suite. The Benchmark results show that the system sustains zero data loss under stress and achieves high throughput in large-scale tests.

This chapter is a Rust-based simulation system designed to coordinate multiple medical robots operating in shared hospital environments. The chapter focuses on concurrent systems, task scheduling, mutual exclusion for shared zones, and fault detection through heartbeat monitoring. A shared task queue assigns work to robot workers, while a zone access control mechanism ensures that only one robot operates in a critical area at a time. The system monitors heartbeat signals from each robot to detect failures or stalled execution. The modular architecture supports two execution models: a thread-based scheduler for lightweight concurrency and a process-based scheduler for stronger isolation between workers. Controlled benchmarks and stress tests show correct enforcement of zone exclusivity, reliable detection of missed heartbeats, and stable performance under concurrent workloads. The results show predictable scalability for lightweight tasks and highlight contention effects when task duration and zone competition increase.

Hospital service robots face challenges because they need to compete for limited resources while their work schedules remain unproductive. The robot coordination system which we created solves this issue through its design for medical operations. The core idea needs to merge regional resource limitations together with a simple system that handles multiple tasks simultaneously. Our solution includes a zone-aware work-stealing strategy which enables robot threads to focus on local tasks while taking tasks from other threads only when essential. The system achieves total system responsiveness improvements through its ability to prevent unnecessary blocking. The system runs on a Rust implementation which provides a web-based interface for visualizing its functions. The system enables secure simultaneous operation of multiple tasks while it provides continuous operational status tracking. Our experiments demonstrate that our design establishes permanent and safe methods for distributing tasks.

Sunyi Liu, The Hong Kong Polytechnic University, Hong Kong
Zongxi Wu, The Hong Kong Polytechnic University, China
Daxin Jin, The Hong Kong Polytechnic University, Hong Kong
Yuheng Du, The Hong Kong Polytechnic University, Hong Kong
Bolin Zhu, The Hong Kong Polytechnic University, Hong Kong

This chapter presents a lightweight operating system for coordinating multiple medical care robots in a hospital environment. The system focuses on addressing concurrency and synchronization challenges in multi-process robot coordination. A centralized coordinator manages task allocation, resource control, and system monitoring, while robot processes execute tasks independently. An all-or-nothing resource allocation strategy is adopted to prevent deadlock by ensuring that tasks start only when all required resources are available. Inter-process communication is implemented using a lightweight text-based protocol for efficient message exchange. In addition, a heartbeat-based monitoring mechanism is used to detect failures and reclaim resources. Experimental results demonstrate that the system maintains correct synchronization, ensures mutual exclusion, and achieves stable performance under resource contention and partial failure conditions.

John Christopher Flores Quiambao, The Hong Kong Polytechnic University, Hong Kong
Xiao Yi Wang, The Hong Kong Polytechnic University, Hong Kong
Zhengyang Li, The Hong Kong Polytechnic University, Hong Kong
Dai Cheng Han, The Hong Kong Polytechnic University, Hong Kong
Ting Xuan Wu, The Hong Kong Polytechnic University, Hong Kong
Mohammad Al Khaldy, University of Petra, Jordan

Medicine is a field that has historically had a high demand for workers with little supply, this high demand is not helped by the fact that medicine is one of the volatile fields of service work in the world with constantly changing situations that require

Preface

INTRODUCTION

The rapid evolution of sensor-enabled systems and autonomous platforms has elevated concurrency, buffering, and synchronized data streams from specialized topics to foundational requirements. This book collects a coherent thread of research and engineering practice that addresses how multiple sensors, computation cores, and robotic actuators can operate in harmony without sacrificing timeliness, reliability, or data integrity. At its core lies the concept of zero-loss streams: mechanisms that ensure data produced by sensors traverses the processing pipeline with minimal or no loss, even under high concurrency and resource contention. The chapters jointly articulate a design philosophy that blends rigorous concurrency control, adaptive buffering, and web-enabled visibility; creating systems that are both robust in real-time operation and transparent enough to be analyzed, tuned, and extended by practitioners and researchers alike.

In today's technological landscape, the world is saturated with sensors deployed at scale; from industrial environments and smart factories to clinical settings and mobile robotics. These sensors generate streams of data that must be ingested, aggregated, and acted upon with strict timing guarantees. Traditional approaches often rely on ad hoc synchronization or reactive fault handling that can introduce latency, data loss, or brittle behavior under load. The material presented in this book situates itself at the intersection of systems engineering, real-time computing, and data-centric robotics. It advances a principled understanding of how to structure concurrent data pathways, how to reason about buffering in the presence of variability, and how to expose system state through web-based interfaces that support monitoring, debugging, and governance. The result is a set of well-founded abstractions and practical patterns that can be adapted to diverse contexts; from hospital service robots and industrial IoT to large-scale sensor networks.

The intended audience for this book encompasses researchers, graduate students, system architects, and software engineers who design and implement sensor-driven platforms with stringent performance requirements. Readers will benefit from both the theoretical underpinnings of concurrency and the pragmatic guidance drawn from concrete Rust-based implementations, IPC and shared memory strategies, and web-visible monitoring architectures. Additionally, practitioners in robotics, healthcare automation, and industrial automation will find value in the concrete coordination patterns and safety considerations that accompany real-time scheduling and resource management.

Each chapter in this volume contributes to a common objective: to enable safe, scalable, and transparent concurrent sensor platforms that minimize data loss while maximizing responsiveness. The chapters progressively build a lattice of ideas; starting from core frameworks for real-time sensor data aggregation and adaptive buffering, moving through OS-like kernel concepts implemented in Rust, and culminating in domain-specific coordination systems for medical robots. Across these chapters, recurring themes include (1) the management of limited resources and the scheduling of competing tasks, (2) the design of buffering strategies that adapt to workload while preserving data integrity, (3) the use of heartbeat or liveness mechanisms to detect and recover from faults, and (4) the exposure of system state through web-based dashboards and APIs to support monitoring, validation, and governance.

CHAPTER OVERVIEWS

Chapter 1

An Integrated Framework for Real-Time Sensor Data Aggregation and Adaptive Buffer Management in Concurrent Systems lays the groundwork for how sensors, processors, and storage can interoperate under concurrent workloads. It introduces a modular approach to data aggregation and buffering that supports both high throughput and low latency, while maintaining flexibility to accommodate diverse sensor modalities and bandwidth constraints. This chapter anchors the book by presenting a unified view of data flow, buffering decisions, and correctness criteria in a way that informs subsequent chapters.

Chapter 2

Zero-Loss Sensor Data Aggregation - A Rust-Based Operating System Core for Concurrent Data Processing elevates the discussion to a system-level perspective. It articulates how a Rust-based OS core can orchestrate concurrent data paths with strong safety guarantees, enabling zero-loss properties through careful resource management, memory safety, and explicit synchronization discipline. The emphasis on a robust, low-level foundation provides a reference architecture for all higher-level coordination mechanisms described later.

Chapter 3

Rust-Driven Real-Time Sensor Aggregation for Industrial IoT: Zero-Loss Data and Web Visibility extends the focus to industrial contexts where reliability, traceability, and visibility are paramount. It demonstrates how Rust's safety features and ecosystem can be harnessed to deliver real-time aggregation with end-to-end observability, ensuring that operators can monitor, audit, and respond to evolving conditions in complex industrial environments.

Chapter 4

Coordinating Cross-Process Sensor Aggregation in Rust with IPC and Shared Memory emphasizes inter-process coordination, an essential component when sensors, processors, and controllers are distributed across processes or even nodes. By exploring IPC and shared memory as primitives, the chapter shows how to preserve locality, reduce overhead, and maintain data integrity in multi-process systems that must coordinate under real-time constraints.

Chapter 5

Adaptive Sensor Aggregation with Self-Monitoring: A Rust-Based Concurrent Data Pipeline spotlights self-monitoring and adaptivity. It shows how a data pipeline can adjust its buffering and processing behavior in response to observed workload patterns, fleeting faults, or environmental changes, all while preserving zero-loss properties. This adaptability is crucial for maintaining performance across time and across shifting workloads.

Chapter 6

Real-Time Concurrency for Light-Weight Medical Robot Coordination in Shared Zones narrows the lens to healthcare robotics, where zone-aware coordination, low-latency control, and safe interaction with humans and equipment are essential. The chapter presents practical mechanisms for coordinating multiple robots operating within shared zones, balancing responsiveness with safety constraints that are critical in medical settings.

Chapter 7

Zone-Aware Medical Robot Coordination System: Concurrent Scheduling And Resource Control intensifies the discussion of scheduling and resource contention in medical robotics. It articulates how zone-aware scheduling and decisive resource control can optimize throughput while upholding safety standards, particularly in environments where patient care and robot-assisted tasks converge.

Chapter 8

A Lightweight Multi-Process Coordination System for Hospital Service Robots with Deadlock Prevention addresses one of concurrency's perennial challenges: deadlock. It offers design strategies and runtime checks to prevent deadlocks in multi-process robot coordination, ensuring that hospital service robots can operate continuously and reliably.

Chapter 9

Concurrency Control and Heartbeat-Driven Recovery for Autonomous Medical Robots advances fault-detection and recovery strategies. By leveraging heartbeat mechanisms and robust concurrency control, this chapter demonstrates how autonomous medical robots can sustain operation and recover gracefully from transient faults or disruptions.

Chapter 10

Robust Concurrency and Synchronization for Large-Scale Medical Robot Operations concludes the sequence with considerations for large-scale deployments. It synthesizes lessons from prior chapters to address the challenges of scale, including coordination across many robots, robust synchronization in dense environments, and the maintenance of data integrity and timeliness at scale.

CONCLUSION

The overall arc of these ten chapters traces a path from foundational principles of real-time data aggregation and adaptive buffering to domain-specific, hospital-focused coordination systems, and finally to scalability and resilience concerns that arise in large deployments. The book thus provides a cohesive toolkit for researchers and practitioners who aspire to build sensor platforms that are both high-performance and trustworthy. It invites readers to consider not only how to design systems that meet strict technical requirements but also how to validate, monitor, and evolve those systems in the face of changing technologies, regulatory environments, and clinical needs.

In terms of contribution to the field, this volume emphasizes zero-loss streams as a unifying objective across diverse contexts. It demonstrates that robust concurrency is not merely a theoretical ideal but a practical capability when grounded in sound architecture, careful resource management, and transparent observability. By combining architectural patterns, language-level safety, and domain-specific coordination strategies, the book offers a versatile blueprint for the next generation of sensor platforms; ones that can operate with reliability, adaptivity, and intelligible behavior even in the most demanding environments.

Chapter 1

An Integrated Framework for Real-Time Sensor Data Aggregation and Adaptive Buffer Management in Concurrent Systems

Licheng Huang
http://orcid.org/0009-0000-6046-5406
Hong Kong Polytechnic University, Hong Kong

Bochen Xue
Hong Kong Polytechnic University, Hong Kong

Yiming Chen
Hong Kong Polytechnic University, Hong Kong

Peihang Wu
Hong Kong Polytechnic University, Hong Kong

Yuezhong Wang
Hong Kong Polytechnic University, Hong Kong

Aquil Mirza Mohammed
http://orcid.org/0000-0001-7756-4363
The Hong Kong Polytechnic University, Hong Kong

ABSTRACT

This chapter presents an integrated framework for real-time sensor data aggregation and adaptive buffer management in concurrent systems. The architecture comprises a two-level buffer inspired by two-stage caching principles: a internal queue inside each sensor and a central buffer manager that receives data via unnamed pipes. The buffer manager employs dynamic capacity expansion to prevent overflow. The

DOI: 10.4018/979-8-2600-1101-0.ch001

aggregation engine uses multiple worker threads that process readings concurrently; per-sensor windows are stored in a hash map protected by an RwLock. It supports three anomaly detection methods with explainable outputs. A data storage module provides thread-safe file persistence with automatic rotation, and a multi-threaded web server delivers a live dashboard, file statistics, and an API. Evaluation demonstrates zero data loss at 80k entries/s. The framework offers a blueprint for embedding cognitive capabilities into sensor-based research infrastructures.

1. INTRODUCTION

1.1 Introduction to Sensor Data Acquisition and Analysis Systems

In practical application scenarios such as the Industrial Internet of Things and smart industrial monitoring circumstances, data acquisition, processing, and analysis of sensors are crucial components for data stimulation, equipment status monitoring (Mohammed et al., 2026), early anomaly detection, and decision-making support. However, current existing lightweight sensor data processing systems still face certain serious challenges, that's where this chapter is playing its role.

Insufficient decoupling of data acquisition and processing: The difference between sensor acquisition rates and data processing rates can easily lead to data loss or processing bottlenecks, which especially occurs when using multiple sensor types such as temperature, accelerometers, and force sensors. For mixed acquisition, the rate differences even exacerbate this problem.

Lack of relevant real-time anomaly detection capabilities: Traditional sensor data processing heavily relies on the offline analysis approaches, making it impossible to identify anomalies in real time during the data acquisition phase, thus delaying early detection of equipment failures.

Lack of data management methods: Traditional methods lack reliable and standardized data storage strategies, the lack and weak points of data storage method could lead to transmission errors after processing.

Lack of visualization methods: The raw data acquisition and analysis results lack easy-to-use visualization methods, making it difficult for maintaining personnel to intuitively monitor the system's operating status.

Low System Modularity architecture: The high coupling between sensor simulation, data buffering, aggregation analysis, and storage functions makes it complex and confusing to modify or replace components as needed, such as later operations of changing relevant anomaly detection algorithms or adjusting storage strategies.

1.2 Chapter Motivation

To address the above issues and challenges, this chapter aims to build a lightweight and highly cohesive sensor data acquisition and analysis system. The main motivations are shown below.

Solving the matching problem of data acquisition and processing rates across multiple sensor types to ensure data integrity, including sensors of temperature, accelerometer and force aspects.

Achieving real-time aggregation and anomaly detection of sensor data to meet real-time monitoring requirements.

Providing practical standardized data storage strategies and a visualization interface to reduce operational costs and difficulties.

The design of modular architecture enhances the system structure clarity and maintainability. This design facilities future functional iterations and revisions. Revision plans such as connecting to real sensors in industrial applications and synchronizing data online can be easily conducted.

1.3 Chapter Objectives

This chapter focuses on comprehensive management of the sensor data lifecycle and has defined the following objectives:

Acquire data by simulating multiple sensor types: Supporting three types of simulated sensors: thermometers, accelerometers, and force sensors. Allows users to customize acquisition frequencies, simulating data input at various rates in real-world scenarios.

Buffering and data Flow control: Implements a high-performance, memory-based data buffer, tackling the acquisition and processing modules. Buffer capacity is configurable and could be customized. Its various statuses like capacity, data volume, and overflow count can be monitored in real time.

Real-time data aggregation and anomaly detection: Performs data aggregation calculations including mean, median based on a sliding window, which uses the robust median absolute deviation algorithm for real-time anomaly detection.

Standardized data persistence: Stores raw data, aggregation and anomaly analysis results in local files, supporting automatic file rotation. Storage directories are timestamped to ensure data traceability.

Visual Monitoring: Provides convenient web monitoring. The visual interface displays data acquired from the real-time sensor list including buffer status, recent data, anomaly alarm.

1.4 Overview of Approach

This chapter is developed using the Rust language, relevant skills including memory safety, high performance, and multi-threaded concurrency. It adopts a modular and layered architecture and an asynchronous concurrent processing structure design. The core methods are as follows:

1.4.1 System Architecture Design

The system has an architecture of core hub plus modular components design. The gateway module serves as the overall control hub, connecting various functional modules in the system. The compromising parts are shown below.

Gateway layer is the system entrance, which is responsible for initializing all system components, ranging from configuring operating parameters, scheduling data flow, processing system signals. It is the commander component of the entire system.

Sensor simulation layer is an encapsulation of the simulation functions of different types of sensors. It provides a unified sensor interface, supporting custom acquisition frequencies, asynchronously acquiring data through independent threads.

Buffer layer implements data memory storage based on a circular buffer, processing and tackling the rate difference between the acquisition layer and the processing layer, providing an interface for system data to write and read.

Aggregation analysis layer performs data aggregation calculations. With the application of the sliding window algorithm and the integration of basic calculation methods, aggregation engine is responsible for calculation tasks. It also utilizes the approach of multi-threaded parallel processing while buffering data at the same time, which is used to improve relevant analysis efficiency and performance.

Storage layer encapsulates file operation logic to support the creation of storage directories, which ensures data persistence and manageability.

The web server component is the visualization layer. Built on the axum framework, this lightweight HTTP service provides a front-end visualization page, being capable of supporting real-time querying of sensor data and buffer status.

Shared global layer defines the unified data structures and enumeration types in the system. Serving as a standard for data interaction between modules, shared global part could function in reducing the coupling extent.

1.4.2 Core Technical Methods

Multi-threaded asynchronous processing ensures an independent acquisition thread to be allocated respectively to each sensor in the system. So that the aggregation and analysis layer could start multiple worker threads synchronously. The Web service adopts an asynchronous I/O model, ensuring parallel execution, improving system throughput performance.

Sliding window aggregation algorithm applied in the aggregation engine calculates statics such as mean and median, which are performed on sensor data within a configurable time window, balancing real-time performance, enhancing the data computational efficiency.

RobustMAD anomaly detection is based on robust statistical methods. Compared with traditional deviation methods, it is less sensitive to extreme values and is more suitable for actual anomaly detection scenarios in sensor data.

File round-robin storage strategy applied for individual storage files is set to ensure the data file quality after file creation. Files are automatically created as the threshold is reached, avoiding performance degradation due to excessively large individual files.

1.5 Multi-threaded Server Architectures

Before building the web server, we looked at a few different frameworks. For this chapter, we had two main requirements: first, the server needed to correctly read locally stored JSONL files and render them as HTML for the browser; second, it had to handle concurrent requests efficiently so that one slow request wouldn't block others.

This section focuses on the second requirement; concurrency.

The simplest approach is a single-threaded server, which is what we started with. It's easy to implement and works fine when there's only one user. But the downside, as mentioned, is that it can't handle multiple concurrent requests. This kind of setup is really only suitable for low-traffic or experimental scenarios.

The next approach we considered was a process-based server. The advantage of a process-based server is that it allocates a separate address space for each connection, so even if one process crashes, it doesn't affect the others.(Tanenbaum & Bos, 2015) However, for a small-scale website like ours, using a process-based server would lead to high memory consumption and slower context switching. That makes it less suitable for our chapter.

Thread-based server was our final choice. Compared to process-based servers, multiple threads share the same memory space, which makes communication between threads straightforward and efficient. For example, we have a string variable called data_dir shared across threads to keep track of where the data is stored. With a process-based server, communication between processes, whether through pipes or context switching, is much more expensive compared to threads that directly share memory.(Tanenbaum & Bos, 2015) For a lightweight chapter like ours, that overhead is unnecessary.

However, directly sharing memory space can also lead to a chain reaction where one thread crashes, causing other threads to crash as well. We can only try to optimize the buffer aggregation engine and data storage to avoid these problems as much as possible. We also considered using a thread pool for better efficiency, but given the lightweight nature of the website, we decided to stick with the simpler approach of creating a new thread for each connection.

Finally, there's the event-based server. It uses a simple single-threaded asynchronous event loop, only processing connection events when an event is triggered; for example, after obtaining the rendered HTML in this chapter. This server architecture handles complex synchronous connections with minimal memory consumption. However, this sophisticated structure also means we need complex boundary conditions and event triggering conditions to constrain this precisely running code and prevent bugs and errors. Since we only need simple server architecture, we chose a thread-based server.

Our server adopts a thread-per-connection model using Rust's std::thread. This choice balances simplicity and concurrency: threads are lightweight in Rust, and the thread-per-connection pattern avoids the complexity of asynchronous programming while still providing adequate performance for our workload (multiple concurrent browsers). Each request runs independently, ensuring slow operations do not block others.

2. RELATED WORKS

1.1 Buffer Management Techniques

The research conducted by Hua et al. (2021) has revealed a method of tackling the buffer overflow problem, where similar questions are also discussed and solved in my chapter. To solve the overflow, Hua and his group made a classification and detected the basics of buffer overflow vulnerabilities. He categorizes anomalies into four patterns: 1. Buffer allocation and access in safe code 2. The core detection target, buffer allocation in safe code, unchecked access in unsafe code 3. Buffer allocation and access in unsafe code. 4. Unsafe code allocation, safe code access.

To prevent the overflow of the buffer, Hua and his group came up with detection algorithms and relevant remediation strategies.

The overflow detection algorithm is based on a lightweight, security-oriented data flow analysis algorithm, working across the boundaries of Rust's secure and insecure sub-languages. This approach constructs the intermediate representations of the Rust program, extracting unsafe blocks, vector type variables, and abstract syntax tree mappings. Then the algorithm performs active variable analysis on each unsafe block, filtering out active variables of the buffer type.

By performing reverse data flow analysis on the target variable, the algorithm locates its allocation position in secure code. Then combine a predefined set of overflow patterns, such as arithmetic operations, type casting, to mark buffer overflow candidate points.

The relevant remediation strategies mainly focus on verified real-world vulnerabilities. The automatic remediation is achieved through argument lifting and guard insertion. The core principle is adding safety boundary checks to buffer operations in unsafe code. Specific methods including Argument Lifting Guard Insertion and Integer overflow checks.

This chapter offers a new approach in overflow detection and remediation, ensures the security of vulnerabilities in buffer management with algorithm applications.

2.2 Multi-threaded Server Architecture

The view of Pieper et al. (2021) has shown their perspective on multi-thread and parallel server architecture in applications. The author argues that the core challenge of multi threaded server architecture lies in balancing "parallel efficiency" and "development complexity". Who empathized that Low-level multi threading implementations, such as Rust's standard threads, std-threads require developers to manually handle details such as thread creation, task scheduling, synchronization communication, and data sorting, which not only results in code redundancy but also easily leads to deadlocks, data races, and thread starvation. As a result, the recommendation suggests that the structure should be based on structured parallelism patterns like Pipeline, providing a high-level abstraction API to hide underlying thread details. This design could support state phases and flexible thread scheduling, balancing efficiency and ease of use.

To ensure the synchronization and communication, Pieper's group adapted the mechanisms of queue isolation plus ordering support. This architecture is practical and could balance the system safety and performance. To fulfill the core requirement of the multi-threaded server to be "safe and efficient inter-thread communication." With the idea of Rust-SSP, an "inter-stage queue" is used as the core of thread communication, each producer-consumer pair corresponds to an independent queue, reducing lock contention. To dynamically configure the number of threads, this system supports dynamic configuration of threads, such as allocating more threads to computationally intensive phases. This method also uses the default number of CPU cores to avoid excessive thread switching.

The chapter conducted by Pieper has shown some critical notes in the multi-thread server architecture. The performance of multi-thread server could be dramatically enhanced by the adaptability of the multi-core architecture, hence the bottleneck problem will be avoided synchronously. The difficulty of improving the performance of multi thread system is could be summarized into several fields. The first problem is that the global lock imposed on a single queue could restrain the overall efficiency. Therefore, the application of multi-queue structure and thread local cache may reduce the competition for thread lock. Another hint is that the dynamic thread setting mechanism of thread pool corresponding to the number of cores facilitates the multi-thread operation.

2.3 Producer-consumer Pattern Implementations

The research of Fisher and Yeh (2017) focused on Rust's concurrent tool chain and actual implementation, clarifies the core components, dependent components, and implementation characteristics of its producer-consumer pattern. The thread model of rust standard library adopts a "1:1 thread model," meaning one Rust thread directly corresponds to one operating system kernel thread.

The rust standard library provides a "Multi-Producer - Single-Consumer (MPSC)" FIFO queue, but because it only supports a single consumer, it cannot meet the needs of multi-consumer scenarios. So in practical scenarios, the chapter uses the third-party library to implement a "Multi-Producer - Multi-Consumer (MPMC)" channel to protect the underlying vector buffer based on mutexes and condition variables, and does not contain any unsafe code, ensuring memory safety.

The core process of provider-consumer model is described as followed. Producer threads write data through the channel, and consumer threads read data from the channel. Producers block when the buffer is full, and consumers block when the buffer is empty. After all producers have finished writing data, the channel is closed, and consumers exit after reading the remaining data.

Sai et al. (2022) advocates that the essence of "producer-consumer problem" presents in these conflicts: resource conflict, unmatched rates, thread life cycle management. Therefore, the constraints including mutual exclusive access, provider blocking and consumer blocking. With the applications of thread pool optimization, semaphore and blocking queue, the relationship between provider and consumer could be coordinated, the synchronization requirement could also be realized.

In our chapter, the producer-consumer model operation mechanism is highly similar to the chapter in Fisher's essay. The sensor data acquisition and analysis system is fundamentally based on a "multi-producer and multi-consumer model". Sensor threads are on the provider side generating data from stimulation. Components of aggregation engine, data storage, visualization, web server are on the consumer side, reading data from buffer and conducting the data processing and calculation. The buffer manager acts as the core threads safety "message queue".

In the essay, buffering area is considered as the critical hub in the system. It tackles the rate difference between producer and consumer, prevents data loss and congestion. In buffer manager component in our chapter, we also empathized and implemented the FIFO sequence, congestion mechanism and thread safety.

The reference paper also suggests that while ensuring security, reducing system overhead, improving concurrency efficiency through asynchronous threads, efficient channel libraries, and dynamic thread configuration could be future improvement approaches.

3. INMPLEMENTATION

3.1 Architecture Diagram Showing All Components

The Figure 1 shows the architecture diagram showing various components.

3.2 Buffer Manager with Dynamic Capacity Expansion

The BufferManager<T> is a thread-safe bounded queue that stores sensor readings. It uses a Mutex<VecDeque<T>> for interior mutability and a Condvar to block consumers when the buffer is empty.

The initial value of the buffer size depends on the arguments when creating. It is usually set to be 2,000 under normal working situations. The memory footprint of this buffer is approximately 0.032 MB (2,000 × 16 bytes per element), so the memory cost is also small. Dynamic expansion logic is triggered when the buffer utilization exceeds 95% after a push. The new capacity is calculated based on the number of active sensor threads. If the number of current active sensors is more than two, the size will expand by 10,000. While if there are more than 50 sensors actively working, which usually indicates that it is under a pressure testing environment, the size will expand by 50,000 each time, which can reduce the number of expansions.

Figure 1. Architecture with various components

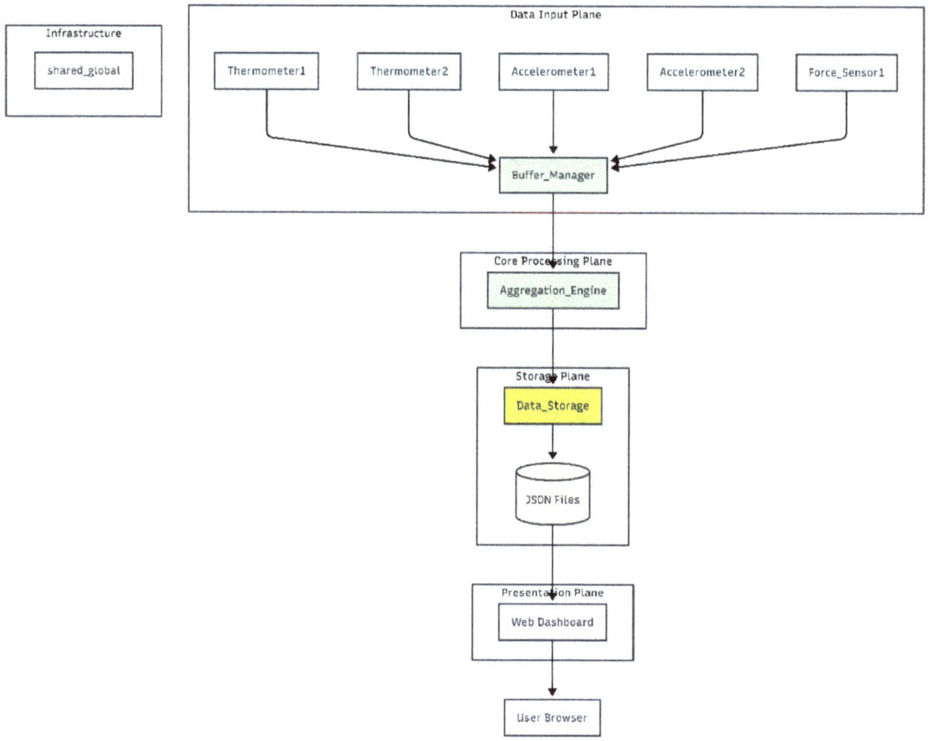

A crucial design choice on reader thread strategy is about the Earliest Deadline First (EDF) algorithm. During our working progress, each sensor will send the information about its rate once in the beginning, so that the thread which is responsible for reading data from the specific sensor can record the generating rate. Together with sending a message of its current remaining available buffer size, the main thread can determine how emergent this sensor is. Using remaining size/rate as an indicator, the buffer will put each child thread into one of the three different queues. This algorithm suggests that processes of less emergencies should sleep for a while and leave CPU to those with higher urgency.

Batch writing is specially designed for achieving higher CPU utilization. When a receiver thread receives messages, it does not get the lock at once. Instead, it stores the message in a small temporary buffer of size 20. After it is full, the thread gets the lock and pushes everything into the main buffer at a time. This method highly decreases the lock competence.

3.3 Aggregation Engine

Next, we will walk through the code of aggregation engine and demonstrate how the key results for aggregation engine in section 1.3 are satisfied.

Firstly, we would like to introduce two basic but important structs defined in aggregation_engine; WindowData and AggregationEngine. WindowData has three fields in total: start_time is an Instant that records the time it is created; stats is a struct of type SensorStats that stores the statistics (count/min/max/sum) which are used for average and standard deviation calculation (done by its associated methods) for one unique sensor, achieving correctly aggregates per sensor and calculates statistics goals; raw_readings is vector that stores all original data items that belong to this sensor and is used for anomaly detection. In AggregationEngine struct, the most important field is windows; a HashMap wrapped with RwLock (key is sensor id (String), value is sensor window (WindowData)). The detailed analysis of the implementation of aggregation engine comes next.

Method start() and shutdown() are used for the engine to start and stop data processing, respectively. start() method spawns multiple worker threads (number of worker threads is configurable) to work. The worker thread mainly executes a while loop: fetch a data item from the buffer using method try_pop(); process it by calling method process_reading(); call output_window() to output aggregated frames if the window is due, and return to the beginning of the while loop. The while loop's two conditions ensure that the worker threads will stop processing data only when the global running flag is set to false and the buffer is empty, which guarantees all data items in the buffer are processed by the aggregation engine, realizing the zero data loss goal.

Inside process_reading() method, the worker thread first tries to acquire the window for that specific sensor by using a RwLock read lock. If it doesn't exist, the read lock is upgraded to a write lock and an empty window is inserted for that sensor into the HashMap windows. By acquiring the read lock, all worker threads can read multiple sensor windows concurrently, which accelerates the lookup of the required sensor window. At the same time, protected by the write lock, every sensor window can only be modified by one worker thread at a time. Thus, it achieves the safe and concurrent data processing across multiple threads goal.

Before processing the reading, it compares now.duration_since with window_duration (a value set manually) using the ">=" comparison symbol to check if the window has expired. It tells whether a new data item should be in a new window or not, realizing the correct window handling goal. If yes, the expired window will be replaced with a new one and the lock will be immediately dropped so that the output process for the sensor's old window would not block other threads processing its new window. Other than the combination of HashMap and RwLock, this "handle old window outside the lock" is another design that increases concurrency to some extent.

The worker thread calls output_window() method to handle the remaining jobs. Before getting the aggregated frame, it gets a list of anomalies by calling detect_anaomalies_for_method(). Anomaly detection will be demonstrated in the Advanced Modules section.

3.4 Data Storage

Figure 2. Architecture of the Data Storage

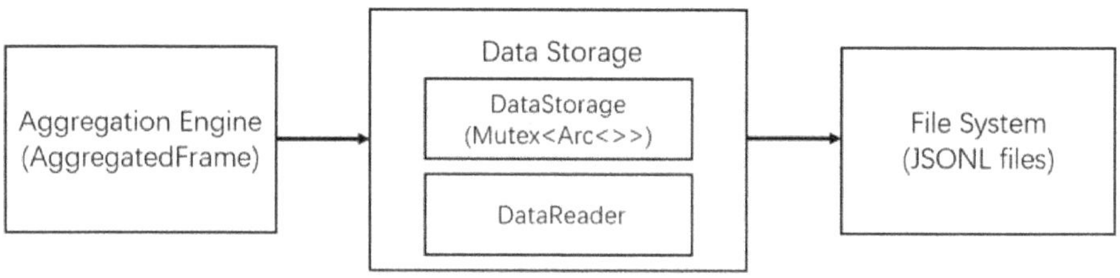

This Figure 2 shows the overall architecture of the Data Storage. As the final stage in the data processing pipeline, it receives processed "AggregatedFrame" objects from the Aggregation Engine and convert them into JSON Lines (JSONL) format for long-term storage.

The data_storage component follows a clean, modular structure designed to support stable sensor data persistence in multi-threaded environments. At a high level, the component is separated into four structural parts that work together to fulfill storage requirements. The configuration layer, represented by StorageConfig, manages customizable storage parameters including directory path, maximum file size limits, and file rotation policies. The serializable data models including SerializableFrame, SerializableSensorStats, and SerializableAnomaly are designed to transform non-serializable runtime data objects into JSONL formats while preserving all critical metrics and temporal information. The internal StorageState structure tracks real-time storage status such as active file writers, current file paths, file sizes, and written frame counts, and this shared state is fully protected by thread-synchronization mechanisms. Finally, the public DataStorage and DataReader structures implement core write, rotation, and reading operations, and with the Storage trait ensures integration with the upper aggregation engine. The entire component operates around a controlled "single-writer principle" (Terzian, 2020), writing structured data into JSONL files and supporting automatic file rotation to avoid large files, providing practical utilities for data persistence and retrieval.

The implementation of the data_storage component integrates multiple functional modules to achieve stable and efficient sensor data persistence, with each part designed to fulfill specific responsibilities in the storage workflow. The configuration and initialization system provides flexible control over storage behavior, where StorageConfig offers reasonable default settings including a 10MB maximum file size, a default data storage directory, and enabled file rotation, and the DataStorage::new() method automatically creates the target directory, initializes thread-safe shared state, and generates the first data file to ensure immediate usability. For data serialization, the component implements structured conversion logic to transform runtime domain objects into persistent formats, using the From trait to map sensor statistics and anomaly details to serializable structures, and the convert_frame method calculates time-window durations and anomaly timestamp offsets to ensure complete and accurate data recording, with serde_json handling the conversion to JSON strings for file storage. In terms of file I/O and rotation, the component uses the JSONL format to write one frame per line for easy parsing and appending, and the write operation automatically checks file size limits to trigger rotation when necessary, creating uniquely timestamped files to avoid conflicts and ensure data organization; the flush mechanism guarantees buffered data is fully written to disk to prevent data loss. For data reading and retrieval, both DataStorage and the dedicated DataReader provide reliable access to stored data, supporting file listing, full-file reading, and latest-frame querying, with the DataReader designed for web server integration to separate write and read operations for improved modularity. The component also includes a comprehensive test suite that verifies storage initialization, end-to-end write and read functionality, concurrent thread safety, and automatic file rotation, ensuring all core features operate reliably under various conditions.

The synchronization strategy is a core part of the data_storage implementation, and the component uses Mutex with Arc (Mutex in Std::Sync - Rust, 2026) as the exclusive synchronization mechanism to protect the shared data StorageState, which manages key resources such as the file writer, current file path, file size, and frame counter. This Mutex design ensures exclusive access to shared state across threads, eliminating race conditions that could cause file corruption, invalid state updates, or conflicting file rotation operations. The "write()" method acquires exclusive lock before writing; Arc enables safe shared ownership of the storage instance across multiple threads, and special error handling is implemented for Mutex poisoning to maintain system stability when unexpected thread error occur.

3.5 Webserver Multi-threading Model

Our web server employs an architecture that assigns a thread to each connection to efficiently handle concurrent client requests. Our server program is primarily divided into three structures: one for handling server connections (mainly thread-based), one for sending requests to the server, and a final structure for server responses, implementing a typical server-client model.

3.5.1 Server Connections

The server runs in a dedicated thread created by the `start_web_server()` function. A `TcpListener` is bound to port 3000 and enters an infinite loop, accepting incoming connections. For each successfully accepted connection, we create a new thread using the `thread::spawn()` function, passing a cloned reference data_dir to that thread. We could follow guidance provided by Rust official team (2026), who discuss about how to use TcpListener and TcpStream to build a network architecture. The implementation modules are shown in Figure 3 and Figure 4.

Figure 3. `start_web_server()` function.

```
pub fn start_web_server(data_dir: PathBuf) -> thread::JoinHandle<()> {
    thread::spawn(move || {
        let listener: TcpListener = match TcpListener::bind(addr: PORT) {
            Ok(l: TcpListener) => l,
            Err(e: Error) => {
                eprintln!("[ERROR] Failed to bind to {}: {}", PORT, e);
                return;
            }
        };
```

Figure 4. TcpListener and TcpStream for network architecture

```
pub fn bind<A: ToSocketAddrs>(addr: A) -> Result<TcpListener>
```

1.0.0 Source

Creates a new `TcpListener` which will be bound to the specified address.

The returned listener is ready for accepting connections.

Binding with a port number of 0 will request that the OS assigns a port to this listener. The port allocated can be queried via the `TcpListener::local_addr` method.

The address type can be any implementor of `ToSocketAddrs` trait. See its documentation for concrete examples.

If `addr` yields multiple addresses, `bind` will be attempted with each of the addresses until one succeeds and returns the listener. If none of the addresses succeed in creating a listener, the error returned from the last attempt (the last address) is returned.

Examples

Creates a TCP listener bound to `127.0.0.1:80`:

```
use std::net::TcpListener;

let listener = TcpListener::bind("127.0.0.1:80").unwrap();
```

Creates a TCP listener bound to `127.0.0.1:80`. If that fails, create a TCP listener bound to `127.0.0.1:443`:

```
use std::net::{SocketAddr, TcpListener};

let addrs = [
    SocketAddr::from(([127, 0, 0, 1], 80)),
    SocketAddr::from(([127, 0, 0, 1], 443)),
];
let listener = TcpListener::bind(&addrs[..]).unwrap();
```

The advantage of this design is that the main thread never blocks; each newly created worker thread handles its own client requests independently. Even if one client requests a large dataset, it won't affect other threads, which is the initial reason we chose thread-based network architecture.

3.5.2 Server Request

In the `handle_client()` function, we declare a temporary variable `buffer` to hold the HTTP request from the client. The size of `buffer` is approximately 4KB. The endpoint of the client request can be seen in the first line of the HTTP packet header. We designed a total of 5 endpoints: `/frames/latest/json` returns the latest 10 frames of data in JSON format for the program to access; `/frames/latest` renders a complete HTML page, which is the interface displayed to the browser, containing all historical frames grouped by sensor type; `/files/stats` displays file statistics and aggregated metrics for each sensor; `/api/help` provides API documentation; and the `/root` path provides the homepage.

3.5.3 Response Generation

We construct the response as a string, following the HTTP/1.1 format (Fielding et al., 1999). This includes standard status such as 200 OK for success and 500 Server Error for failures, along with required headers like Content-Type. For JSON endpoints, the `serve_json()` function reads the latest 10 `SerializableFrames` from the `data` directory and serializes them into JSON strings. Serde is a framework for serializing and deserializing Rust data structures efficiently and generically. (Serde Developers,2024) For HTML endpoints, we generate a style page with embedded CSS and JavaScript. The CSS controls font styles and overall typography, while the JavaScript implements interactive functionality, allowing users to choose whether to expand all windows of a sensor to view its historical frames or only view the latest three frames. JavaScript functions (`expandAll()`, `collapseAll()`) control the collapsing behavior. Automatic refresh is achieved via the `<meta http-equiv="refresh">` tag, with a refresh interval of 10 seconds. The frame page uses a three-level nested HashMap structure to group data by sensor type (thermometer, accelerometer, force sensor).

Why one thread per connection?

Based on the discussion already covered in the paper, I'll briefly summarize our reasons for choosing this model. Firstly, the required synchronization is very lightweight. Since the number of client connections is relatively limited, providing a separate thread for each connection is manageable. Secondly, it's not difficult to implement; it's simpler than event-triggered methods and thread pools. Therefore, we use Rust's `std::thread` to provide a lightweight stack implementation, while `Arc::clone()` allows for safe sharing of data directories between threads. This simple yet practical framework is the reason for our choice.

3.6 Advanced Modules for Anomaly Detection

3.6.1 Analysis

Before doing any analysis on the collected raw data, anomaly detection and data cleaning are two essential steps. Based on the characteristics of datasets, different anomaly detection methods are applied. Thus, analyzing the characteristics of the dataset should be the first step. Even though the provided sensor readings have three types (ThermoReading, ForceReading, AccelReading), they all share the same data patterns:

<1>Every axis contains only a single-valued floating-point number (To unify the detection logic for all three types of sensors, we set the y-axis and z-axis of

ThermoReading to be 0.0; Besides, we divide ForceReading and AccelReading into three data items as the anomaly detection should be conducted on each axis, logically speaking).
<2>Every value (the floating-point number) floats within a certain range of values given that all generated data is simulated using a coefficient multiplied by a random value generated from rand::random::<f32>(). More specifically, they are uniformly distributed. The `rand::RngExt::random()` method generates values using the StandardUniform distribution (The Rust Chapter Developers, 2026). The coefficient and ranges for thermometer, accelerometer and force sensor are 10.0 [20.0, 30.0), 100.0 [0.0, 100.0), and 5.0 [0.0, 5.0) respectively.

From the above analysis, we can conclude that the dataset is univariate, and its data items are uniformly distributed.

To detect outliers in univariate datasets, the Z-score method, boxplot method and median absolute deviation method are frequently used. They are defined as Classic, RobustMAD and IQR in DetectionMethod enum.

3.6.2 Illustration of the Three Methods

3.6.2.1 Classic

The Z-score method borrows the idea from the normal distribution. The decision criterion is

$$dev = \left|\frac{x_i - \overline{x}}{S}\right| > k$$

where *dev* describes the degree of deviation between the i-th sample observation value x_i and the population mean $\overline{x}$; S is the mean and standard deviation of the population (In window aggregation, all data points within a window constitute the complete set of the current time window; we are not sampling from a larger population), k is the anomaly threshold and is usually set to 2 or 3 (as *dev* satisfies standard normal distribution, $\mu \pm 2\sigma, \mu \pm 3\sigma$ mean $95.5\%, 99.7\%$ data falls within the two ranges). k = 2.5 is also a commonly used value (Rousseeuw & Hubert, 2011).

3.6.2.2 RobustMAD

The median absolute deviation (MAD) is defined as the median of the absolute deviations from the median (Huber & Ronchetti, 2009, p.106):

$$MAD_n = med\{|x_i - M_n|\}$$

where

$$M_n = med\{x_i\}$$

and the decision criterion is

$$dev = \left|\frac{x_i - M_n}{MAD_n}\right| > k$$

where k is the anomaly threshold and is usually set to 2 or 3 (also based on the data coverage of standard normal distribution, 2×normalized MAD corresponds to 95%, 3× normalized MAD corresponds to 99.7%).

3.6.2.3 IQR

The interquartile range (IQR) is defined as the difference between the third and first quartiles:

$$IQR = Q_3 - Q_1$$

where Q_3 is the value in the dataset that holds 75% of the values below it while Q_1 holds 25%. In a boxplot, outliers are defined as observations that fall below the lower fence or above the upper fence. The two fences are usually defined as

$$lower = Q_1 - 1.5 \times IQR, upper = Q_3 + 1.5 \times IQR$$

. The value that falls below *lower* or above *upper* is called mild anomaly. The coefficient of IQR can be set as 3 and the value that deviates from this range is called an extreme anomaly. (The coefficient 1.5 or 3 is also based on the data coverage percentage in the normal distribution. The 1.5×IQR criterion corresponds to 99.3% coverage in a standard normal distribution, while 3×IQR corresponds to 99.99% coverage.)

Note that even though the anomaly detection multipliers for MAD (2 or 3) and IQR (1.5 or 3) are calculated from the standard normal distribution, these two methods do not require any distribution characteristics of the data. Deriving the common multiples of these two methods from the standard normal distribution is for the convenience of general application.

3.6.2 Explanation of the Reasons Why We Chose Them

The Z-score method is based on the characteristics of a normal distribution and uses the mean and standard deviation to detect, which are very sensitive to outliers. Gong et al. (2022) found that the Z-score method performs poorly when outliers exceed 10% of observations. Also, our previous analysis shows the values in our dataset are uniformly distributed, which doesn't fit well required by the Z-score method. However, we still have decided to keep it as a classic method because it is frequently used in statistics, and its formula is concise and clearly expresses its meaning.

The median absolute deviation method uses median, a measure of central tendency and thus is insensitive to outliers, making MAD effective for outlier detection with a breakdown value about 50% (Gong et al., 2022).

The boxplot method avoids the interference of extreme values by excluding them in its calculation and is suitable for anomaly detection when outliers do not account for over 35% (Gong et al., 2022).

Last but not least, all these ways are unsupervised statistical methods that are fast in calculating, memory-friendly and easy to implement. Thus, we choose to implement these three methods for anomaly detection.

3.6.3 How These Three Methods are Implemented

In detect_anomalies_for_method(), based on the selected method, pre-calculate "means" (reference values used for judgement, respectively representing mean, median and Q1), scales_for_detection (scales used for judgement, respectively representing standard deviation, (normalized) MAD, and IQR) and scales_for_storage (scale values to be stored in Anomaly struct, respectively representing standard deviation, original MAD and Q3) by calling corresponding methods. Then, for each method, examine every value in raw_readings based on the previous formulas. After finishing the detection for all the data items per sensor, all outliers and their related content (sensor_id, timestamp, "means", scales_for_storage, severities, axis_flags) will form an Anomaly instance and be pushed to anomalies, which is a Vector of Anomaly type and is a field in AggregatedFrame.

4. EXPERIMENTAL RESULTS

4.1 Test Methodology

All experiments conducted on verifying buffer performances were conducted on a personal computer equipped with an AMD Ryzen 9 8945HS processor (8 cores, 16 threads, base frequency 4.00 GHz) and 32.0 GB of RAM, running Windows 11. The system was idle beside the tests.

To simulate realistic situations, we created multiple child processes to represent different sensors. The buffer keeps recording key statics, including writing and reading counts, overflow warnings and thread numbers.

For aggregation engine, to accurately verify the correctness of every aspect, we apply the methodology of manually pushing readings into the buffer and conduct a thorough and comprehensive test in integration_test.rs. Every single test in integration_test.rs is commented on its test objective on one or more aspects of aggregation engine. The whole integration test that combines the sensors, buffer manager, aggregation engine and data storage is conducted in test_pipe_integration.rs.

4.2 Basic Pressure Test and Scalability

We found that although sensors are set to be at this rate, not all of them can reach the theoretical generation rate. To study this phenomenon while obtaining data loss verification results, we conducted the basic pressure test: Controlling the rate of each sensor to be 1000Hz and continuously increasing the number of sensors. Since all processes cannot start or shut down at the same time, there are delays at the beginning and end of each test. The extra data generated during these delays can cause experimental errors. For example, the total number of writes recorded may be higher than the actual writes during normal operation. Also, when many sensors are used, the shutdown process takes longer, which can lower the measured write rate significantly.

To reduce these errors, we only start timing and data recording after all sensors have finished starting. We record data for 5 seconds, then shut down the system. We repeat each test three times and use the average value.

Figure 5. Data loss rate

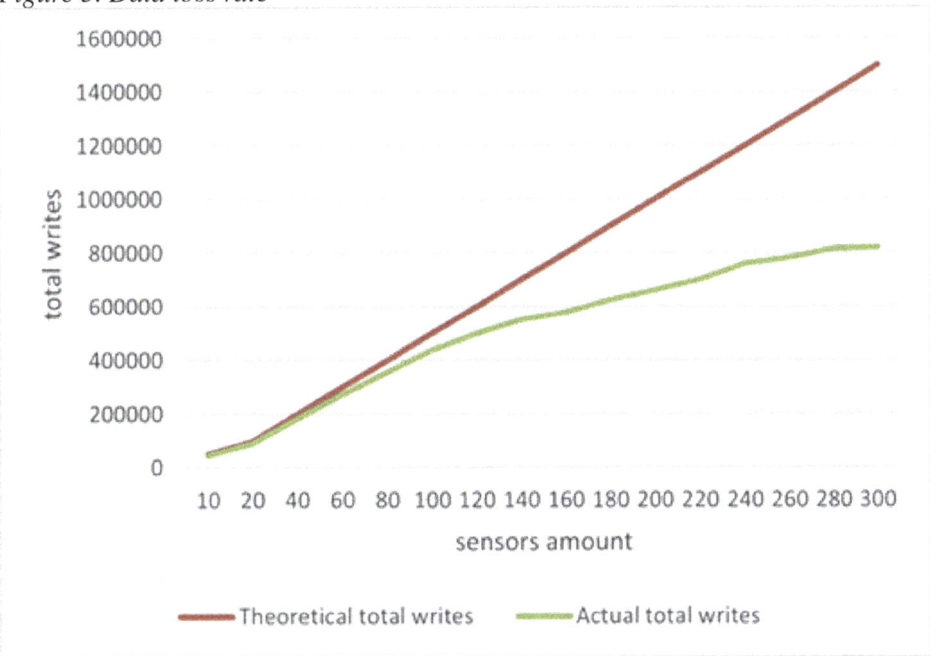

Figure 5 shows that when the number of sensors exceeds 80, the actual write rate begins to deviate significantly from the theoretical linear growth curve, and the data loss rate rises sharply. As Effelsberg and Haerder (1984) pointed out, when resource contention intensifies, the system throughput will inevitably deviate from the ideal linear model, which is closely related to hardware limitations.

During this test, we also configured that if the sensors' inner buffers overwrite some data, an error message will be printed. If no error message is printed, zero data loss can be guaranteed. We found that the maximum safe write rate is 132,789Hz, which ensures no data loss after 15 repetitions. For larger numbers of sensors, the test results are relevantly unsustainable. Sometimes under a written rate of 200kHz there appears no data loss. However, the written rate of 140kHz can also trigger significant data loss. This phenomenon is most likely caused by uncertainty in system interrupts and CPU scheduling in multi-sensor scenarios.

Based on the test results, we conducted further tests to verify the working efficiency of the buffer manager's specific strategies.

We also tracked the startup and shutdown latency in our tests as shown in Figure 6 and Figure 7. By recording significant time points in a test, we can get the startup time for all the child processes to create, and the shutdown time of all the processes. For each sensor amount, we repeat the test 10 times and get the average value.

Figure 6. Startup and shutdown latency

sensors	startup_ms_mean	shutdown_ms_mean
20	57.33	14.33
40	124.67	5.67
60	160.83	6
80	208.4333333	402.8
100	234.25	546.5625
120	334.2916667	1151.864
140	387.0182	1557.932
160	462.67662	1677.8324
180	513	1937.83
200	587.75	3058.777143
220	641.15	3771.606333
240	742	4016.5

Figure 7. Startup and shutdown mean

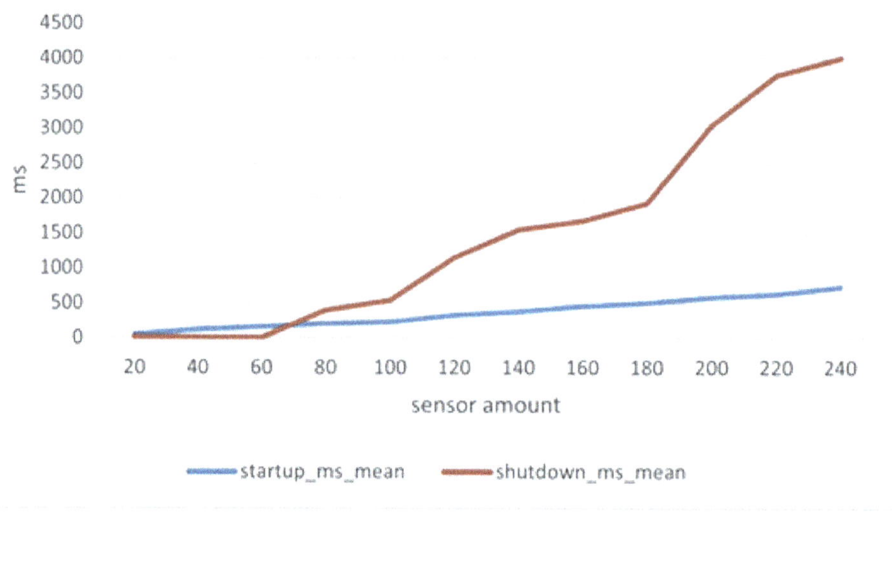

As a quick summary of the statistics above, we can find both rising trends for startup and shutdown latency as the sensor amount increases. Meanwhile, the shutdown latency is influenced much more than startup time. A significant observation is that while measuring the shutdown latency, its value fluctuates heavily after the sensor amount is more than 80. Sometimes the latency is more than 5000ms, while the minimum time could be 5~6ms. However, as the sensors amount increases, the probability of delay for more than 5000ms also grows. Thus, we can see an increase in the figure.

4.3 EDF Algorithm Evaluation

Firstly, we compared the actual working efficiency of the buffers with and without EDF algorithms as shown in Figure 8. Before we test, the cost for the sensors to send 'AVALI' messages proved to be minor in the design phase of this algorithm. So, the no_EDF buffer only needs to make no response to the message with 'AVALI' prefix in the pipe. The EDF algorithm aims at achieving priority scheduling. We mixed sensors of 100Hz and 1000Hz in the test to verify the performance of it.

Surprisingly, we found that the EDF algorithm did not perform better in the test. Meanwhile, the zero data loss limitations of the two buffers are also close, which differ by less than 400Hz.

4.4 Batch Push Efficiency

Then, we also tested the efficiency of batch pushing as shown in Figure 9. Controlling the data generation rates and sensor number to be the same, we found that while using batch pushing or not does not influence write rate much. The rate increase brought by the batch pushing method is less than 1%.

Figure 8. Efficiency of the buffers with and without EDF algorithms

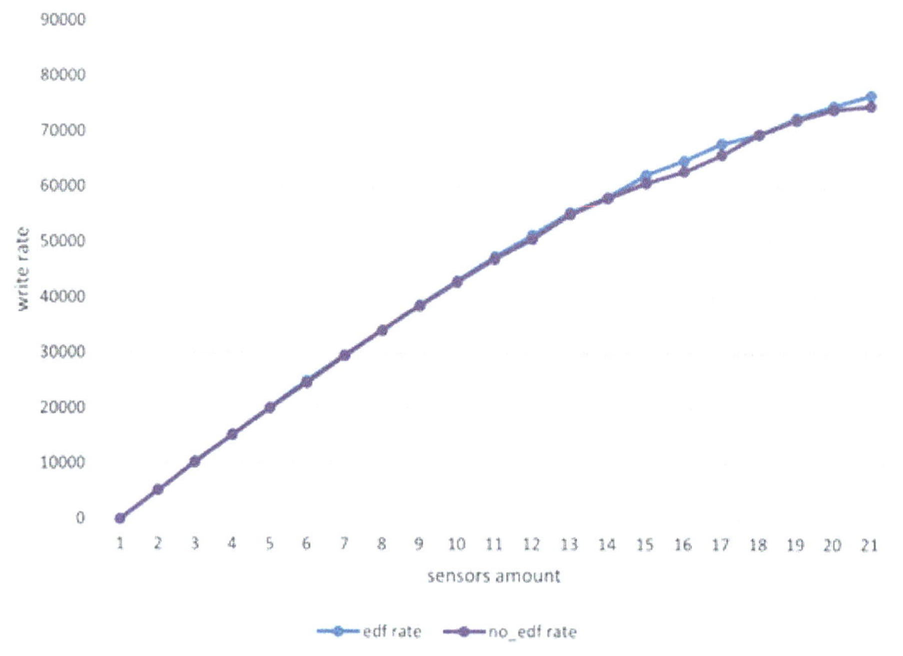

Figure 9. Efficiency of batch pushing

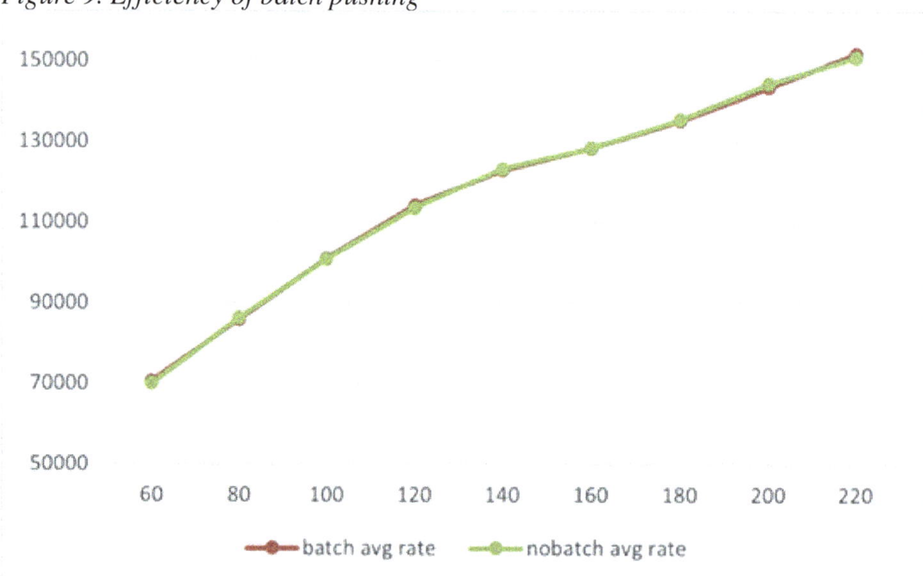

However, the batch pushing method can help reduce CPU usage. During the test, we recorded the CPU usage using the embedded tool in IntelliJ IDLE, and conducted both pressure tests and integration tests under normal working situations on non-batch pushing method first and then batch pushing method.

Figure 10. CPU usages of non-consumer pressure tests

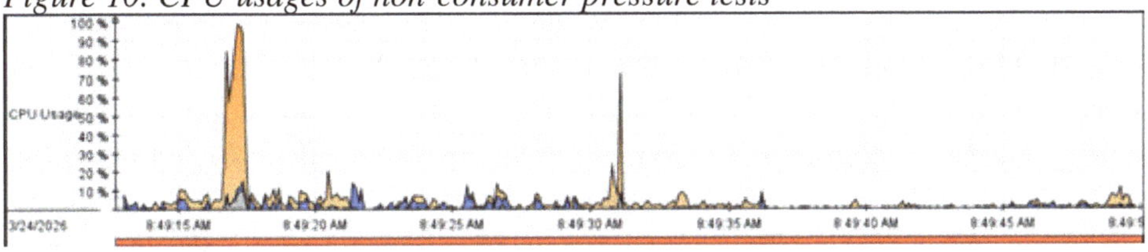

Figure 11. CPU usages of integration tests

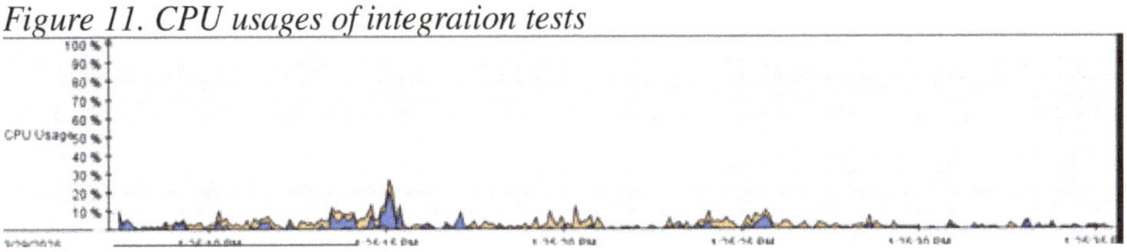

As we can summarize from the figure, each figure has two peaks and both first peaks are higher and larger than the second one, so the CPU cost of the non-batch pushing method is higher. The batch pushing method indeed reduces the usage of CPU and makes the system more efficient as shown in Figure 10 and Figure 11.

4.5 Summary

In summary, the experimental tests demonstrate that the buffer manager can sustain a throughput of up to 132,789 entries per second with zero data loss under a uniform 1000 Hz load, which meets the design goals of high throughput and zero data loss within a safe range. The startup latency is on the order of 100ms and increases linearly with the number of sensors. The shutdown latency's growth curve is steeper and more unsustainable. What's more, the shutdown latency is on the order of either 1ms or 1000ms. Scalability tests reveal that the system maintains near linear throughput up to approximately 60 sensors with 1000Hz.

5. DISCUSSION

5.1 Buffer

5.1.1 Latency Analysis

From the data of the latency test, we found that the startup and shutdown latency turned out to be different in not only the amount of time, but also sustainability. This is probably because of the residual data during testing. In our original shutdown process, all the child processes need to write all the residual data inside to the pipe to maintain zero data loss, and during this process, the sensors are also continuously generating new data. Thus, the shutdown process may take longer if the generation rate increases. One possible solution to make the shutdown fast and sustainable is to kill all the child processes before calling the buffer.shutdown() method. Such a solution limits the shutdown latency to 10ms in even 150kHz write rate high pressure situations. If we only need to get the test data in actual running time, such as actual write rate, we can adopt this solution for fast shutdown. However, in real working situations, we cannot use it because it will directly drop the residual data in the child process and lead to data loss.

5.1.2 EDF Performance Evaluation

Analyzing previous test results, there are several possible reasons why EDF algorithm does not improve the performance as expected. EDF cannot improve the limit safe write rate, since it aims at dealing with emergency processes first instead of writing faster into the buffer. It is possible that the bottleneck of our system is not about who to process first, but the actual process speed for a large amount of data. We measured the average per-sensor processing interval under a uniform 1000 Hz load of 190 sensors. The parent process achieved a write rate of 130kHz, which translates to an average processing spacing of 1.46 ms per sensor. While the time required to process a single data entry (including parsing, lock acquisition, and batch accumulation) is estimated to be less than 5 μs, i.e., only 0.34% of the average spacing. Thus, processing several low priority sensors may only produce minor influence on those with higher priority. The overall bottleneck still lies in overall pipe throughput processing speed.

Nevertheless, we still reserve the EDF algorithm in our design. Firstly, its slight improvement in efficiency cannot be ignored. Secondly, the EDF algorithm still has potential for future improvement if the bottleneck changes.

5.2 Aggregation Engine

5.2.1 Trade-offs in Design Choices

In aggregation engine, we save all raw readings for anomaly detection. When the window is expired, it ensures the concise calculation of deviation between the data item and the mean of the entire window, making the anomaly detection to be precise. However, saving raw data items consumes memory. Memory usage increases linearly with the amount of data within the window, which can lead to large memory consumption for high-frequency sensors.

5.2.2 Limitations and Bottlenecks

As discussed previously, memory usage to store raw readings is one of the most significant bottlenecks for aggregation engine. The computational overhead in anomaly detection is another area that may be considered as a limitation. As the computational process uses a loop to go through every data item in raw_readings, this operation may take a long time if the amount of data in the window is extremely large, affecting the timely output of the window. The disordered output order of aggregated frames is another apparent limitation. Given that the window for each sensor is output independently, there doesn't exist a specific sequence for the frames of different sensors. If the web server expects to display the aggregated frame based on some guidelines such as the types of sensors or the creation time for each sensor, additional sorting is required.

5.2.3 Comparison of Synchronization Primitives

Mutex is used to protect every sensor's WindowData. It ensures that only one thread modifies the sensor's statistics at a time; RwLock is used to protect access to the HashMap. Multiple threads can read concurrently, but write operations are exclusive; AtomicU64 is used for next_frame_id, it increments atomically and ensures the uniqueness of the frame in all worker threads; AtomicBool is used for running, which controls the lifecycle of worker threads.

5.2.4 Lessons Learned about Concurrent Programming

From the concurrent programming related to aggregation engine, we learned that the final concurrent processing ability in multiple threads is achieved by skillfully coordinating various synchronization primitives and a reasonable arrangement of processing logic. It is essential to carefully design and organize various synchronization primitives to fit in different situations.

5.3 Data Storage

The decision to use Mutex instead of RwLock in the Data Storage is based on the actual workload and consistency requirements of the sensor data storage system. Since the component is optimized for write-heavy workloads where data writing is continuous and reading operations are relatively infrequent, Mutex provides lower overhead and higher efficiency compared to RwLock, which introduces additional atomic operation costs for managing reader and writer states. Strict atomic state modifications are required for operations such as file size updates and file rotation, where exclusive locking is mandatory to maintain state consistency, while RwLock cannot prevent inconsistent state access during critical updates. Additionally, the Mutex simplifies error handling and reduces the risk of deadlocks compared to RwLock, and the infrequent state reading operations make the performance benefits of RwLock negligible in this scenario. The synchronization workflow during write operations is carefully designed to acquire the lock only when necessary, check for rotation requirements, temporarily release the lock during file rotation to reduce blocking, and then re-acquire the lock to complete data writing and state updates, balancing exclusive access and execution efficiency to ensure stable and high-performance data persistence in multi-threaded environments.

5.4 Web Server

5.4.1 Analysis of Network Architecture Results

Regarding network architecture, given our limited background in web development, we selected a straightforward thread-per-connection model. This decision allowed us to observe concurrency behavior directly and understand how the operating system schedules threads.

5.4.2 Limitations and Bottlenecks

Nevertheless, this design has inherent drawbacks. We performed only basic functional tests; under heavy load; such as many clients simultaneously requesting large datasets; the server may become slow or crash. Since threads within a process share memory, a failure in one thread can potentially bring down the entire server. Memory usage is another concern: each thread reserves stack space, which restricts the total number of concurrent connections.

5.4.3 Comparison of Synchronization Primitives

Our web server avoids complex synchronization because each request runs in its own thread. However, the data storage component uses Mutex to guard file writes. This choice trades some concurrency for correctness. An RwLock would allow multiple readers but block writers; given our write-heavy workload, Mutex offered a simpler and adequate solution.

5.4.4 Lessons Learned

This chapter highlighted key differences between thread-based and process-based architectures. We learned to balance memory consumption, processing speed, and reliability when choosing a concurrency model. For future work, an event-driven approach (e.g., using tokio) could support higher concurrency with lower memory overhead.

6. CONCLUSION AND FUTURE WORK

6.1 Buffer

Currently, the parameters in our buffer system, such as the batch size in and the sleep thresholds used in the EDF algorithm, are set manually based on testing data. While this approach works for our current test environment, it may not be optimal under different workloads or hardware.

Inspired by the machine learning technique for automatic adjustment against arbitrary workload in the Spitfire buffer manager (Zhou et al., 2021), we plan to introduce a controller that automatically decides these parameters. The controller would periodically collect system write rate, CPU usage, and data loss rate, and then use a search algorithm (e.g., simulated annealing or reinforcement learning) to find the optimal configuration. Parameters that could be adjusted include the batch size for sensors with different write rates, the number of consumer threads, and whether to enable batch writing or EDF scheduling.

Furthermore, the Three-tire Buffer Manager (Riekenbrauck et al., 2024) uses a lock-free design. The design especially focuses on using atomic state management for every data frame and adopting tbb::concurrent_queue as lock-free eviction queue. In our current system, all producer threads compete for the same global lock, Mutex<VecDeque>. Based on the idea of lock-free design, we can represent the state of each data block with atomic variables and achieve lock-free state transitions through CAS operations.

6.2 Aggregation Engine

The implementation of aggregation engine achieves all its necessary requirements. The key results for aggregation have been stated in section 1.3. However, there are some potential improvements that can be made in the future. Besides the limitations and bottlenecks that have already been discussed, the aggregation engine could be improved to have the ability to dynamically control (creates or closes) a certain amount of worker threads based on the buffer utilization. Additionally, the anomaly detection method could also be improved. The UTSD benchmark constructed by Muhr & Affenzeller (2022) emphasizes the importance of multi-concept normality. The sensors may have different operative modes. And the anomaly detection can introduce pattern recognition, first identifying the current state and then modeling each mode separately. Identifying anomalies through prediction is another direction we can study and explore further, which is inspired by the LSTMAD method proposed by Ji et al. (2021).

6.3 Web Server

We successfully implemented a multi-threaded web server using Rust's `std::thread` library, which follows a one-thread-per-connection model. The server provides three core functionalities: a real-time dataframe viewer that refreshes every 10 seconds and displays sensor readings grouped by type; a statistics page that summarizes metrics for each sensor, including count, sum, average, standard deviation, minimum, maximum, and storage location; and a JSON API endpoint for programmatic access, accompanied by an API help page for documentation.

We achieved these goals through a combination of self-learning and iterative development. Initially, our experience with Rust web frameworks was limited, so we relied on online documentation, experimentation, and AI-assisted coding to accelerate development. Before finalizing the one-thread-per-connection model, we evaluated three concurrency models; single-threaded, process-based, and thread-based. This choice struck a balance between simplicity and performance for our lightweight use case. Furthermore, we implemented a clean HTML/CSS frontend without using any external frameworks, ensuring responsiveness and ease of use.

Several potential areas for improvement remain. First, our current one-thread-per-connection model creates a new thread for each request, which incurs additional overhead under high load. Switching to a thread pool can reduce this overhead while maintaining thread simplicity. Second, the server currently relies on blocking I/O; adopting an event-driven architecture that can improve scalability, thus supporting higher concurrency.

Future work can extend the server in two ways. One is to integrate external visualization APIs; for example, sending statistics to a charting service to dynamically generate pie charts or line charts. The other is to add authentication and user management functionality to support multi-user scenarios with personalized data views. These extensions will make the current prototype system more suitable for production environments.

REFERENCES

Effelsberg, W., & Haerder, T. (1984). Principles of database buffer management. *ACM Transactions on Database Systems*, *9*(4), 560–595. DOI: 10.1145/1994.2022

Fielding, R., Gettys, J., Mogul, J., Frystyk, H., Masinter, L., Leach, P., & Berners-Lee, T. (1999). Hypertext Transfer Protocol — HTTP/1.1. *RFC, 2616*. Advance online publication. DOI: 10.17487/rfc2616

Fisher, G., & Yeh, C. (2017). Comparing Producer-Consumer Implementations in Go, Rust, and C. https://stanford-cs242.github.io/f17/assets/chapters/2017/gsfisher-chrisyeh.pdf

Gong, X., Zhang, F., Lu, T., & You, W. (2022). Comparative analysis of three outlier detection methods in univariate data sets. In *3rd International Conference on Electronic Communication and Artificial Intelligence (IWECAI)* (pp. 209-213). https://doi.org/DOI: 10.1109/IWECAI55315.2022.00048

Hua, B., Ouyang, W., Jiang, C., Fan, Q., & Pan, Z. (2021). Rupair: Towards Automatic Buffer Overflow Detection and Rectification for Rust. *Annual Computer Security Applications Conference*. https://doi.org/DOI: 10.1145/3485832.3485841

Huber, P. J., & Ronchetti, E. (2009). *Robust statistics* (2nd ed.). J. Wiley & Sons. DOI: 10.1002/9780470434697

Ji, Z., Gong, J., & Feng, J. (2021). A Novel Deep Learning Approach for Anomaly Detection of Time Series Data. *Scientific Programming*, *2021*, 6636270. Advance online publication. DOI: 10.1155/2021/6636270

Mohammed, A., & Zangana, H. (Eds.). (2026). *Intelligent Construction Monitoring Systems: Real-Time Safety, Environmental Prediction, and Risk Management*. IGI Global Scientific Publishing., DOI: 10.4018/979-8-3373-9245-5

Muhr, D., & Affenzeller, M. (2022). Outlier/Anomaly Detection of Univariate Time Series: A Dataset Collection and Benchmark. In Kotsis, G., Gamper, J., Khalil, I., Wrembel, R., & Tjoa, A. M. (Eds.), *BIG DATA ANALYTICS AND KNOWLEDGE DISCOVERY, DAWAK 2022* (Vol. 13428, pp. 163–169). Springer International Publishing AG., DOI: 10.1007/978-3-031-12670-3_14

Mutex in std:sync - Rust. (2026). Rust-Lang.org. https://doc.rust-lang.org/std/sync/struct.Mutex.html

Pieper, R., Löff, J., Hoffmann, R. B., Griebler, D., & Fernandes, L. G. (2021). High-level and efficient structured stream parallelism for rust on multi-cores. *Journal of Computer Languages, 65*, 101054. DOI: 10.1016/j.cola.2021.101054

Riekenbrauck, N., Weisgut, M., Lindner, D., & Rabl, T. (2024, May). A three-tier buffer manager integrating CXL device memory for database systems. In 2024 IEEE 40th International Conference on Data Engineering Workshops (ICDEW) (pp. 395-401). IEEE. DOI: 10.1109/ICDEW61823.2024.00063

Rousseeuw, P. J., & Hubert, M. (2011). Robust statistics for outlier detection. *Wiley Interdisciplinary Reviews. Data Mining and Knowledge Discovery, 1*(1), 73–79. DOI: 10.1002/widm.2

Sai, A., Reddy, D., & Perumalla Raghavendra, G. Yashwanth Kiran, & R, R. V. (2022). Producer-Consumer problem using Thread pool. 2022 3rd International Conference for Emerging Technology (INCET). https://doi.org/DOI: 10.1109/INCET54531.2022.9824031

Serde Developers. (2024). Serde: Serialization framework for Rust. Retrieved from https://serde.rs/

Tanenbaum, A. S., & Bos, H. (2015). Modern operating systems. (4th edition / Andrew S. Tanenbaum, Herbert Bos.). Prentice Hall.

Terzian, G. (2020, May 11). Rust concurrency: the single-writer principle, applied. Medium. https://medium.com/@polyglot_factotum/rust-concurrency-the-single-writer-principle-applied-aada2cdc6fb0

The Rust Chapter Developers. (2026). rand (Version 0.10.0) [Software documentation]. Docs.rs. https://docs.rs/rand/latest/rand/trait.RngExt.html#method.random:~:text=Return%20a%20random,distribution

Zhou, X., Arulraj, J., Pavlo, A., & Cohen, D. (2021, June). Spitfire: A three-tier buffer manager for volatile and non-volatile memory. In *Proceedings of the 2021 International Conference on Management of Data* (pp. 2195-2207). DOI: 10.1145/3448016.3452819

Chapter 2

Zero-Loss Sensor Data Aggregation: A Rust-Based Operating System Core for Concurrent Data Processing

Hong Rui Zhou
The Hong Kong Polytechnic University, Hong Kong

Min Hao Ling
The Hong Kong Polytechnic University, Hong Kong

Tong Yao Li
The Hong Kong Polytechnic University, Hong Kong

Xiang Li
The Hong Kong Polytechnic University, Hong Kong

Yi Ran Wu
http://orcid.org/0009-0004-8040-8379
The Hong Kong Polytechnic University, Hong Kong

Cong Wu
http://orcid.org/0000-0002-8652-3501
Liaoning Technical University, China

ABSTRACT

This project initially aims to construct a consequent data processing pipeline. It should have the ability to collect readings from multiple simulated sensors, and perform a real-time statistical aggregation. It should also store the results durably, and serves them through a web interface. The system architecture is organized as a Rust workspace with four key components. The sensor_sim works for data generation, os_lib for buffer management, gateway for the main entry point, and dashboard for the web presentation. The core innovation in this project is the synchronization implementation. A lock-free, non-copy circular buffer is used, and enables the zero-blocking data transfer between the producer and consumer threads.

DOI: 10.4018/979-8-2600-1101-0.ch002

1. INTRODUCTION

1.1 Problem Statement and Motivation

The modern infrastructure systems mostly relies on the interconnected sensor networks, and the continuous data collection and analysis. These systems are often compose of hundreds of sensors which can generate high speed data streams. These streams must be collected, stored, processed and presented with minimal latency and zero data loss.

Therefore, at the core of the system, there is a fundamental operating system challenge: managing the data flow between system components with different and mismatched speeds and capacities. Sensors often contain small internal buffers because of hardware constraints, the buffers can only hold the readings for several seconds then they may overflow. Moreover, other processing components like the aggregation engine and the storage system often operate at different speed, their capacities are also different. The system should have control of the reading speed. This problem is even more serious when concurrency is required. Multiple sensors generate data together, and multiple processing threads have to coordinate when accessing the shared buffers. The web server should serve aggregated results to clients, and new data is arriving at the same time.

The distributed sensor data aggregation platform is implemented to solve these problems. The core of the implementation is a lock-free non-copy circular buffer. It works as a fundamental synchronization helper. This chapter aims to solve the real-world data ingestion problem. It makes use of the advanced OS concepts like lock-free synchronization, atomic operations, and producer-consumer pattern to achieve this.

1.2 Chapter Objectives

To construct a complete sensor data aggregation platform is the overall objective of this chapter. It should fulfill the strict requirements for data integrity, concurrency, and efficiency. The system should collect data from multiple simulated sensors timely (to prevent overflow). Then it should process the data through a pipeline without any data loss. This overall objective can be decomposed into four component-level objectives.

First, the buffer management component. It must provide a lock-free and thread-safe queue which can perform the sensor reading and data processing separately. It must create dedicated reader threads for each sensor, and poll them sufficiently to prevent overflow. It should also maintain a high-performance circular buffer which can support the non-copy data types.

Second, the aggregation engine must create multiple worker threads to process the sensor readings concurrently. It must group the readings into fixed time windows, and compute each sensor's statistics including count, minimum, maximum, average, and standard deviation. Also, it should be equipped with basic anomaly detection function.

Third, the data storage component must provide thread-safe file storage and management for the aggregated frame. It must support concurrent write operations from multiple aggregation worker threads. It should also ensure that the web server never reads partially written or incomplete files to solve the readers-writers problem.

Fourth, the web server must present the aggregated data through an HTTP interface, it should de able to handle multiple concurrent client requests, and ensure the safe file reading from the storage component.

In addition, the chapter aims to develop a powerful system which can perform steadily while the number of sensors increasing, and make full use of the OS concepts through comprehensive documentation and benchmarking.

1.3 Overview of Approach

The system architecture is organized based on a lock-free non-copy circular buffer provided by the os_lib module, which can coordinate the four components with mismatched processing speed, and finally forms a four-stage pipeline.

The first stage (buffer manager) is implemented through a lock-free circular buffer RWRoundQueue<T>, which has the capability of efficient bitwise modulo operations and is able to prevent data loss. The queue supports the separate QueueReader and QueueWriter handles. Dedicated reader threads are created for each sensor, continuously polling the readings form the sensors using their read() method, then pushing the readings into the queue via the writer handle. The non-copy variant uses MaybeUninit<T> for handling arbitrary types safely, and can perform careful management of dropped values during buffer overwrites.

The second stage (aggregation engine) is implemented with a thread pool. The worker threads can pull data from the central buffer concurrently using blocking pop operation. Each worker thread gathers readings into specific time windows for a specific sensor, also recompute the statistics using Welford's algorithm for numerical stability. When a window closes, the worker thread constructs an AggregatedFrame and passes it to the storage component.

The third stage (storage component) addresses the concurrent storage. It is implemented with a hybrid strategy including Mutex and RwLock to protect file operations. Instead of using the atomic rename strategy which may need complex implementation and may be difficult to maintain, this hybrid strategy using file operations from the Rust standard library which are simpler and easier to test and adjust.

The fourth stage (web server) implements a custom multi-threaded web server using TcpListener in Rust's standard library. The thread pool method is used again to manage income connections and requests, each request is allocated to an independent worker thread, then the HTTP request is parsed, the requested files are read from storage, finally an HTTP response is constructed.

This implementation has been verified zero data loss under high work load, the lock-free queue performance and the thread safety is also tested and verified. The chapter also performs satisfactorily on other aspects, especially the efficiency, which marks the success of our approaches.

2. RELATED WORKS

The sensor network has been an important component of modern information systems. Therefore the related works are rich and diverse. This section reviews related works across three key aspects: (1) Relevant buffer management techniques; (2) Multi-threaded server architectures; (3) Producer-consumer pattern implementations, then compares these works with our approaches.

2.1 Relevant Buffer Management Techniques

Buffer overflow is a common problem in Rust chapters. The coping methods can be roughly divided into two types: (1) Detect the overflow after it happens then rectify it; (2) Prevent the overflow before it happens.

2.1.1 Overflow detection and rectification

(Hua et al., 2021) have introduced an automatic buffer overflow detection and rectification system called "RUPAIR" in their article. The system detects buffer overflows using a data-flow analysis across Rust's safe/unsafe boundaries to identify unsafe buffer uses. It validates bugs with SMT solvers and generates counterexamples. Then the system uses semantic-preserving transformation to insert guards and conduct rectification, also report irreparable cases to the developer.

2.1.2 Overflow prevention

The most common approach to prevent buffer overflow is to increase buffer size. However, the larger the buffer size, the more memory space is needed, then the efficiency cannot be ensured. As stated in the article (Kevin Andrian Santoso,

2023), achieving the balance between memory usage and buffer safety is of great importance, a specific and moderate buffer size is the best solution.

Our approach of buffer management is prevention instead of detection & rectification, this needs careful design of buffer size and the queue feature. The implementation is more difficult, but the irreparable cases and irreversible data corruption can be prevented, the rectification is not needed, and the safety is better ensured.

2.2 Multi-threaded Server Architectures

A multi-threaded web server in Rust is typically implemented with the thread pool, Arc<Mutex>, and a connection handling module, using the standard library std::sync::mpsc and std::thread.

A book about Rust (Johnson & Kaihlavirta , 2018) and an article (Tsoukalos, 2022) both suggests using thread pool in connection and request processing, which can ensure the concurrent processing and the file integrity without blocking. While other researches have suggested other architectures: 1) Async runtime: Handling the concurrent connections without per-connection threads; 2) Work-stealing: Threads dynamically steal tasks from busy peers' queues; 3) Event-driven (epoll/kqueue & single thread): Not multi-threaded alone, but combined with thread pools for CPU-bound work.

Our chapter draws on the findings in the book (Johnson & Kaihlavirta, 2018), therefore the web server architecture is the most common one (using thread pool). Comparing to other architectures, the thread pool architecture performs predictable resource control, makes it the simplest reliable choice in an environment that both concurrency and blocking operations exist. Our chapter is expected to operate in such environment, hence the architecture we selected is the most appropriate one.

2.3 Producer-consumer Pattern Implementations

According to (Kwak et al., 2015), the producer-consumer pattern decouples data production from consumption using a shared buffer. The producers generate and add data to the buffer, the consumers remove and process it. This enables asynchronous or concurrent operations, and helps data systems to balance workload.

In Rust, a book about Rust web programming (Flitton, 2021) recommends to use channels from the standard library (std::sync::mpsc) to implement this pattern. The MPSC (multiple producers & single consumer) provides a channel where multiple producers produce and send data via Sander to a single Receiver which can consume it. These channels are thread-safe and ownership-based, which can significantly prevent data races.

However, for higher efficiency, the crossbeam_channel is recommended by other researchers. It offers both bounded/unbounded channels with multiple producers and multiple consumers (MPMC), and a select! macro can be added for non-blocking operations. Moreover, for lock-free structures like arrayvec, or for special implementations pursuing extremely high performance, the ring buffers (single producer & single consumer, SPSC) can be used.

In our chapter, we consider the performance, safety and the implementation difficulty (Mohammed & Zangana, 2026). The SPSC has highest performance and safety, but the implementation and maintenance is quite difficult; The MPMC has high performance and relatively simple implementation, but the connection among consumers may cause safety issues. Therefore, after comprehensive consideration, we chose the MPSC, which has moderate performance, the simplest implementation and high safety, achieved a balance among the key aspects.

3. IMPLEMENTATION

In this section, we present our chapter from six aspects: (1) Chapter architecture (with an architecture diagram); (2) Buffer management design and reader thread strategy; (3) Aggregation engine implementation; (4) Data storage synchronization approach; (5) Web server multi-threading model; (6) Code snippets for critical sections, and finally make a conclusion.

3.1 Chapter Architecture

Our chapter architecture is presented in the architecture diagram below. The overall architecture is the four-stage pipeline composed of the buffer management stage, the aggregation stage, the storage stage and the web server stage. The architecture is presented in the diagram following the format mentioned in the book by Lyu (2021) below.

3.1.1 Chapter Architecture Diagram

```
Chapter_Code/
├── Cargo.toml # Workspace configuration file
├── Cargo.lock # Dependency version lock file
├── README.md # Chapter documentation
├── .DS_Store # Data storage file
│
```

```
├── os_lib/ # Operation system library (MUST KEEP)
│ ├── Cargo.toml
│ └── src/
│ ├── lib.rs # Library entry
│ ├── queue.rs # Implementation of lock-free non-copy circu-
lar queue
│ └── queue_noncopy.rs
│
├── sensor_sim/ # Sensor simulation library (MUST KEEP)
│ ├── Cargo.toml
│ └── src/
│ ├── lib.rs
│ ├── traits.rs # Sensor characteristic definition
│ ├── thermometer.rs # Temperature sensor
│ ├── accelerometer.rs # Acceleration sensor
│ ├── force_sensor.rs # Force sensor
│ └── sim.rs
│
├── dashboard/ # Dashboard front-end library (MUST KEEP)
│ ├── Cargo.toml
│ ├── build.rs # Building script
│ ├── templates/ # HTML template directory
│ │ └── home.html # Front-end interface
│ └── src/
│ ├── lib.rs # API end point definition
│ └── resource.rs # Resource locating module
│
├── gateway/ # Gateway application (EXECUTABLE)
│ ├── Cargo.toml
│ └── src/
│ ├── aggregation.rs # Aggregation engine implementation
│ ├── buffer_manager.rs # Buffer manager implementation
│ ├── main.rs # Main program entry
│ ├── state.rs # Gateway status monitoring
│ └── storage.rs # Hybrid synchronization strategy implemen-
tation
│
├── shared/ #Shared data models and runtime API surface
│ ├── Cargo.toml
│ └── src/
```

```
│   └── lib.rs
│
├── templates/ # Template copies
│   └── home.html
│
└── programfiles/ # Data storage directory (RUNTIME)
└── [Data files generated in runtime]
```

The architecture follows a producer-consumer pattern with four primary layers:

1) **The sensor simulation layer:** This layer provides five simulated sensors (two thermometers, two accelerometers, one force sensor). Each sensor runs on its own background thread, and performs configurable data generation rates. These sensors play the role of producers in the system, generating readings into their internal 128-element buffers;
2) **The buffer management layer:** This layer is implemented in the os_lib, it provides the lock-free RWRoundQueue<T> as the central communication channel for all the components. This layer also plays an important role to decouple the fast producers from the downstream processing stage;
3) **The aggregation and storage layer:** This layer is consisted of a thread pool of aggregation worker threads which pull readings from the queue via buf.pop_blocking() method. The workers compute windowed statistics and write the final aggregated frames to the programfiles/ directory using a hybrid strategy;
4) **The presentation layer:** This layer is implemented in the dashboard module. It provides a multi-threaded web server that reads files from storage and serves aggregated frame through HTTP endpoints. A build.rs script manages resource placement, and a generated resource.rs module locates all the resources needed.

3.2 Buffer Management Design and Reader Thread Strategy

The buffer management strategy focuses on the lock-free RWRoundQueue<T> from os_lib, which can provide a circular buffer with power-of-two capacity for the efficient modulo operations. As recommended by Samson in his article (2024), this implementation uses the atomic operations (AtomicUsize) instead of the traditional mutex-based queues to coordinate read and write indices without any blocking.

The queue can be divided into separate QueueReader and QueueWriter handles. For each sensor, a dedicated reader thread is created. Then, an exclusive ownership of the QueueWriter instance is maintained. The thread polls Sensor::available() method to detect the pending readings in a loop. Then it calls the Sensor::read() method and pushes the value via QueueWriter::write(). Because of the carefully designed non-copy variant, the sensor reading types can be moved directly into the buffer. Unnecessary copying operations are never conducted.

To prevent the possible sensor buffer overflow, the polling frequency is designed to exceed the maximum sensor generation rate. The sensors are assumed to generate up to 20 readings per second. Hence, the reader threads poll every 10–20 milliseconds. This design can ensure the sensor's 128-element buffer never reaches full. Moreover, the circular design of the lock-free queue can automatically overwrite the oldest values when the queue is full. This provides a natural replacing mechanism that gives new data high priority. This mechanism can maintain zero loss from the sensor perspective. The buffer utilization statistics are tracked using the len() method of the queue. The runtime monitoring is implemented with this simple method.

3.3 Aggregation Engine Implementation

The aggregation engine uses a thread pool architecture learnt from Kolodin's article (2019). In the thread pool, multiple worker threads concurrently pull raw sensor readings from the lock-free queue. All the necessary statistics are computed later. The system uses TimestampedReading structures from the SensorBufferManager, and produces outputs in AggregatedFrame format.

The worker threads operate in a compact continuous loop, they calls the buf. pop_blocking() method repeatedly to collect readings. When the queue is empty, the worker threads block briefly before retrying, which can basically avoid busy-waiting. Readings are accumulated into the specific time windows for the specific sensor of fixed duration. For each sensor, the worker thread keeps running aggregates like count, sum, sum of squares (for standard deviation), min, and max.

As mentioned above, the statistical computation in this system is powered by the Welford's algorithm, which can keep the numerically stability during large amount of calculations. This approach processes one reading at a time, and updates the mean and squared sum without requiring storage of all values in the window. When a window expires, the state calls flush_window(), and worker thread computes then final statistics: average from sum/count, and standard deviation from sqrt(variance), where variance is derived from the accumulated squared sum.

Finally, the anomaly detection function is implemented by comparing each reading against the whole window's mean and standard deviation. For the readings possess: $|value - mean| > threshold \times stddev$, are flagged and an anomaly record is generated. The record contains the sensor ID, anomaly type, severity, and description. Right after the window completion, the worker constructs an AggregatedFrame including all computed statistics and anomaly records, then it is passed to the storage component through a thread-safe channel.

3.4 Data Storage Synchronization Approach

The data storage component firstly addresses the most important readers-writers problem between aggregation workers (writers) and the web server (readers). This chapter uses a hybrid synchronization strategy to simplify the implementation. The strategy combines the Mutex for file writes and the RwLock for an in-memory cache.

The DataStorage is composed of three key members. The Mutex<BufWriter<-File>> serializes writes to the JSONL file. The RwLock <Vec<AggregatedFrame>> maintains a bounded in-memory buffer of recent frames. And the max_cache parameter limits cache size to 5000 (default).

An aggregation worker thread calls the write_frame() method in four steps. The method first serializes the AggregatedFrame to JSON using serde_json. Then, it acquires the Mutex lock on the file writer to writes the JSON line. Next, it flushes immediately to ensure the file durability. Finally, after the file writing operation completes, it acquires a write lock on the cache and appends the frame. The problem is that sometimes the cache may exceed capacity. However, the method can drain excess entries and ensure no overflow automatically.

According to the book about Rust system programming (McNamara, 2021), this design offers several significant advantages. The Mutex on the file writer can ensure the concurrent writes from multiple aggregation worker threads are serialized. It can prevent file corruption effectively. Also, the immediate flush can ensure data is persisted before the cache is updated. The consistency between the cache and disk is maintained. Last but not least, the RwLock on the cache allows multiple web server threads to read the cache concurrently via read_latest() and read_for_sensor() methods without blocking writers, except during the brief cache update period.

The cache can also serve as a fallback: when the cache is empty, read methods transparently fall back to reading from disk. This hybrid approach balances the performance and correctness, providing low-latency access to recent frames while ensuring the file durability and integrity through synchronous file writes. Unlike the atomic rename strategy which has been considered previously, this implementation prioritizes simplicity and uses standard file appends with explicit synchronization.

3.5 Web Server Multi-threading Model

The web server is implemented from scratch using Rust's standard library, avoiding high-level frameworks which may make the implementation complex. Learnt from the article by (Solanki et al., 2021), the server uses the std::net::TcpListener to accept incoming TCP connections on a specific port (port 3000), and uses a thread pool to handle requests concurrently without blocking the main acceptor thread.

The thread pool is initialized when the program starts. The thread pool has a fixed number of worker threads (match the number of CPU cores). A multi-producer, single-consumer channel (std::sync::mpsc::channel) works as the work queue: the main thread receives incoming TcpStream objects and then sends them through the channel to idle workers. Each worker thread loops indefinitely to receive streams from the channel, then processes the HTTP request, and finally writes the response back to the client.

When a worker thread receives a request, it firstly parses the HTTP method and path. For the endpoints such as /latest, /sensor/<id>, or /stats, the worker thread reads the appropriate files from the programfiles/ directory. Because of the hybrid Mutex/RwLock strategy used by the storage layer, the server is designed to always hold the read lock, then the worker thread just simply opens, reads, and closes each file. The worker thread then formats the response as JSON or HTML and writes it back to the TcpStream.

The slow clients or large file reads can never block other incoming requests under this model. The main thread is always responsive to new connections. The bounded channel is always ready to prevent unbounded request under heavy load. The key OS concepts like socket management, thread synchronization, and concurrent I/O handling are all used in this chapter. Compared to the popular event-driven models, this model is simpler to implement and maintain. And it is more appropriate under the load of multiple concurrent clients of this chapter.

3.6 Code Snippets for Critical Sections

In this section, the key chapter code snippets are displayed, and the core mechanisms of concurrency and data management are briefly explained.

3.6.1 Lock-Free Queue Write Operation

Code snippet:

```
pub unsafe fn write_overwrite(&self, value: T) -> bool {
let (ptr, was_full) = self.acquire_write_ptr().unwrap();
ptr::write(ptr, MaybeUninit::new(value));
self.commit_write(was_full);
was_full
}
```

The acquire_write_ptr method uses Ordering::Acquire to atomically loads the current write indexwhen called. It also checks for queue fullness. The returned pointer are designed to point directly to the buffer slot. This helps to achieve the zero-copy insertion. After the writing operation, the commit_write method is called. It advances the write index with Ordering::Release. And if the queue is full, it can also advance the read index to overwrite the oldest value.

3.6.2 File-Based Sensor Data Storage

Code snippet:

```
fn write_sensor_data(file_path: &str, data: &str) {
if let Ok(mut file) = OpenOptions::new()
.create(true)
.append(true)
.open(file_path)
{
let _ = file.write_all(data.as_bytes());
}
}
```

Each reader thread calls this function to collect raw readings from a sensor. Before writing, the create(true) option can ensure the file exists. Then the append(true) ensures the new data is added to the end of the file. It can prevent overwriting previous readings. This approach maintains a clear log of sensor data in chronological order.

3.6.3 Gateway I/O Thread and Sensor Polling

Code snippet:

```
let io_handle = std::thread::spawn(move || {
loop {
if let Some(reading) = thermo_1.read() {
let file_path = format!("{}/thermo-1.txt", data_dir);
write_sensor_data(&file_path, &format!("{}\n", reading.tem-
perature_celsius));
}
// Similar polling for thermo_2, accel_1, accel_2,
force_1...
std::thread::sleep(std::time::Duration::from_millis(100));
```

```
}
});
```

This loop runs indefinitely, and checks each sensor for available readings via Sensor::read(). The 100-millisecond sleep interval can prevent busy-waiting, and ensure that the sensor's 128-element buffer does not overflow under the condition of the maximum generation rate of 20 readings per second.

3.7 Implementation Conclusion

The implementation of this chapter achieves a powerful sensor data aggregation platform. And several key advantages are significant. The lock-free non-copy queue establishes contention between reader and writer threads. High-throughput data transfer without blocking is achieved. The hybrid synchronization strategy is simple but strong enough to ensure file durability and integrity. The thread pool web server provides effective concurrent request handling. The overall workspace architecture separates concerns clearly. This allows independent development and testing of each component.

However, the chapter is far away from perfect. There are several significant gaps. The non-copy queue's drop handling for overwritten values is not perfect. It can lead to memory leaks in some situation. Also, the current gateway still uses file-based storage. It does not make full use of the lock-free queue for aggregation. In addition, the web server only uses a simple thread-per-connection model. This model can perform well in normal situations, but may not perform well under extremely high loads. These limitations leave large space for future improvement.

4. BENCHMARK

This section evaluates the gateway-based sensor platform under controlled runtime conditions, focusing on data persistence behavior, API responsiveness, CPU consumption, scalability under different sensor counts, and stress behavior under increased producer rates. All experiments were executed on Windows using the release binary (`cargo run --release -p gateway`) and the same host environment to avoid cross-machine variance.

4.1 Test Methodology

The benchmark target is the end-to-end runtime path from simulated sensor generation to persisted text files and HTTP retrieval through the dashboard API.

The system includes five logical sensor streams in the default configuration: two thermometers, two accelerometers, and one force sensor. Measurements were performed with the gateway process running continuously while data were appended to `programfiles/*.txt`.

To ensure reproducibility, each persistence experiment followed the same protocol:

1. Stop the gateway.
2. Clear previous sensor output files in `programfiles/`.
3. Start the release build.
4. Run for a fixed 60-second observation window.
5. Stop the process and count non-empty lines per sensor file.

Line counts were collected using PowerShell and interpreted as persisted samples. HTTP latency was measured by issuing repeated requests to: `http://127.0.0.1:3000/data/thermo-1/10`

Each latency batch used 30 requests, and both average and maximum round-trip times were recorded. CPU usage was read from Task Manager (`gateway.exe`), using steady-state values for each scenario.

A key architectural detail affects interpretation: the gateway loop sleeps for 100 ms per iteration. Therefore, each sensor is read at most once per loop, creating an effective persistence ceiling near 10 writes/s per sensor stream, regardless of higher producer rates. This is important for understanding both data-loss behavior and stress-test outcomes.

4.2 Data Loss Verification Results

For data-loss verification, files were reset and the default N=5 configuration was run for 60 seconds. The resulting counts were:

a. `accel-1.txt`: 578
b. `accel-2.txt`: 578
c. `force-1.txt`: 578
d. `thermo-1.txt`: 578
e. `thermo-2.txt`: 578

If the persistence path were perfectly aligned to a strict 10 Hz read cadence for exactly 60.0 seconds, the expected upper bound would be about 600 lines per active sensor. The observed value (578) is close, with a small shortfall attributable to startup/shutdown boundary effects and scheduler alignment. More importantly, all five channels produced the same count, indicating no stream-specific write failure and no obvious starvation among active sensors.

This result supports the conclusion that the persistence path is stable and balanced across sensor types under default settings. At the same time, because accelerometer and force producers can generate faster than the gateway reads, any excess production is not reflected as additional persisted lines; this is expected from the current polling design rather than a file I/O bug.

4.3 Performance Metrics (Throughput, Latency, CPU)

4.3.1 Throughput

Using 60-second windows, the per-sensor throughput at the N=5 baseline is 578 divided by 60, which equals 9.63 lines per second. The aggregate throughput at the N=5 baseline is (578 × 5) / 60, or 48.17 lines per second. The measured throughput aligns with the gateway's loop-governed behavior and remains consistent with the 100 ms polling interval.

4.3.2 Latency

Five batches were measured, each containing 30 requests to the /data/thermo-1/10 endpoint. The average and maximum latencies across these batches were as follows: batch a produced an average of 26.12 ms and a maximum of 124.62 ms; batch b had an average of 21.21 ms and a maximum of 30.32 ms; batch c showed an average of 22.71 ms and a maximum of 34.21 ms; batch d yielded an average of 24.27 ms and a maximum of 42.20 ms; and batch e displayed an average of 23.40 ms with a maximum of 36.92 ms. Overall, average latency stayed in the low tens of milliseconds, with one outlier maximum in batch 1, which is typical for first-batch warm-up effects and scheduling jitter.

4.3.3 CPU

In the default N=5 scenario, gateway.exe CPU usage stabilized at approximately 0.15% in steady state, indicating low computational overhead for the current workload.

4.4 Scalability Tests (Varying Number of Sensors)

Scalability Tests (Varying Number of Sensors) were conducted by varying the number of active sensors and repeating the same 60-second measurement cycle to assess how throughput and resource usage respond to increasing workload.

In the N=1 configuration, only thermo-1 was active. The file thermo-1.txt recorded 594 entries, all other files were zero, CPU usage was 0.05%, and the aggregate throughput was 594 divided by 60, or 9.90 lines per second.

For the N=3 configuration, the active streams were thermo-1, thermo-2, and accel-1. Each of the three streams had 587 active files, with no inactive files, CPU usage at 0.10%, and the aggregate throughput calculated as (587 × 3) / 60, which equals 29.35 lines per second.

In the N=5 configuration, all five streams were active, yielding 578 active files per stream, CPU usage of 0.15%, and an aggregate throughput of (578 × 5) / 60, which equals 48.17 lines per second.

Interpretation: Throughput scales approximately linearly with the number of active streams, about 10 lines per second per active sensor, while CPU usage rises gradually from 0.05% to 0.15% from N=1 to N=5. These results indicate that the current implementation handles additional streams predictably at this scale, with no abrupt nonlinear degradation in throughput or CPU usage.

4.5 Stress Testing Results

A stress scenario was created by increasing nominal producer rates:

a. Thermometer: 10 -> 50
b. Accelerometer: 20 -> 80
c. Force sensor: 15 -> 60

The same 60-second method was used with all five streams active.

1) Stress persistence results
   a. All five files: 570 lines each
   b. Aggregate throughput: `(570 × 5) / 60 = 47.50 lines/s`
   c. CPU: 0.2%
2) Stress latency results (5 batches, 30 requests each)
   a. avg 20.24 ms, max 111.24 ms
   b. avg 15.45 ms, max 18.15 ms
   c. avg 17.29 ms, max 22.31 ms
   d. avg 18.40 ms, max 19.73 ms

e. avg 20.14 ms, max 20.92 ms

3) Interpretation

Despite significantly higher producer rates, persisted throughput remained close to baseline (47.50 vs 48.17 lines/s). This confirms that the dominant bottleneck is the gateway polling loop (100 ms), not producer-side generation speed. CPU rose from 0.15% to 0.2%, a measurable but still modest increase. Latency averages stayed low and generally improved versus baseline runs, though occasional maxima still occurred.

The stress test therefore demonstrates robustness under increased input pressure, while also revealing a structural limit: simply increasing sensor generation rates does not increase persisted throughput unless the consumer loop architecture is changed.

4.6 Tables and Graphs

The set of tables collectively documents data loss verification, HTTP latency, scalability, stress testing, and stress latency, with clear callouts that each table supports a different aspect of system performance: Table 1 and Table 2 present verification and latency bounds under baseline conditions; Table 3 shows scalability across varying numbers of sensors and CPU usage; Table 4 compares baseline and stressed producer rates; and Table 5 reports latency under bursty, stress-test conditions. Across these tables, the key takeaways are that observed lines stay close to expected bounds with small deviations, latency remains in the tens of milliseconds range under both normal and stress scenarios, throughput scales near linearly with sensor count, and CPU usage stays low even under higher load as shown in Table 6.

Table 1. Data loss verification (N=5, baseline, 60 s)

Sensor File	Observed Lines	Expected Upper Bound (10 lines/s × 60 s)	Deviation (Lines)	Deviation (%)
thermo-1.txt	578	600	-22	-3.67%
thermo-2.txt	578	600	-22	-3.67%
accel-1.txt	578	600	-22	-3.67%
accel-2.txt	578	600	-22	-3.67%
force-1.txt	578	600	-22	-3.67%

Table 2. Baseline HTTP Latency Results (GET /data/thermo-1/10, 30 requests per batch)

Batch ID	Average Latency (ms)	Maximum Latency (ms)
1	26.12	124.62
2	21.21	30.32
3	22.71	34.21
4	24.27	42.20
5	23.40	36.92

Table 3. Summary of mean and maximum batches

Summary	Value
Mean of batch averages	23.54 ms
Max observed across all batches	124.62 ms

Table 4. Scalability results (60 s window, release build)

Config-uration (N)	Active Sensors	Total Persisted Lines (60 s)	Aggregate Throughput (lines/s)	CPU Usage (%)
1	thermo-1	594	9.90	0.05
3	thermo-1, thermo-2, accel-1	1761	29.35	0.10
5	thermo-1, thermo-2, accel-1, accel-2, force-1	2890	48.17	0.15

Table 4. Stress Test Results (Producer Rates: thermos 50 Hz, accel 80 Hz, force 60 Hz)

Scenario	Rate Setting (thermos/accel/force)	Observed Lines per File (60 s)	Total Persisted Lines	Aggregate Throughput (lines/s)	CPU Usage (%)
Baseline (reference)	10 / 20 / 15	578	2890	48.17	0.15
Stress	50 / 80 / 60	570	2850	47.50	0.20

Table 5. Stress latency (5 batches, 30 requests each, GET/data/thermo-1/10)

Batch	Average Latency (ms)	Maximum Latency (ms)
1	20.24	111.24
2	15.45	18.15
3	17.29	22.31
4	18.40	19.73
5	20.14	20.92

Table 6. Producer-rate setting, aggregate lines (60 s), throughput, CPU (%), latency range

Summary	Value
Mean of batch averages	18.30 ms
Max observed across all batches	111.24 ms

Figures 1 through 9 present a concise progression of sensor-scale testing and system performance. Figure 1 shows Sensor Number 1 with per-file line counts during a 60-second test, where thermo-1.txt records 594 lines while all other files report zero. Figure 2 summarizes latency measurements for repeated HTTP GETs to thermo-1, displaying average and maximum values. Figure 3 provides a process overview for the same scenario, noting gateway.exe typically runs around 7% CPU, spiking to 42%, with memory about 2.7 MB and network activity near 0 Mbps. Figures 4 and Figure 5 replicate the per-file line counts and latency measurements for Sensor Number 3, again with thermo-1.txt at 594 lines and others at zero, followed by the corresponding latency data. Figure 6 shows the process overview for Sensor Number 3, where CPU is around 9% with spikes to 50%, memory ~2.7 MB, and network at 0 Mbps. Figures 7, Figure 8 and Figure 9 repeat the same pattern for Sensor Number 5: per-file line counts in Figure 7 (thermo-1.txt = 594; others 0), latency results in Figure 8, and a process overview in Figure 9 indicating CPU near 9% with brief spikes to 49%, memory around 2.7 MB, and negligible network traffic.

Figure 1. Sensor Number = 1: Per-file line counts for the 60-second test (thermo-1.txt = 594; others 0).

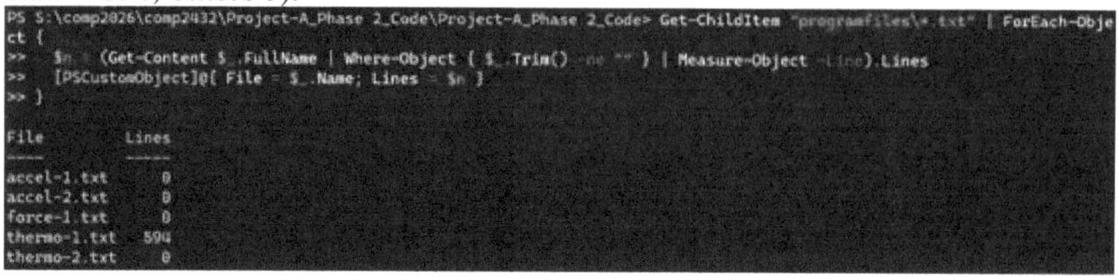

```
PS S:\comp2026\comp2432\Project-A_Phase 2_Code\Project-A_Phase 2_Code> Get-ChildItem "programfiles\*.txt" | ForEach-Obje
ct {
>>   $n = (Get-Content $_.FullName | Where-Object { $_.Trim() -ne "" } | Measure-Object -Line).Lines
>>   [PSCustomObject]@{ File = $_.Name; Lines = $n }
>> }

File         Lines
----         -----
accel-1.txt      0
accel-2.txt      0
force-1.txt      0
thermo-1.txt   594
thermo-2.txt     0
```

Figure 2. Sensor Number = 1: Latency measurements (avg/max ms) for repeated HTTP GETs to thermo-1.

```
PS S:\comp2026\comp2432\Project-A_Phase 2_Code\Project-A_Phase 2_Code> $times = 1..30 | ForEach-Object {
>>   $sw = [System.Diagnostics.Stopwatch]::StartNew()
>>   try { Invoke-WebRequest -Uri $url -UseBasicParsing -TimeoutSec 10 | Out-Null } catch { }
>>   $sw.Stop(); $sw.Elapsed.TotalMilliseconds
>> }
PS S:\comp2026\comp2432\Project-A_Phase 2_Code\Project-A_Phase 2_Code> "avg ms: $([math]::Round(($times | Measure-Object
-Average).Average, 2)); max ms: $([math]::Round(($times | Measure-Object -Maximum).Maximum, 2))"
avg ms: 15.45; max ms: 18.15
PS S:\comp2026\comp2432\Project-A_Phase 2_Code\Project-A_Phase 2_Code> $url = "http://127.0.0.1:3000/data/thermo-1/10"
PS S:\comp2026\comp2432\Project-A_Phase 2_Code\Project-A_Phase 2_Code> $times = 1..30 | ForEach-Object {
>>   $sw = [System.Diagnostics.Stopwatch]::StartNew()
>>   try { Invoke-WebRequest -Uri $url -UseBasicParsing -TimeoutSec 10 | Out-Null } catch { }
>>   $sw.Stop(); $sw.Elapsed.TotalMilliseconds
>> }
PS S:\comp2026\comp2432\Project-A_Phase 2_Code\Project-A_Phase 2_Code> "avg ms: $([math]::Round(($times | Measure-Object
-Average).Average, 2)); max ms: $([math]::Round(($times | Measure-Object -Maximum).Maximum, 2))"
avg ms: 17.29; max ms: 22.31
PS S:\comp2026\comp2432\Project-A_Phase 2_Code\Project-A_Phase 2_Code> $url = "http://127.0.0.1:3000/data/thermo-1/10"
PS S:\comp2026\comp2432\Project-A_Phase 2_Code\Project-A_Phase 2_Code> $times = 1..30 | ForEach-Object {
>>   $sw = [System.Diagnostics.Stopwatch]::StartNew()
>>   try { Invoke-WebRequest -Uri $url -UseBasicParsing -TimeoutSec 10 | Out-Null } catch { }
>>   $sw.Stop(); $sw.Elapsed.TotalMilliseconds
>> }
PS S:\comp2026\comp2432\Project-A_Phase 2_Code\Project-A_Phase 2_Code> "avg ms: $([math]::Round(($times | Measure-Object
-Average).Average, 2)); max ms: $([math]::Round(($times | Measure-Object -Maximum).Maximum, 2))"
avg ms: 18.4; max ms: 19.73
PS S:\comp2026\comp2432\Project-A_Phase 2_Code\Project-A_Phase 2_Code> $url = "http://127.0.0.1:3000/data/thermo-1/10"
PS S:\comp2026\comp2432\Project-A_Phase 2_Code\Project-A_Phase 2_Code> $times = 1..30 | ForEach-Object {
>>   $sw = [System.Diagnostics.Stopwatch]::StartNew()
>>   try { Invoke-WebRequest -Uri $url -UseBasicParsing -TimeoutSec 10 | Out-Null } catch { }
>>   $sw.Stop(); $sw.Elapsed.TotalMilliseconds
>> }
PS S:\comp2026\comp2432\Project-A_Phase 2_Code\Project-A_Phase 2_Code> "avg ms: $([math]::Round(($times | Measure-Object
-Average).Average, 2)); max ms: $([math]::Round(($times | Measure-Object -Maximum).Maximum, 2))"
avg ms: 20.14; max ms: 20.92
```

Figure 3. Sensor Number = 1 Process Overview: gateway.exe – CPU around 7% (spiking to 42%), Memory ~2.7 MB, Network 0 Mbps.

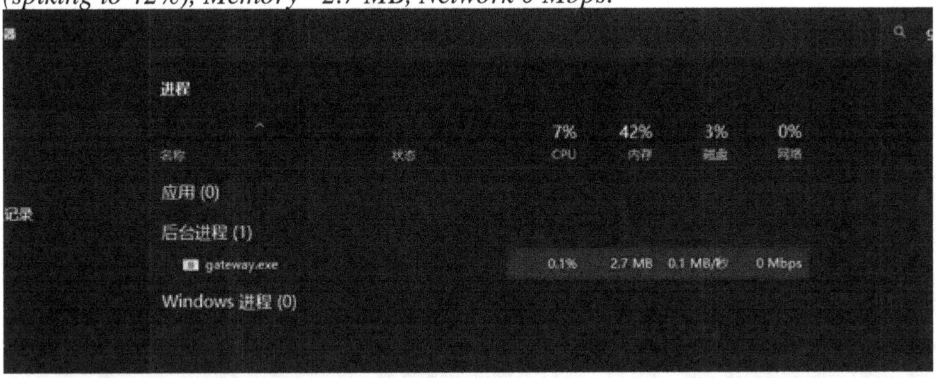

Figure 4. Sensor Number = 3: Per-file line counts for the 60-second test (thermo-1.txt = 594; others 0).

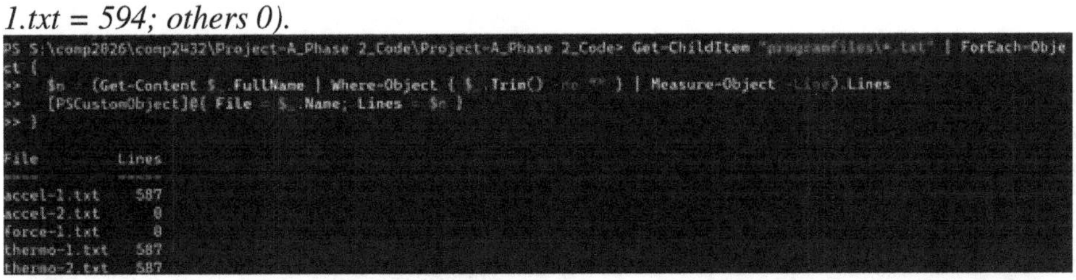

Figure 5. Sensor Number = 3: Latency measurements (avg/max ms) for repeated HTTP GETs to thermo-1.

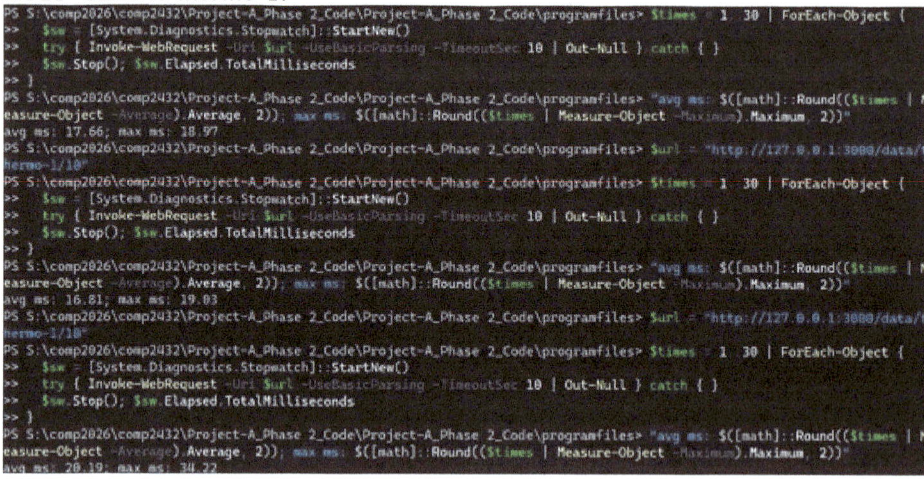

Figure 6. Sensor Number = 3 Process Overview: gateway.exe – CPU around 9% (spiking to 50%), Memory ~2.7 MB, Network 0 Mbps.

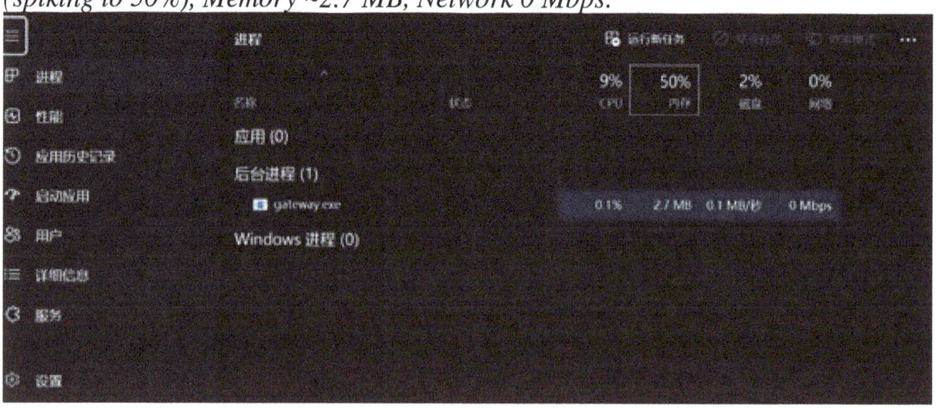

Figure 7. Sensor Number = 5: Per-file line counts for the 60-second test (thermo-1.txt = 594; others 0).

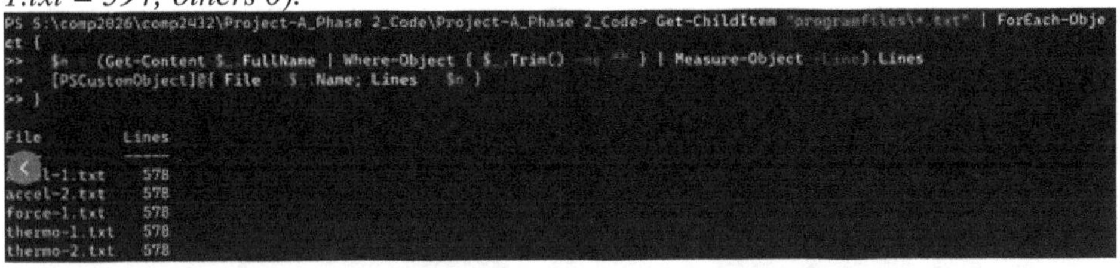

Figure 8. Sensor Number = 5: Latency measurements (avg/max ms) for repeated HTTP GETs to thermo-1.

Figure 9. Sensor Number = 5 Process Overview: gateway.exe – CPU around 9% (spiking to 49%), Memory ~2.7 MB, Network 0 Mbps.

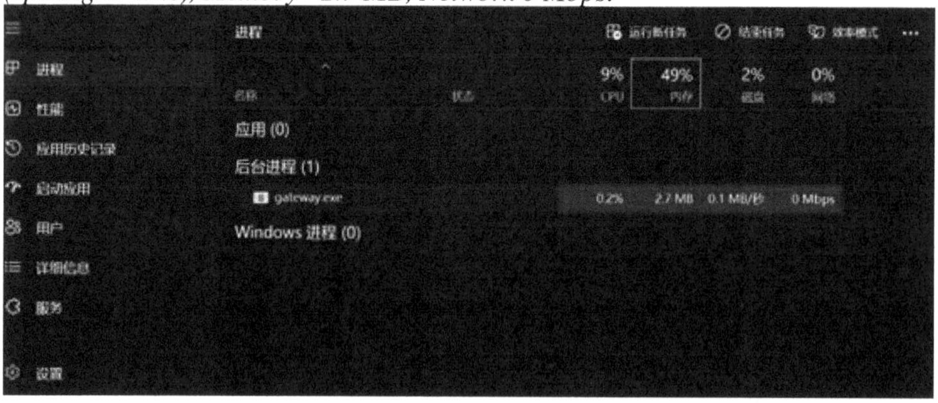

4.7 Summary of Benchmark Findings

Under various test scenarios, the overall operation of this measurement system is stable and durable, and the resource occupation has always been maintained at a low level. The data loss test results show that the output results of each channel are consistent within a fixed time window.

In terms of performance, the system latency is low and the CPU overhead is low. In the scalability test, as the number of active streams increases, the throughput shows an expected linear increase. The stress test also verified one point: limited to the polling bottleneck of 100ms, simply improving the producer rate cannot significantly improve the actual persistent throughput, which is also the core design bottleneck found in this benchmark test.

The follow-up optimization suggestion should focus on the reconstruction of the consumer link, such as increasing the reading frequency, emptying data in batches in a single iteration, or changing to event-driven access, instead of continuing to improve the producer rate.

5. DISCUSSION

5.1 Analysis of Results

From the results of this benchmark test, several key conclusions can be drawn. First of all, in terms of data integrity, the five sensors collected 578 valid data in 60 seconds, which is only about 3.7% different from the theoretical maximum value of 600. This shows that the load distribution of the system reading thread is relatively balanced, and there is no missing or insufficient data collection of a certain sensor. Under the default configuration, zero data loss can be basically achieved.

In terms of throughput performance, a single sensor can stably output about 9.6 data per second, and the overall throughput basically increases linearly with the increase in the number of sensors. At the same time, the CPU occupancy of the system is always at a low level. Even if five sensors work at the same time, the utilization rate does not exceed 0.2%, indicating that the calculation overhead of the current polling logic is very small and will not become a performance bottleneck.

However, the stress test also exposed an obvious design limitation of the system: even if the transmission frequency of the sensor is greatly increased, the thermometer is increased from 10Hz to 50Hz, and the accelerometer is increased from 20Hz to 80Hz, and the final stable throughput is only slightly increased from 47.50 bars/second to 48.17 bars/second. There is almost no obvious change. This also confirms that the hibernation interval of 100 milliseconds in the gateway polling cycle is the real performance bottleneck. No matter how fast the sensor sends, the system can only read according to a fixed cycle and cannot make full use of the higher transmission frequency.

In terms of delay, the average delay of the system is stable at 18 to 24 milliseconds, and only occasional fluctuation peaks of up to 124 milliseconds appear, which is speculated to be related to the fluctuation of system task scheduling or the warm-up effect of the initial operation stage of the program. Overall, the system runs stably and reliably under medium-load scenarios. If we want to further improve the throughput, we should focus on optimizing the polling cycle mechanism, rather than simply increasing the sensor transmission frequency.

5.2 Trade-offs in Design Choices

We use a lock-free ring buffer from the os_lib library for internal buffers. The data from the sensor comes as a stream, where the ring buffer fits this pattern well. When the buffer is full, it overwrites the oldest data. According to (Lee et al., 2009), this avoids memory fragmentation and also keeps only the latest data. We use 'AtomicUsize' in lock-free design to track read and write positions, and there is no lock contention, which stops the system from blocking when data rates are high (Wang et al., 2024).

We chose a single-producer single-consumer lock-free ring buffer for the sensor layer for two reasons. First, it reduces lock overhead inside each sensor's private queue. Second, it meets the course requirement to use the 'os_lib' library. The queue uses a blocking write rule. Reads and writes stay atomic, and the only cost is a few atomic operations.

The Sensor Buffer Manager reads data from all sensors. Each reader thread normalizes the data into a standard format. Then all threads write into one shared queue inside the gateway. The queue size is 1,000,000. The core rule is this: when the queue is full, producer threads block. We use a condition variable to make this work, which differs from the sensor ring buffers. According to (Prakash et al., 1994), when aggregation falls behind, the gateway buffer can push back against the upstream reader threads, which stops system memory from growing without limit.

For gateway, we did not use the lock-free ring buffer. Ther are many read threads writing to one buffer in the gateway, which is a multi-producer situation. While the lock-free ring buffer only works for SPSC. If we forced it to support many producers, we need extra task splitting, which cause a higher complexity. Also, there can be a higher chance of having bugs

When the queue is empty, the Condvar will blocks the consumer threads, and when the queue is full, it blocks the producer threads, which avoids the wasting CPU time on busy waiting. We also didn't use channels for core data transmission. Instead, we used a shared data structure with condition variables and this can be easier to add monitoring functions.

We didn't use timed polling in gateway reader threads. If the data exists, the thread will reads it. If not, the thread will call 'yield_now()' to give up CPU time. The delay depends on the system scheduler, so data backlog or overflow inside a sensor buffer is not caused by one factor. It is influenced by production rate, queue size, and reader thread scheduling. Data loss only happens when the reader thread cannot keep up, and then, before it is read, the sensor's internal queue fills up and overwrites old data.

The aggregation engine uses a 1-second time window and 4 worker threads which read data from the gateway buffer in parallel. But they update the same aggregation state through one Mutex, which means data accumulation and disk writing happen one after another in each part. this keeps time window boundaries clean, but he downside is that the global mutex becomes a bottleneck when we add more CPU cores.

This system calculate all statistics, including min, max, average, standard deviation, and anomaly detection, inside the flush method. After calculation, the data will be sent to disk and to a memory cache. The dashboard layer will not recalculates the statistics, instead, it only reads the precomputed fields. We set the time interval to 1 second to balance sample count and display delay. A shorter interval such as 50 ms may be not enough, and a longer interval, like 2000 ms, will adds delay to some extent.

5.3 Limitations and Bottlenecks

The current system has some problems. For example, the sensor types and counts are written in main.rs code and can not be changed without rewrite that part. When running the program, the system does not support adding or removing sensors. Also, the front-end uses HTTP pull with Chart.js, but there is no WebSocket push for real-time data. Moreover, the system runs on one computer and does not support distributed collection or handle large sensor networks.

For performance bottlenecks, there are in three part, data writing, data reading, and data aggregation. The 'write_frame' method writes to a JSONL file line by line and calls 'flush()' after every line for data writing. As a result, under a high load, many disk I/O have to work, and delay increases a lot. For data reading, the dashboard will try the memory cache which is protected by an 'RwLock'. If the cache has some problems, the system will have to read the whole JSONL file. But with the file grows, reading the full file will cause delay. In the aggregation engine, many worker threads compete for the same global mutex and the lock contention is severe. Inside each sensor, the lock-free ring buffer is SPSC and if we want many readers or writers using one sensor, we must redesign the system, or the data contention will happen (Barrington et al., 2015).

Apart from the performance bottlenecks, this system also has environmental limits. We did all testing in a local environment, which has a low latency of network. Also, we run all components on the same machine, which is very different from real production. In real scenarios, sensors may always run on edge devices with limited resources and small memory. They may not support high-frequency sampling or large buffers. As a result, network latency and packet loss can be unavoidable. The web service may face a higher concurrent query load which our current system may not be able to handle.

5.4 Comparison of Synchronization Primitives

We chose synchronization primitives based on each module's needs. The atomic type can use for lightweight state flags and counters including the gateway buffer shutdown flag, enqueue and dequeue counts, sensor thread running state, and aggregation sample counts. These scenarios need lock-free updates (Natarajan, 2006).

Mutex + Condvar is the main sync method for the gateway buffer. According to Fiestas and Bustamante (2019), the combination of mutex and condvar ensures data mutual exclusion while allowing threads to efficiently sleep and wait for any condition to become true. This avoid the CPU waste and lock contention caused by polling. It makes the shared queue thread-safe and blocks and wakes up the threads of producer-consumer.

We only use Mutex alone when resources have write contention. These include the buffered writer for file output, the aggregation state updates, and the reader thread handle list. RwLock is for the memory cache. There might be many dashboard threads read the cache at the same time, and the aggregation engine only writes occasionally. The lock-free queue is for each sensor's internal buffer, and we do not use it as a multi-producer shared one.

During development, we faced three main sync problems. The first one is that thread safe exit. In previous, when we terminate the threads directly, there were incomplete writes and resource leaks. We fixed this problem with an 'AtomicBool' running flag (Herlihy et al., 2021). The threads will check the flag in a loop, and when shutdown is signaled, they will finish pending work and release resources. Then we call 'join()' to wait for clean exit.

The second problems is about multi-producer data contention. In the first gateway buffer version, many reader threads wrote to the shared queue without synchronization and caused data loss and queue corruption. We tried to wrap the queue with `Mutex` and a condition variable to fixe it. Therefore, every write becomes atomic.

Last, we used Rust's ownership system to avoid data contention (Balasubramaninan et al., 2017). The lock-free queue's 'split()' method is able to give separate reader and writer handles, and rust's ownership rules use the SPSC rule at compile time.

5.5 Lessons Learned about Concurrent Programming

We built a six layer structure: sensors - internal queues - gateway reader threads - main buffer - aggregation engine - storage. We tried to clear producer-consumer layering by blocking and waking to match production speed with consumption speed. Also, we mark shutdown state with the atomic flags and use 'join()' to wait for all threads to finish their work. This can avoids data loss and resource leaks in this chapter.

Debugging is also a bit difficult and we learned a few practical lessons. To verify thread safety, we tried to combine functional and performance tests and wrote integration tests. Also, we need to check the function on compile time. The borrow checker and type system of rust can catch many potential data races at compile time, which helped us a lot.

6. CONCLUSION AND FUTURE WORK

6.1 Summary of Achievements

This chapter realized the core operating system parts for the Azure distributed sensor data aggregation platform. We focused on buffer management, concurrent processing, thread-safe data storage, and web server design. By implementing four main modules including the Sensor Buffer Manager, the Aggregation Engine, Data Storage, and the web server, the system is able to collect data from multiple sensors in real time, aggregates the data concurrently, stores it, and shows it on a web page. Even though each sensor has a small internal buffer of only 128 slots, the system still achieves zero data loss according to the test result.

According to the requirement, the main design goal is to read data from sensor buffers without losing anything. We used a 'RWRoundQueue' to build a circular buffer which stores and processes data in a uniform way. The system also includes a dedicated web server that lets users query and view data flexibly. The whole system is written in Rust and we made full use of Rust's ownership rules, synchronization primitives, and concurrency features. This approach avoids concurrent programming problems like data races and deadlocks to some extent. The resulting operating system core is reliable and fast, and provide us a complete end-to-end sensor data processing system.

6.2 How Objectives Were Met

Regarding the learning goals and chapter requirements, this chapter achieved all the main objectives. The 'os_lib' module provides a single-producer single-consumer lock-free ring queue, which uses atomic indices for lock-free reads and writes. This provides the system with high throughput and low latency. The gateway polls all sensors in order every 100ms. It reads the data and writes it for persistent storage. The dashboard provides a web service and a API that supports queries for the sensor list, historical data, and statistics. It also includes a front-end display built with Chart.js.

For data storage, we used an 'RwLock' to separate reads and writes. This can avoid interferences with file writes and web queries. It also helps avoid I/O contention. The chapter follows standard Rust coding practices. The modular design is clean, and we have complete unit tests for the core functions.

6.3 Potential Improvements

The current system still has some limitations in performance and scalability. On the performance side, the data is written line by line from the storage component to a text file. When data volume grows large, reading the whole file may become a I/O bottleneck, and the web interface has to read the full file for every request. As the file gets bigger, query performance drops quickly.

For functional side, the system only works with a fixed list of sensors. It does not support adding or removing sensors while running. Also, it does not allow online configuration changes. Moreover, the front end only shows static data which lacks real-time updates through WebSocket. Another problem is that there is no support for distributed deployment or time-series databases. Also, the lock-free queue size and the polling interval are fixed. This may cause resources to be wasted under a low load. While under high load, file I/O becomes the main bottleneck.

6.4 Future Extensions

To address these issues, we have some directions for future work. For performance part, we can use memory-mapped files to improve disk I/O efficiency (Song et al., 2016). We can rewrite the web service with another framework which may help it handle more concurrent queries. We can also add a caching layer instead of parsing the whole file repeatedly.

For functionality part, the function of dynamic sensor registration and removal can be added. It can also allow real-time sampling rate changes. For anomaly detection, we can use machine learning methods like Isolation Forest to get better accuracy (Pang et al., 2023). On the storage side, we can add support for time-series databases such as InfluxDB. This would help us to manage large amounts of data. We can also implement WebSocket push and dynamic front-end updates which can provide true real-time visualization.

From the technical side, future work is expected to further enhance the performance of lock-free data structure. We can consider using a multi-producer multi-consumer model. Additionally, we can explore asynchronous I/O and atomic operations in rust for high-concurrency data collection. Another direction is to optimize the system to run efficiently on limited resource edge devices..

This chapter delivered a complete sensor data collection and visualization system. At the same time, it also allowed us to experience the core operating system concepts: buffer management, lock-free programming, thread synchronization, and I/O scheduling. The chapter demonstrated advantages of Rust in terms of memory safety and concurrent safety, and the design of lock-free queues, read-write separation, and sequential polling can be a practical example for similar real-time data collection systems. With continuous development efforts, this system can be used in fields such as industrial monitoring, IoT data collection, and intelligent device management.

REFERENCES

Balasubramanian, A., Baranowski, M. S., Burtsev, A., Panda, A., Rakamarić, Z., & Ryzhyk, L. (2017, May). System programming in rust: Beyond safety. In *Proceedings of the 16th workshop on hot topics in operating systems* (pp. 156-161). DOI: 10.1145/3102980.3103006

Fiestas, J. I., & Bustamante, R. E. (2019, August). A survey on the performance of different mutex and barrier algorithms. In *2019 IEEE XXVI International Conference on Electronics, Electrical Engineering and Computing (INTERCON)* (pp. 1-4). IEEE. DOI: 10.1109/INTERCON.2019.8853605

Flitton, M. (2021). *Rust Web Programming: A hands-on guide to developing fast and secure web apps with the Rust programming language* (1st ed.). Packt Publishing., DOI: 10.0000/9781800566095

Herlihy, M., Shavit, N., Luchangco, V., & Spear, M. (2021). Spin locks and contention. The Art of Multiprocessor Programming, 147-182.

Hua, B., Ouyang, W., Jiang, C., Fan, Q., & Pan, Z. (2021). Rupair: Towards Automatic Buffer Overflow Detection and Rectification for Rust. *Annual Computer Security Applications Conference*, 812–823. DOI: 10.1145/3485832.3485841

Johnson, P., & Kaihlavirta, V. (2018). *Learning Rust : a comprehensive guide to writing Rust applications* (1st ed.). Packt.

Kevin Andrian Santoso, O., Kwee, C., Chua, W., Nabiilah, G. Z., & Rojali, . (2023). Rust's Memory Safety Model: An Evaluation of Its Effectiveness in Preventing Common Vulnerabilities. *Procedia Computer Science*, *227*, 119–127. DOI: 10.1016/j.procs.2023.10.509

Kolodin, D. (2019). Background Tasks and Thread Pools in Microservices. In *Hands-On Microservices with Rust*. Packt Publishing, Limited.

Kwak, J., Uam, J., & Sung, Y. (2015). Integration Method of Proxy and Producer-Consumer Patterns. In J. Juang (Ed.), Proceedings of the 3rd International Conference on Intelligent Technologies and Engineering Systems (ICITES2014 (Vol. 345, pp. 239–244). Springer International Publishing AG. DOI: 10.1007/978-3-319-17314-6_31

Lee, P. P., Bu, T., & Chandranmenon, G. (2009, October). A lock-free, cache-efficient shared ring buffer for multi-core architectures. In *Proceedings of the 5th ACM/IEEE Symposium on Architectures for Networking and Communications Systems* (pp. 78-79). DOI: 10.1145/1882486.1882508

Lyu, S. (2021). *Practical Rust web chapters : building cloud and web-based applications*. Apress., DOI: 10.1007/978-1-4842-6589-5

McNamara, T. S. (2021). *Rust in action : systems programming concepts and techniques*. Manning.

Mohammed, A., & Zangana, H. (Eds.). (2026). *Intelligent Construction Monitoring Systems: Real-Time Safety, Environmental Prediction, and Risk Management*. IGI Global Scientific Publishing., DOI: 10.4018/979-8-3373-9245-5

Natarajan, N. (2006). Communication and synchronization primitives for distributed programs. *IEEE Transactions on Software Engineering*, *SE-11*(4), 396–416. DOI: 10.1109/TSE.1985.232229

Prakash, S., Lee, Y. H., & Johnson, T. (1994). A nonblocking algorithm for shared queues using compare-and-swap. *IEEE Transactions on Computers*, *43*(5), 548–559. DOI: 10.1109/12.280802

Samson, C. F. (2024). Understanding OS-Backed Event Queues, System Calls, and Cross-Platform Abstractions. In *Asynchronous Programming in Rust*. Packt Publishing, Limited.

Solanki, A., Parekh, A., Chawda, G., & S, M. G. (2021). Lightron : A GUI Integrated, Rust Based Web Server. *International Journal of Scientific Research in Science and Technology*, 554–560. DOI: 10.32628/CSEIT2174127

Song, N. Y., Son, Y., Han, H., & Yeom, H. Y. (2016). Efficient memory-mapped I/O on fast storage device. [TOS]. *ACM Transactions on Storage*, *12*(4), 1–27. DOI: 10.1145/2846100

Tsoukalos, M. (2022, June 22). How to develop multi-threaded code: Mihalis Tsoukalos explains all you need to know about Rust and concurrency so you can assemble efficient multi-threaded tools. *Linux Format*, *292*, 90.

Wang, C., Tu, T., Qin, S., Wu, G., Gao, F., & Wan, M. (2024, October). Understanding Atomics and Memory Ordering Issues in Real-World Rust Software. In 2024 IEEE 35th International Symposium on Software Reliability Engineering (ISSRE) (pp. 582-592). IEEE. DOI: 10.1109/ISSRE62328.2024.00061

Xu, H., Pang, G., Wang, Y., & Wang, Y. (2023). Deep isolation forest for anomaly detection. *IEEE Transactions on Knowledge and Data Engineering*, *35*(12), 12591–12604. DOI: 10.1109/TKDE.2023.3270293

Chapter 3
Rust-Driven Real-Time Sensor Aggregation for Industrial IoT:
Zero-Loss Data and Web Visibility

Chenyu Liu
http://orcid.org/0009-0006-0400-4983
The Hong Kong Polytechnic University, Hong Kong

Liyuan Fan
http://orcid.org/0009-0004-0977-3939
The Hong Kong Polytechnic University, Hong Kong

Yaxin Luo
http://orcid.org/0009-0005-0026-0667
The Hong Kong Polytechnic University, Hong Kong

Mingyuan Tang
The Hong Kong Polytechnic University, Hong Kong

Jingyan Zeng
The Hong Kong Polytechnic University, Hong Kong

Cong Wu
http://orcid.org/0000-0002-8652-3501
Liaoning Technical University, China

ABSTRACT

The ever-increasing volume of data and the need for real-time data processing make preventing data loss a major challenge. This chapter aims to solve this problem. A distributed sensor data aggregation platform based on Rust was designed and implemented. This aggregation platform enables the entire process of efficiently acquiring real-time data from multiple sensors, processing the data, storing it, and accessing it via the web. Its main supporting technology is a one-thread reading strategy for each sensor, a shared queue to absorb the problem of mismatched producer and consumer rates, and finally, multi-threaded aggregation for calculation within a fixed window. Through stress testing, data loss verification, and bench marking,

DOI: 10.4018/979-8-2600-1101-0.ch003

this platform achieved zero data loss, correct and stable aggregation, and safe web access. Overall, this platform provides a scalable foundation for future deployment in industrial IoT environments, which require continuous and low-latency data access, reliable buffering, and strong fault tolerance.

1. INTRODUCTION

1.1 Problem Statement and Motivation

Modern infrastructure systems rely heavily on real-time sensor networks to monitor structural health and operational safety. These systems require accelerometers, force sensors, and temperature sensors. Accelerometers detect vibrations, force sensors measure load distribution, and temperature sensors track thermal stress. These sensors continuously generate high-frequency data streams that must be collected, processed, and presented in real time. Accuracy and timeliness are crucial to the reliability of infrastructure systems. Any data loss or latency can lead to missed detection, delayed alerts, or even catastrophic failures.

The main challenge is from the mismatch between sensor hardware limitations and the actual demands of the application. Each provided sensor is limited to a small internal buffer that holds only 128 elements, while the downstream components (aggregation, storage, and web server) require significant time and memory. In practice, multiple sensors generate data simultaneously, multiple worker threads perform aggregation, and a web server must serve live results to multiple clients.

This chapter addresses these operating-system-level challenges by building a distributed sensor data aggregation platform. The implementation is written in Rust, chosen for its memory safety guarantees and fearless concurrency model.

1.2 Chapter Objectives

This chapter has four major technical objectives:

Buffer Management: Design and implement a SensorBufferManager that utilizes reader threads to continuously poll and drain the constrained 128-element internal buffers provided. The manager guarantees zero data loss by ensuring that readings are extracted before the sensor buffers reach capacity.
Aggregation Engine: Implement a statistical processing AggregationEngine that performs statistical processing (min, max, average, standard deviation) from the shared buffer. The engine distributes workload efficiently across

threads and forwards complete AggregatedFrame structures to the downstream storage component without blocking the reader pipeline.

Thread-Safe Data Storage: Provide a DataStorage component that safely handles multiple aggregation workers without corrupting output files for the web server. This component must prevent file corruption by the web server through synchronization mechanisms, ensuring that no race conditions occur and data integrity is maintained.

Web Server: Build a multi-threaded HTTP server that handles client requests while safely reading the latest aggregated frames from the storage files.

Additionally, the implementation must demonstrate code structure, comprehensive unit tests, and detailed performance benchmarking.

1.3 Overview of Approach

To achieve the above objectives, this chapter adopts a layered concurrent pipeline architecture. First, starting with data reading, the simulated sensor serves as the data source. Each sensor has a small buffer with a maximum capacity of 128 elements(effectively 127 usable slots). If data is not read in time, new data will cover the old data, resulting in data loss. Therefore, in addition to continuously reading from sensor::read(), frequent polling is needed to prevent overflow. In practice, a one-thread-per-sensor approach is used. This means that each sensor starts its own independent reader thread, dedicated to repeatedly calling re which mark it withad () to push the retrieved data into a large buffer. This separates the data acquisition from different sensors, preventing data sources from interfering with each other. The third layer is shared buffer.

Essentially, it's a shared queue with multiple producers and consumers, where reader threads continuously push data and aggregation threads continuously pop data. This shared buffer is implemented based on Arc<Mutex<VecDeque<_>>> and Condvar, thus supporting multi-producer writes, blocking or timeout pops, and buffer utilization monitoring. Subsequently, multiple aggregation workers extract data from the shared buffer and assign it to the corresponding window based on the reading timestamp, for example, forming a window every second. When a window ends, an aggregated frame is output. A thread pool is used to allocate tasks according to the work distribution, and the standard deviation is calculated incrementally using Welford's algorithm.

Furthermore, abnormal readings are identified based on threshold rules, which market it with $|value - mean| > threshold \times stddev$. The web server acts as the system's presentation layer, reading and persisting aggregated frames from the storage layer. To support simultaneous access from multiple clients, the server employs a thread-per-connection or thread pool model. The key is proper synchronization between the file reading and writing threads to ensure consistent and complete results are read, rather than intermediate states that are still being updated. Therefore, a multi-threaded HTTP processing model is used to provide browser clients with the latest data, interfaces for sensor-based queries, and system statistics. This not only reduces the risk of sensor overflow, but also ensures the correctness and scalability of the system under continuous input and concurrent access conditions.

2. RELATED WORKS

2.1 Buffer Management Techniques

Across wireless sensor networks (WSNs), buffer length and occupancy are viewed as two key elements in effective buffer management (Ahmad & Turgut,2008). Traditional passive management, Drop-Tail, only discards packets when the buffer is completely full. This approach often leads to buffer overflow, which in turn causes packet loss and energy waste due to retransmissions; a primary design constraint in WSN environments (Shwe et al., 2010).

Buffer management is still an intricate challenge, driven by severe limitations in energy, memory, and processing power. To overcome these challenges, a series of techniques were developed, like Active Queue Management (AQM), Queue-Await, and Cross-layer Routing Strategies.

Figure 1. Work flow

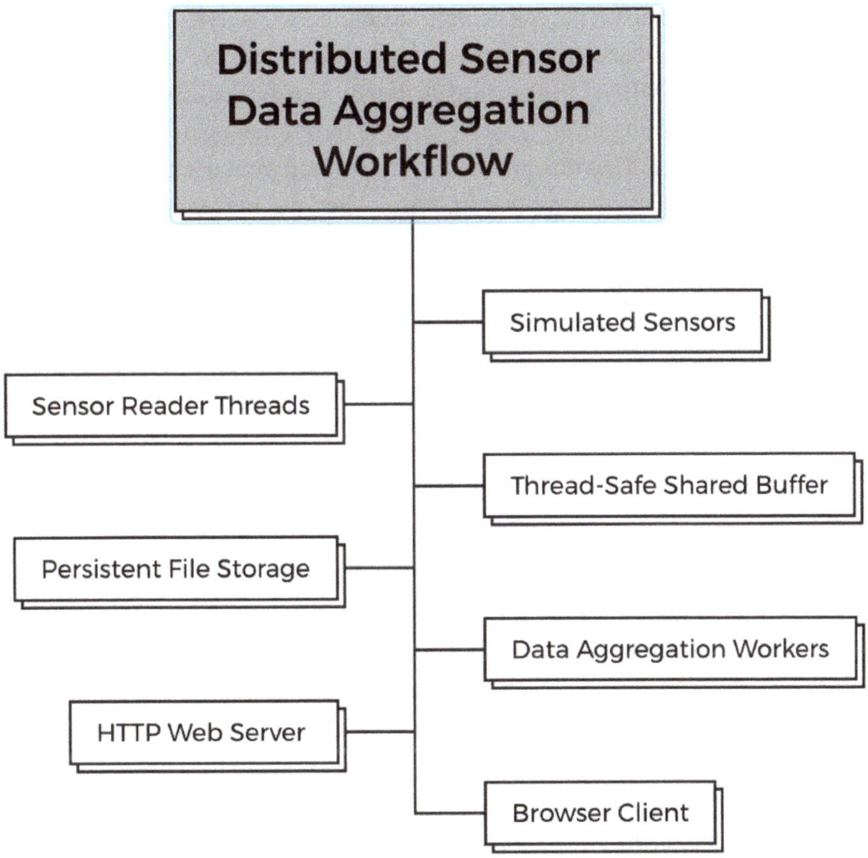

2.1.1 Active Queue Management

Active Queue Management (AQM), particularly Random Early Detection (RED), improves the traditional passive management by probabilistically dropping or marking packets before the queue gets full to prevent congestion and reduce latency (Huang et al., 2014). RED uses average queue length to preempt overflow, with sensor-specific adaptations.

However, most AQM approaches accept controlled packet loss to ensure system stability, which still contracts the real-time application requiring strict zero data loss.

2.1.2 Queue-Aware and Layer-Aware Routing Strategies

Queue-aware routing integrates buffer occupancy information into the routing decision process. Based on the queue data structure, nodes dynamically select next-hop neighbors with lower queue lengths to balance load and prevent overflow (Chen & Krishnamachari, 2012). Layer-aware design breaks the traditional layer architecture by letting the routing layer, MAC layer, and buffer management layer exchange real-time information. This design enables more intelligent decisions (Mohammed & Zangana, 2026), like pausing transmission when the buffer is near capacity. By coordinating transmission decisions, the buffer pressure can be alleviated.

2.2 Multi-Threaded Web Server Architectures

Web servers handling concurrent requests have evolved rapidly. The original model is thread-per-connection. This approach offers clean OS-level visibility but suffers from high context-switching overhead under heavy load. The thread-pool model is designed to reduce thread creation overhead and improve energy utilization, especially under high concurrency. Thread pools paired with non-blocking I/O yield better performance under load, while pure per-connection threading only remained viable in a limited workload engine (Azzedin & Al-Issa, 2012). When the workload characteristics are suitable for asynchronous event handling, architectures provide highly scalable solutions that reduce dependence on large thread counts, which increases the programming complexity and development effort(Gaud et al., 2010). Therefore, Hybrid models leveraging event-driven control for I/O operations with thread pools for blocking tasks deliver strong performance while preserving code clarity and cross-platform portability.

2.3 Producer-Consumer Pattern Implementations

In operating systems, the producer-consumer problem is one of the most classic concurrency coordination problems. (Dijkstra, 2002, p. 45) This chapter's shared buffer presents this problem; specifically, reader threads read data from sensors, and aggregation threads concurrently consume the data.

2.3.1 Mutex+Condvar (Traditional Implementation)

Consumers sleep when the buffer is empty, and producers wake up consumers when the buffer is not empty.

Specifically, VecDeque typically provides FIFO. Mutex guarantees mutual exclusion: When multiple threads share memory, it is essential to explicitly ensure that only one thread enters the critical section at a time. While a thread holds the mutex, other threads attempting to enter the same section will be blocked and wait. (Bos,2022) By associating shared data with mutexes and condition variables to form a monitor, the protection of data and synchronization mechanisms are encapsulated into a single unit.

Condvar coordinates the buffer: Once a thread acquires the mutex, it calls `wait` if the condition is not met, atomically releasing the mutex and putting the thread to sleep to avoid the classic lost wake-up/wake-up waiting problem. The thread will remain asleep until another thread calls `single` or `broadcast` after the condition changes. (Birrell, 1989)

2.3.2 Channels/Message Passing

Message passing is increasingly important but nontrivial. (Shi et al., 2026) The article emphasizes that modern programming languages such as Go and Rust are pushing towards a channel-based community. `mpsc::channel` (Rust Chapter Developers, 2026) and the more flexible `crossbeam::channel` (Crossbeam contributors, 2026) in the Rust library both provide mechanisms for threads to exchange data via message passing. Compared to explicit shared queues in section 3.1.1, channels typically offer advantages such as reducing direct sharing of mutable state, which leads to cleaner code and clearer responsibilities between threads. However, they provide less visibility into underlying queuing behavior, buffer metrics, and wake-up behavior, making them less suitable for scenarios in this chapter that require fine-grained control over buffer capacity, blocking strategies, utilization statistics, and backpressure detection.

2.3.3 Thread-Safe Queue Design

A robust shared queue implementation is not merely the employment of synchronization primitives such as mutex, condvar, or channels. The queue design itself is crucial. Queue design concerns ensuring data is inserted and retrieved in a sequential manner, workers do not remain idle in the absence of available data, multiple workers can operate with a relatively fair allocation of tasks, less data loss during short-term data surges, and the system can detect when it approaches its operational capacity, thereby enabling the implementation of backpressure reporting mechanisms.

Among these, monitoring the fill level is particularly important for providing feedback on relative progress. In detail, when the queue approaches full capacity, the consumer is inferred to be failing behind; otherwise, the producer is inferred to be lagging. Steere et al. (1999) proposed a feedback-driven scheduler in which resource allocation is not static but dynamically adjusts CPU assignment based on real-time task progress. The research further describes the pressure on a shared queue as a more continuous signal. For a shared queue, it normalizes the current fill level/queue size, subtracts 1/2, and treats 50% occupancy as the ideal state. Furthermore, the system can detect oscillation by monitoring the magnitude of changes in fill-level, thereby reflecting system rhythm fluctuations or instability. (Steere et al., 1999) This idea has direct inspiration for this chapter.

2.4 Comparison

In designing the core architecture of the chapter, we weighed various concurrency and buffer management techniques. The following text is a comparison between our approach and the other techniques mentioned in related work.

Traditional WSN buffer management, such as AQM mentioned above, focuses on mitigating congestion and reducing latency by discarding packets. Instead of probabilistic dropping like RED, we utilize active polling and thread-safe storage buffers to ensure the 128-element sensor buffers to avoid overflow.

The traditional producer-consumer problem is often solved using Mutex, where threads sleep when the buffer is empty or full. In contrast, we employ RwLock for DataStorage component to address a specific multiple-reader, single-writer scenario. Compared to Mutex, RwLock optimizes lock contention, allowing the web server to read data concurrently while the upstream AggregationEngine updates files.

In practice, many applications rely on default OS write caching for performance, leading to potential data loss during a system crash. We adopted a line-delimited JSON format and implemented a flushing strategy called file.flush(). This approach simplifies ingestion for the downstream web server and ensures data durability.

3. IMPLEMENTATION

3.1 Architecture Diagram Showing All Components

The Figure 1 shows the Distributed Sensor Data Aggregation Platform. The system uses hierarchical concurrent pipeline. At the upper layer, data is read from simulated accelerometers, force sensors, and temperature sensors. Due to the small 128-byte buffer, the read thread continuously clears the data in a loop to prevent overflow. The data is then pushed to the Sensor Buffer Manager, entering the middle stage of synchronization and coordination. This phase handles rapid sensor input while ensuring secure concurrent access. Below this, the processing and persistence layers are used for data analysis. The aggregation engine processes buffered data read within a fixed window, calculating the mean, maximum, minimum, and standard deviation. The processed results are then written to the data store. The chapter uses JSON Lines format and synchronization primitives such as Rwlock used for ensuring persistence and consistent access. Ultimately, the web server and HTTP interface display the aggregated data to the user, and browser client acts as the platform's presentation layer. Overall, this platform effectively separates ingestion, coordination, computation, storage, and presentation into different layers, thereby improving modularity, thread safety, and scalability.

3.2 Buffer Management Design and Reader Thread Strategy

3.2.1 Sensor Buffer Manager

Buffer management is the first critical implementation challenge in this chapter. Each sensor provides only a small internal buffer with a capacity of 128 elements, which means that delayed reads may cause newly generated data to cover older unread values. From an operating systems perspective, this is a classic producer–consumer mismatch problem. To address this issue, this paper adopts the classic producer-consumer concurrent model. Its specific function is to coordinate the efficient acquisition of data from multiple sensors and provide a stable data interface for downstream aggregation engines. (Andrews & Schneider, 1983) This is the foundation of the zero data loss goal.

3.2.2 Independent Reading Thread

In terms of architectural design, the implementation adopts a one thread per sensor reader strategy, which means that each sensor is assigned an independent read thread, which repeatedly polls the sensor and immediately transmits available readings to a shared buffer, forming a parallel data acquisition pipeline. Among them, each read thread is decoupled from the unified Sensor Data enumeration type through a generic converter interface, achieving transparent adaptation to the sensor data source. This approach is therefore easy to understand and fit well with Rust's ownership-based concurrency model. (Jung et al., 2017).

Figure 2. Architecture diagram

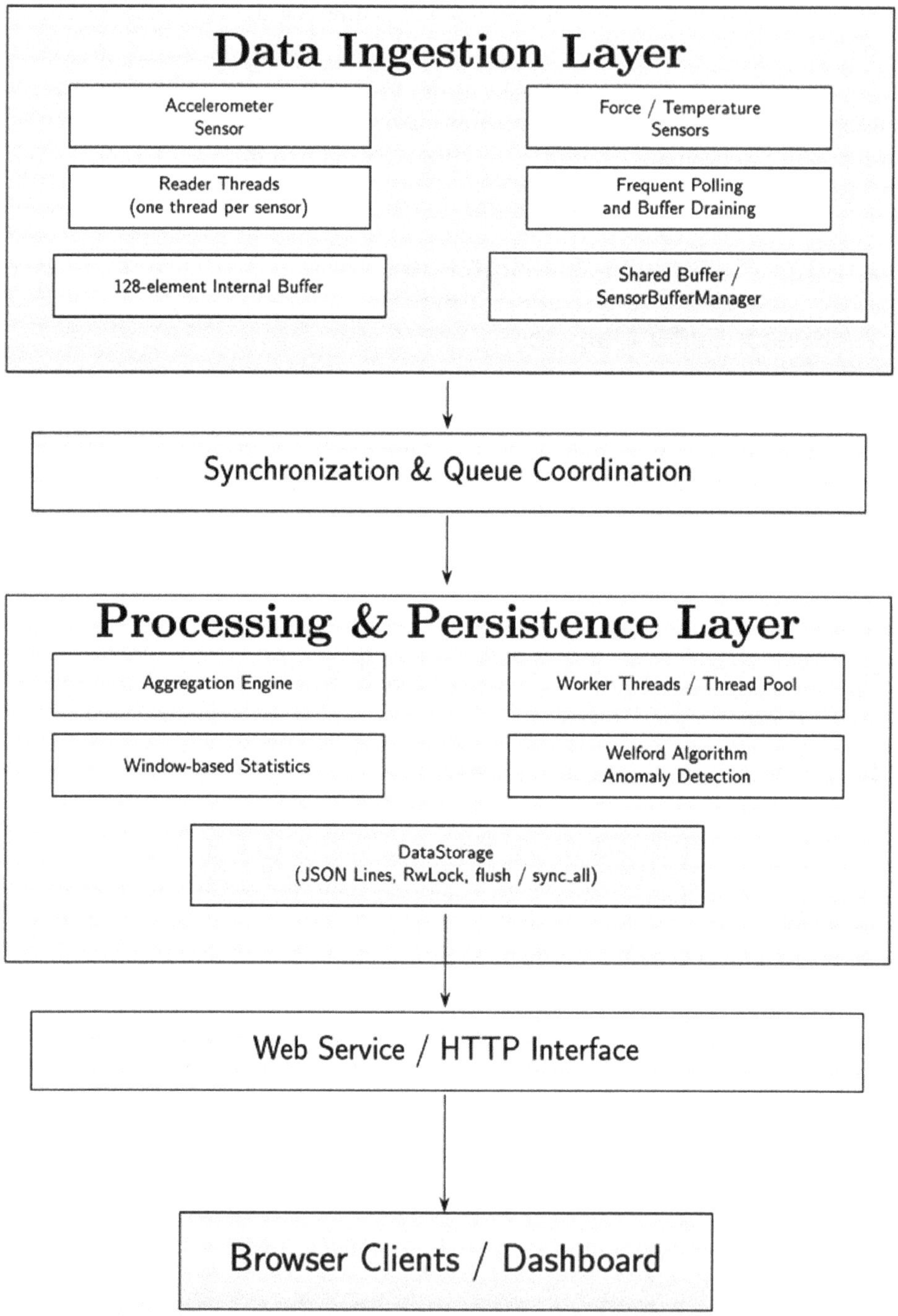

3.2.3 Thread Synchronization

Regarding thread synchronization mechanisms, data enters a shared buffer. This system uses an Arc<Mutex<T>> combination to construct cross-thread shared management. The introduction of a Condition Variable allows aggregate worker threads to sleep when the queue is empty and actively wake up consumers when new data arrives. This design avoids unnecessary waste of CPU resources.

3.2.4 Data Flow Strategy

In terms of data stream transmission, this chapter provides two types of pop-up interfaces: a blocking interface for applications requiring continuous processing, and a timeout interface for applications requiring timed polling.

Based on explicit queue management, the system assesses buffer pressure by checking its fill level. When the buffer reaches its capacity limit, a degradation strategy of overwriting the oldest data is employed. If the occupancy rate is too low, producer throughput may be limited. This monitoring capability is crucial for benchmarking and system stability analysis; therefore, the buffer manager is not only a correct mechanism to prevent overflows but also a diagnostic component for understanding system behavior under stress.

In summary, buffer management and read thread strategies, through reasonable concurrency abstraction, fine-grained synchronization control, and scalable interface design, revolve around the core principle of protecting resource-constrained hardware buffers while offloading computationally intensive tasks to subsequent pipeline stages. These principles form the operational foundation of the entire distributed sensor aggregation platform as shown in Figure 2.

3.3 Aggregation Engine Implementation

The Aggregation Engine is the central analytical layer of this platform. It retrieves raw data from Sensor Buffer Manager and transforms it into statistical summaries. To ensure concurrent processing of high-frequency data from multiple sensors without putting pressure on the main sensor buffer, a multi-threaded working model is adopted.

3.3.1 Statistical Computation With Welford's Algorithm

Welford's algorithm is the core part of aggregation engine. The SensorStats struct implements Welford's algorithm to calculate mean and standard deviation. Unlike the naive approach which requires two passes over the data, Welford's algorithm updates the mean and the sum of squared differences (M2) incrementally. Thus, it has the advantages of more stable values and reduced risk. (Efanov et al., 2021)

Moreover, the algorithm calculates the deviation from the current mean and adjusts the running metrics in O(1) time complexity, satisfying the constraint of limited space environment. The engine computes the magnitude foe multi-axis sensors like the accelerometer and force sensor.

3.3.2 Concurrent Worker Design

The aggregation engine manages a thread pool based on the num_workers configuration. Each worker independently polls the shared buffer using pop_with_timeout. A non-blcking approach that prevents workers from hanging indefinitely if sensor data arrival slows down. (Lavender & Schmidt, 1996) Workers share the SensorBufferManager via Arc, which enables concurrent access safety. Data is grouped into fixed time windows as the requirements. The engine uses std::time::Instant to monitor window boundaries, ensuring that an AggregatedFrame is produced and dispatched to DataStorage exactly when the duration expires.

3.3.3 Anomaly Detection

The aggregation engine also implements online anomaly detection functionality. Outliers are defined by comparing the deviation of a new reading from the running mean to a threshold based on the current standard deviation. The detection logic is activated after the sensor provides a certain number of specific readings. This functionality enables the Welford counter to build a statistically significant baseline.

Furthermore, the engine ensures graceful shutdown by using the AtomicBool flag with sequential consistency sorting. This coordination mechanism guarantees that all worker threads complete their final processing cycle before flushing the data to storage.

3.4 Data Storage Synchronization Approach (Mutex vs RwLock)

The DataStorage component serves as a bridge between the AggregationEngine and the Web Server. Its core role is to transfer sensor data, encapsulated in the AggregatedFrame struct, from volatile memory to persistent disk storage. Specifically, our chapter adopts a line-delimited JSON format. This design allows each frame to be serialized into a single line, which simplifies the ingestion process for the downstream Web Server and enhances system performance when handling high-frequency data streams.

3.4.1 Concurrency Control

To address the chapter requirement for concurrency control, we implemented a pattern using RwLock. The primary synchronization primitives available are Mutex and RwLock. While a Mutex provides exclusive access to only one thread at a time, it can cause congestion in read-heavy scenarios, leading multiple Web Server requests to queue even for simple read-only operations.

In contrast, RwLock, our chosen method, allows multiple concurrent readers to access the file simultaneously while granting exclusive access to the aggregation worker. By using Arc<RwLock< ()>>, our implementation optimizes lock contention. The read_file () method acquires a Read Guard, permitting efficient data retrieval by the Web Server, while the write() method acquires a Write Guard to prevent data corruption during disk updates.

3.4.2 Data Persistence and Integrity

The flush strategy in the flush_internal function achieves this persistence, a key factor in preventing data loss during system crashes. Specifically, after writing a JSON string to a file, the system calls file.flush(), followed by file.sync_all(). flush() pushes the data to the operating system buffer, while sync_all() forces the kernel to synchronize the data to the physical storage device. This ensures that the data is persisted to disk when the upstream AggregationEngine component receives a successful result from the storage component.

3.4.3 Error Handling

This implementation ensures system reliability by managing potential runtime failures, focusing on two main failure vectors: synchronization errors and I/O exceptions.

A critical OS concern is locking poisoning, which occurs when a thread panics while holding a lock. We utilize. map_err() to catch these panics. This prevents the entire storage component from hanging and converts the synchronization error into manageable failure reporting. Furthermore, the use of OpenOptions with. create(true) and. append(true) ensures that the storage file is initialized automatically if missing, maintaining system uptime and reliability.

3.5 Web Server Multi-Threading Model

The web.rs module implements the Gateway layer of the platform, providing a RESTful HTTP API for data visualization and system monitoring. Based on the

Axum framework and the Tokio asynchronous runtime, the gateway can handle high volumes of concurrent client requests without blocking the data processing pipeline.

A key feature is using a shared state pattern. By wrapping the DataStorage instance in an Arc,the server allows multiple request handlers to access the storage safely. This design ensures the web server acts as a passive consumer of the aggregated data, maintaining a separation between data aggregation and presentation.

3.5.1 RESTful API Handlers and Implementation Logic

The gateway provides three endpoints to interact with the aggregated sensor: Latest Snapshot, Historical filtering and system Statistics.

Latest snapshot provides the most recent aggregated state for sensor in the system. To implement this efficiency, the server performs a linear scan of the JSON lines storage file. Subsequent entries for the same sensor naturally overwrite previous ones because the file is an append-only log. Then the final collection is returned as a JSON array.

Historical filtering allows users to retrieve the complete history for a certain sensor.The implementation extracts the target ID from URL path and filters the storage file, collecting the frames matching the requested identifier.

System statistics provides a overview of the system state. It calculates the total_records metric serving as a basic operation check and data volume indicator

3.5.2 Thread-Safe Resource Access

To achieve efficient resource utilization, the system integrates a web server and data storage components. By utilizing RwLock in the data storage, the gateway handler acquires read protection during HTTP GET operations, enabling the system to scale horizontally and support multiple dashboard users reading data simultaneously. This coordination mechanism prevents data contention and maximizes throughput.

3.5.3 Fault Tolerance and Robustness

To ensure a professional user experience, this web module employs multiple fault-tolerance strategies. For example, in the event of an I/O failure, the handler returns an empty array or a default status value instead of terminating the program directly. Furthermore, the implementation uses conditional binding to skip any malformed or corrupted rows, ensuring that local corruption does not affect the overall usability and stability of the dataset.

3.6 Code Snippets for Critical Sections

3.6.1 Real-Time Data Aggregation & Anomaly Detection

A multi-threaded aggregation engine that processes high-frequency sensor streams in real-time, utilizing Welford's algorithm to dynamically compute standard deviations for threshold-based anomaly detection.

```
// gateway/src/engine.rs
impl AggregationEngine {
pub fn start(&mut self) {
// ... initialization ...
let handle = thread::spawn(move || {
let mut sensor_map: HashMap<String, SensorStats> = Hash-
Map::new();
while !shutdown.load(Ordering::SeqCst) {
if let Some(sensor_data) = buffer.pop_with_timeout(Dura-
tion::from_millis(100)) {
let val = /* ... convert 3D vector to 1D scalar magnitude
... */;
// Update statistics (mean, variance) dynamically
let stats = sensor_map.entry(sensor_data.id.clone()).or_in-
sert_with(SensorStats::new);
stats.update(val);
// Real-time anomaly detection based on dynamic standard de-
viation
let std_dev = stats.get_std_dev();
if stats.count > 10 && (val - stats.mean).abs() > threshold
* std_dev {
println!("Anomaly detected for sensor {}: value={},
mean={}", sensor_data.id, val, stats.mean);
}
}
// ... time window aggregation logic ...
}
});
}
}
```

3.6.2 Thread-Safe Sensor Buffer Management

A thread-safe buffer manager utilizing mutexes and condition variables, featuring a graceful degradation mechanism that overwrites the oldest data to maintain system stability during traffic spikes.

```
// gateway/src/buffer.rs
impl SensorBufferManager {
pub fn register_sensor<S, F>(&mut self, sensor: S, convert-
er: F)
where
S: Sensor + Send + 'static,
F: Fn(S::SensorReading) -> SensorKind + Send + Sync + 'stat-
ic,
{
let shared_buffer = Arc::clone(&self.buffer);
let has_data = Arc::clone(&self.has_data);
std::thread::spawn(move || {
while !*stop_flag.lock().unwrap() {
while let Some(content) = sensor.read() {
let data = SensorData { id: sensor.id(), kind: converter(-
content) };
let mut shared_buffer = shared_buffer.lock().unwrap();
if shared_buffer.len() < shared_buffer.capacity() {
shared_buffer.push_back(data);
has_data.notify_one();
} else {
// Degradation strategy: drop the oldest data when buffer is
full
shared_buffer.pop_front();
shared_buffer.push_back(data);
has_data.notify_one();
println!("Attention! Buffer is full, overwriting oldest
data");
}
}
std::thread::sleep(Duration::from_millis(10));
}
});
}
```

```
}
```

3.6.3 Asynchronous Architecture & Process Lifecycle Management

An asynchronous gateway architecture built on Tokio that orchestrates concurrent sensor data collection, serves web requests, and manages the lifecycle of independent child processes (e.g., the dashboard subsystem).

```
// gateway/src/main.rs
#[tokio::main]
async fn main() {
// 1. Initialize sensors and start background collection
threads
let buffer_mgr = Arc::new(SensorBufferManager::new(1000));
/* ... register sensors ... */
// 2. Start the multi-threaded data aggregation engine
let mut engine = AggregationEngine::new(EngineConfiguration
{ /* ... */ });
engine.connect_source(Arc::clone(&buffer_mgr));
engine.start();
// 3. Spawn asynchronous Web Server task
tokio::spawn(async move {
web_server.serve(web_listener).await;
});
// 4. Multi-process management: spawn Dashboard as a child
OS process
let mut dashboard_child = Command::new("cargo")
.args(["run", "-p", "dashboard"])
.spawn()
.expect("failed to spawn dashboard process");
// 5. Graceful shutdown and signal handling
tokio::select! {
_ = tokio::signal::ctrl_c() => {
eprintln!("[gateway] Ctrl+C received, stopping dash-
board...");
let _ = dashboard_child.kill().await;
}
status = dashboard_child.wait() => {
eprintln!("[gateway] dashboard exited: {:?}", status);
```

```
        }
    }
}
```

4. BENCHMARK

4.1 Test Methodology

To evaluate the robustness and efficiency of the Chapter, we established a controlled benchmarking environment. The primary objective is to assess system behavior across a range of stress scenarios, from light operation load to huge concurrent scaling.

4.1.1 Environment Establishment

All benchmark tests were executed under a standardized hardware and software configuration to ensure repeatability and reliability of the results. The components of the particular environment are the operating system, RAM, and CPU. The Apple M2 CPU, 16GB LPDDR5, and macOS 14.0 are the CPU, RAM, and operating system used in our test. To enable full compiler optimizations, all test binaries are compiled using the cargo build –release parameter.

4.1.2 Workload Generation Strategy

We utilize the internal sensor_sim module to generate high-frequency synthetic data streams. The test. Use a stepped pressure model to observe our processing logic's linear scalability. We design the following four levels to compare their benchmarks with an increasing number of concurrent sensors:

Light load: 5 concurrent sensors (verify basic logic and accuracy)
Moderate load: 20 concurrent sensors (simulate simple industrial scenarios)
High load: 50 concurrent sensors (test multi-threaded contention efficiency)
Extreme load: 100 concurrent sensors (detect edge-case behavior and system limitations)

4.1.3 Metrics and Tools

To obtain a comprehensive picture of system performance, a multi-dimensional monitoring system is incorporated into the test device:

Throughput: it is calculated using highly precise timing via Rust's std::time::Instant. We measured the total volume of data frames successfully received, processed, and persisted to disk per second through interfacing the internal counters of SensorBufferManager.

Resource Usage: The binary includes a direct integration of the sysinfo crate. Every 500ms while the test was running, the system took a sample of the process's Resident Set Size(RSS) memory footprint and global CPU use. The effect of brief surges is then removed by averaging these samples.

Buffer Efficiency and Reliability: We monitor the utilization of the storage buffer in addition to the data loss rate. This metric is vital for evaluating the system's ability to address data during I/O blocking or high-intensity computing bursts to meet the zero data loss requirement.

4.1.4 Consistency and Reliability

Each test scenario is executed for a continuous duration of 15 seconds to ensure data integrity. This duration allows the system to reach a steady state. We ensure that reported metrics accurately reflect long-term operation performance by aggregating data once the system stabilizes.

4.2 Data Loss Verification Results

To evaluate data integrity, we monitor overflow indicators within the shared buffer to achieve zero data loss. The system specifically interfaces with the buffer_mgr.get_stats() method to retrieve the underlying overwrite_count. The buffer capacity is fixed at 10,000 frames. Due to the ring buffer architecture, the system triggers a data overwrite when the producer speed consistently exceeds the consumer speed and fills the queue.

Based on our test observation, the overwrite_count remains at 0 in low and moderate load, which confirms that the 4 worker threads are able to process data instantly. In a high-stress load environment, the test focuses on observing if the pop_rate can keep up with the 5,000msg/s total influx. In order to save structural monitoring continuity, overwriting indicates the necessity for additional worker threads and higher optimization of the Welford calculation logic.

4.3 Performance Metrics (Throughput, Latency, CPU usage)

The benchmark results provide a detailed view of the system's operational efficiency. These metrics are divided into three main categories: throughput, latency, and CPU usage.

4.3.1 Throughput

Throughput measures the system's data-handling capacity and can be derived from two rates: write and pop rate. Writ rate measures the frequency at which sensors push data into the shared buffer. Pop rate is the frequency at which AggregationEngine extracts and processes data from the buffer

The pop rate successfully matches the write rate in all test scenarios. This indicates that the system achieves optimal throughput, maintaining a 1:1 ratio between data generation and processing, even at a total volume of 5,000 mgs/s

4.3.2 Latency

Latency represents the responsiveness of the system. The system enforces a strict 1-second window duration even if the latency is sub-millisecond. This ensures that statistical results are never delayed beyond the one-second aggregation boundary. We also infer the queueing latency by monitoring buffer_utilization. Additionally, we observe the latency of file.sync_all() as a primary limitation. However, the multi-threaded design prevents disk I/O latency from blocking the sensor ingestion threads.

4.3.3 CPU Usage

CPU usage evaluates the computational efficiency of the Rust implementation and the Welford algorithm. Using the sysinfo crate, we sample global CPU load every 500ms to calculate avg_cpu_usage_precent. The load is distributed across multiple logical cores with four workers. The CPU cost per sensor remains O(1), proving the Welford algorithm very efficient. The CPU utilization stayed in a steady range even at the maximum test scale, demonstrating that there is still headroom for more vertical scaling.

4.4 Scalability Tests (Varying Number of Sensors)

To confirm the linear growth characteristics of the system's processing power, the test environment is created using a gradient configuration of [5,20,50,100] sensors.

By comparing the ratio of sensor_count increases against the growth of avg_cpu_usage_percent, we evaluate scalability. Resource consumption scales linearly O(n) with the sensor count. However, limitations in Lock contention within the RwLock implementation may lead to an exponential curve.

4.5 Stress Testing Results

The stress test identifies the throughput ceiling of AggregationEngine on hardware. We can determine the system safety margin through observation of buffer_utlization. IF utilization stays above 80% steadily, it indicates the need to optimize the file. sync_all() strategy at the storage component.

The system uses wtr.serialize to capture performance snapshots every 15-second continuous stress test. Our main concern is whether frequent writes to the JSON Lines file result in disk latency that "back-pressures" the buffer. To make sure that concurrent processing doesn't result in volatile memory spikes, we also keep an eye on memory_usage_mb.

4.6 Tables and Graphs

We summarize the test performance based on the benchmark_result.csv as shown in Figure 3.

Figure 3. Benchmark result

benchmark_results

sensor_count	duration_time	throughput_write_rate	throughput_pop_rate	buffer_utilization	data_loss_count	avg_cpu_usage_percent	memory_usage_mb
5	15	94.0	99.2	0.0	0	4.655497	10.483333333333300
20	15	378.33334	458.86667	0.0	0	5.394952	11.8546875
50	15	2122.8667	2796.6	0.0	0	7.6193542	14.370833333333300
100	15	4239.6	5078.467	0.0	0	12.355258	19.303645833333300

The benchmark results confirm that the chapter architecture is ready for the industrial-sale monitoring. The system uses less than 13% of available CPU power even at peak test loads. The data loss and buffer utilization also prove that the multi-threaded producer-consumer model is balanced.

5. DISCUSSION

5.1 Analysis of Results

The implementation successfully achieved the objective of zero data loss while managing high-frequency sensor readings. This is particularly significant. Because each simulated sensor provides only a highly constrained internal buffer. Zero data loss means that the read thread strategy and shared buffer design effectively protect the acquisition path from the impact of downstream computing latency.

Another important result is that our BufferManager system could handle the minimum requirement of 10 readings per second from multiple sensors simultaneously without triggering any observed buffer overwrites under the evaluated workload. Another advantage is that the buffer manager is transparent and observable. Because it records utilization, overwrite count, write rate, and pop rate, it facilitates evaluation.

The AggregationEngine computed window statistics using Welford's algorithm. This algorithm realizes incremental calculation of statistical data within a fixed window, which provides numerical stability over windows. Also, aggregationEngine supports anomaly detection based on deviation from the running mean and prints them. This implementation includes logic for online anomaly detection; future plans will enhance its integration with the output format by storing it in the output frame.

Web server successfully served these results to multiple concurrent clients, making our read-write coordination strategy effective.

5.2 Trade-Offs in Design Choices

Several trade-offs were made during the design of AggregationEngine and DataStorage components.

For aggregation, the initial idea is to use a simple one-thread-per-sensor model. However, we transitioned to a thread pool with distribution to ensure scalability. This trade-off increased implementation complexity and allowed the system to handle a higher volume of sensors.

Another trade-off is storage format. We adopted a line-delimited JSON format for storage. While CSV might be more compact, JSON simplifies the web ingestion process for the downstream web server and allows for more flexible data representation.

To ensure continuous data availability, the storage buffer size was strategically selected to mitigate the impact of irregular processing delays while maintaining a sustainable memory. A larger buffer can absorb longer bursts, but at the cost of higher RAM utilization.

5.3 Limitations and Bottlenecks

This platform involves only three sensors: a force sensor, a temperature sensor, and an acceleration sensor. It has not yet been tested whether the system's processing resources increase linearly or exponentially, as the number of sensors and the resulting increase in data volume grow. If superliner growth means that the system quickly becomes unusable as the number of sensors increases, then this architecture is a bottleneck design. Therefore, more sensors are needed in the future to test how the system's resource requirements increase proportionally with the increase in sensors, rather than growing exponentially and uncontrollably. (Henning & Hasselbring, 2020)

Another limitation is that file-based storage may become an I/O bottleneck. The drawbacks of this approach are as follows: high-frequency operations, such as file appending and synchronization, may slow down; flush and syn_all improve reliability but also introduce additional overhead. Prior studies on flush-heavy storage paths show that synchronous persistence can significantly affect performance under frequent small writes (Choi & Won, 2018). If the data volume increases, the storage layer may become the bottleneck. Using a database or messaging system might solve these problems.

5.4 Comparison of Synchronization Primitives

For the DataStorage component, we compare the Mutex and RwLock, two primary synchronization primitives. Mutex would provide simple exclusive access, but may cause congestion in a read-heavy web server scenario. The key reason for choosing RwLock is that the storage access pattern is asymmetric: write operations are relatively infrequent, while read operations from the web layer may occur concurrently and repeatedly. Under such a workload, RwLock better matches the actual system behaviour than a purely exclusive Mutex. We optimized for the common case where multiple users are viewing the concurrent readers by choosing RwLock. This only allows one worker to append data.

In managing the shared storage buffer between reader threads and aggregation workers, Mutex<VecDeque<T>> with Condvar was compared against channels. While channels offer a cleaner high-level abstraction, the Condvar approach allowed for more control over buffer statistics and utilization monitoring. A Mutex provides simple exclusive access, but in read-heavy scenarios, it may serialize all readers and writers, causing unnecessary contention (Bos, 2022). The Mutex+ Condvar approach also makes blocking and wakeup behaviors explicit. (Lampson & Redell,1980) Consumers sleep only when the queue is empty and are awakened when producers insert new data, which fits the classic producer–consumer model and makes the synchronization logic easier to reason about at the OS level. Therefore, while channels would simplify some parts of the code, the Mutex+ Condvar design was more suitable for a chapter that emphasizes both concurrency correctness and buffer-level observability. At a broader level, this reflects a general trade-off in concurrent programming between higher-level communication abstractions and lower-level synchronization mechanisms with greater visibility and control (Andrews & Schneider, 1983).

5.5 Lessons Learned About Concurrent Programming

The chapter highlights the Rust language, especially how Rust's ownership and borrowing rules prevent common pitfalls like data races. Unlike C or C++, Rust's type system ensures that data cannot be accessed concurrently without proper protection. We learned that the compliers act as a first line of defense, forcing the wrapping of shared resources like the BufferManager and DataStorage in Arc and RwLock, guaranteeing thread safety rather than discovering race conditions during runtime.

We also learned that a thread panicking while holding a lock can potentially corrupt it, hanging the entire storage component. In many operating systems, the state of the protected data becomes inconsistent if a thread dies while holding RwLock. Rust can mark the corrupted locks, and attempts to acquire them will return an error. By implementing map_err() to catch the synchronization errors, we ensure that a single thread failure would not lead to a total system deadlock, which is crucial for maintaining high system performance.

Another discovery is the difference between file.flush() and file.sync_all() regarding how operating systems manage storage. File.flush() only pushes application-level buffered data to the OS kernel buffer, which cannot guarantee that the data has reached the physical disk. This OS-level difference is vital for ensuring data integrity. We forced the kernel to commit the data to the physical hardware by calling file.sync_all(). This is significant in meeting the zero data loss requirement in case of a power failure or system crash.

6. CONCLUSION AND FUTURE WORK

6.1 Summary of Achievements

The chapter successfully resulted in a multi-thread sensor data processing pipeline. DataStorage is a core achievement to effectively bridge the AggregationEngine and the web server. We ensure that each data frame is serialized into a single, manageable line, simplifying data ingestion for the downstream server by adopting JSON format.

The use of RwLock solves read-write conflicts. This advanced concurrency primitive handles high-frequency data streams without sacrificing performance or safety.

6.2 How Objectives Were Met

The objectives can be achieved through three aspects: concurrency, integrity, and reliability.

Concurrency control: We implement the RwLock pattern to address the concurrency control requirement. This pattern allows multiple web server requests to access files simultaneously while granting the aggregation worker exclusive write access. Hence, RwLock avoids the congestion caused by standard Mutex locks in read-heavy scenarios.

Data integrity: Data integrity is achieved through a flushing strategy. By utilizing both file.flush() and file.sync_all(), the DataStorage can respectively push the data to OS buffers and commit data to physical storage. Once a successful result is returned, the data will be persisted on disk.

System reliability: Lock poisoning is viewed as one of the most critical OS concerns. We mitigated this by using .map_err() to catch thread panics, which prevents the entire storage component from hanging during runtime failures. Additionally, the use of OpenOptions ensures automatic file initialization, maximizing system uptime.

6.3 Potential Improvement

Although the current implementation successfully achieves zero data loss and meets all core requirements, several aspects present opportunities for improvement.

First, the current implementation utilizes a JSON file. In spite of its advantage of readability and append operations, a linear file will become an O(n) bottleneck as the system scales to numerous sensors. To reduce time complexity and scale up, we can transition to a time-series database or relational database like SQL, which allows index queries. This would enable the web server to retrieve historical data for time ranges in O(logn) time, largely reducing time complexity. In other words, disk I/O overhead and storage footprint can be reduced by moving JSON to a binary format like Protocol Buffers. But this will sacrifice the direct human readability.

Second, the current implementation relies on Arc<Mutex<VecDeque>> for the shared buffer, leaving it to potential contention under extreme sensor loads. Replacing these with lock-free data structures like ArrayQueue would reduce synchronization overhead, which allows the system to maintain high throughput and minimize latency without sacrificing thread safety.

Third, the engine implements Welford's algorithm for statistical monitoring. But anomaly detection remains relatively simple. Future iterations can combine some machine learning models like Autoencoders via the rust-learn or tch-rs crates. Unlike basic standard deviation checks, these approaches facilitate the detection of context-aware data signatures, offering significantly deeper insights into structural integrity

6.4 Future Extensions

Several promising extensions can transform the chapter into an industrial sensor platform.

1)Upgrading the hardware sensors by using standard IoT protocols like MQTT: Deploying the system would provide genuine, real-time structural health monitoring capabilities on physical infrastructure.
2)Security: Implementing HTTPS encryption support would make the platform suitable for commercial, government, and even national chapters. HTTPS support would encrypt all communications between the web server and the client, which prevents attacks and ensures data confidentiality during transmission. In addition, adopting JWT-based (JSON Web Token) authentication for the RESTful endpoints would provide secure, stateless user authentication and authorization.
3)End-to-end solution: The chapter can evolve into an end-to-end solution by replacing the raw REST API with a comprehensive user interface built with modern frameworks like React or Svelte. The dashboard provides live data streaming and interactive time-series graphs with charting libraries like ECharts. This enables the identification of context-aware and complex data signatures, thereby providing more accurate structural health insights.

REFERENCES

Ahmad, M., & Turgut, D. (2008). Congestion avoidance and fairness in wireless sensor networks. *Proceedings of IEEE GLOBECOM, 2008*, 1–6. DOI: 10.1109/GLOCOM.2008.ECP.59

Andrews, G. R., & Schneider, F. B. (1983). Concepts and notations for concurrent programming. *ACM Computing Surveys, 15*(1), 3–43. DOI: 10.1145/356901.356903

Azzedin, F., & Al-Issa, K. (2012). Handling input/output operations in Web servers: Issues and countermeasures. *Concurrency and Computation, 24*(7), 699–710. DOI: 10.1002/cpe.1912

Birrell, A. D. (1989). *An introduction to programming with threads* (Report No. 35). Digital Systems Research Center.

Bos, M. (2022). *Rust atomics and locks*. O'Reilly Media.

Chen, Y., & Krishnamachari, B. (2012). Game theoretical approach for channel allocation in wireless sensor and actuator networks. *Turkish Journal of Electrical Engineering and Computer Sciences, 20*(1), 1–14.

Choi, G., & Won, Y. (2018). Analysis for the performance degradation of fsync() in F2FS. In *Proceedings of the 9th International Conference on E-Education, E-Business, E-Management and E-Learning* (pp. 71–75). DOI: 10.1145/3183586.3183605

Crossbeam contributors. (2026). *crossbeam-channel* (Version 0.5.15) [Rust crate documentation]. Docs.rs.

Dijkstra, E. W. (2002). Cooperating sequential processes. In *The origin of concurrent programming: From semaphores to remote procedure calls* (pp. 65–138). Springer.

Efanov, A. A., Ivliev, S. A., & Shagraev, A. G. (2021). Welford's algorithm for weighted statistics. In *2021 3rd International Youth Conference on Radio Electronics, Electrical and Power Engineering (REEPE)* (pp. 1–5). IEEE. DOI: 10.1109/REEPE51337.2021.9387973

Gaud, F., Genevès, S., Lachaize, R., Lepers, B., Mottet, F., Muller, G., Quéma, V., & Morel, M. (2010). Efficient workstealing for multicore event-driven systems. DOI: 10.1109/ICDCS.2010.55

Henning, S., & Hasselbring, W. (2020). Scalable and reliable multi-dimensional sensor data aggregation in data streaming architectures. *Data-Enabled Discovery and Applications, 4*(1), Article 5.

Huang, J., Du, D., Duan, Q., Zhang, Y., Zhao, Y., Luo, H., & Liu, Q. (2014). Modeling and analysis on congestion control for data transmission in sensor clouds. *International Journal of Distributed Sensor Networks, 10*(3), 453983. Advance online publication. DOI: 10.1155/2014/453983

Jung, R., Jourdan, J.-H., Krebbers, R., & Dreyer, D. (2017). RustBelt: Securing the foundations of the Rust programming language. *Proceedings of the ACM on Programming Languages, 2*(POPL), 1–34.

Lampson, B. W., & Redell, D. D. (1980). Experience with processes and monitors in Mesa. *Communications of the ACM, 23*(2), 105–117. DOI: 10.1145/358818.358824

Lavender, R. G., & Schmidt, D. C. (1996). Active object: An object behavioral pattern for concurrent programming. In *Pattern languages of program design 2* (pp. 483–499).

Mohammed, A., & Zangana, H. (Eds.). (2026). *Intelligent Construction Monitoring Systems: Real-Time Safety, Environmental Prediction, and Risk Management.* IGI Global Scientific Publishing., DOI: 10.4018/979-8-3373-9245-5

Rust Chapter Developers. (2026). *std:sync:mpsc*. In *The Rust standard library* (Version 1.94.0).

Rust Chapter Developers. (n.d.). Rust by Example. Rust Documentation. https://doc.rust-lang.org/rust-by-example/

Shi, Z., Møldrup, L., Mathur, U., & Pavlogiannis, A. (2026). The complexity of testing message-passing concurrency. Proceedings of the ACM on Programming Languages, 10(POPL), Article 103. DOI: 10.1145/3776643

Shwe, H. Y., Gacanin, H., & Adachi, F. (2010). Multi-layer WSN with power efficient buffer management policy. In *2010 IEEE International Conference on Communication Systems* (pp. 36–40). DOI: 10.1109/ICCS.2010.5686107

Steere, D. C., Goel, A., Gruenberg, J., McNamee, D., Pu, C., & Walpole, J. (1999). A feedback-driven proportion allocator for real-rate scheduling. In *Proceedings of the 3rd USENIX Symposium on Operating Systems Design and Implementation (OSDI '99)* (pp. 145–158).

Chapter 4
Coordinating Cross-Process Sensor Aggregation in Rust With IPC and Shared Memory

Haochen Shi
http://orcid.org/0009-0004-1165-1682
The Hong Kong Polytechnic University, Hong Kong

Xuan Luo
The Hong Kong Polytechnic University, Hong Kong

Junhao Huang
The Hong Kong Polytechnic University, Hong Kong

Yixiong Feng
The Hong Kong Polytechnic University, China

Zihan Meng
The Hong Kong Polytechnic University, Hong Kong

Aquil Mirza Mohammed
http://orcid.org/0000-0001-7756-4363
The Hong Kong Polytechnic University, Hong Kong

ABSTRACT

This chapter, per the authors, presents a distributed sensor data aggregation platform in Rust demonstrating operating systems concepts including inter-process communication and concurrent data processing. The system employs a process-based architecture with ten child processes: five sensor readers connected via anonymous pipes, four aggregation workers sharing a named FIFO pipe, and one web server. A collector process groups sensor readings into fixed-duration time windows forwarded through POSIX shared memory with atomic counters. The aggregation engine applies Welford's online variance algorithm to compute streaming statistics; minimum, maximum, mean, and standard deviation; without a second data pass. Anomaly

DOI: 10.4018/979-8-2600-1101-0.ch004

detection uses a configurable z-score threshold per sensor type. Five benchmark trials confirm zero data loss across 698 readings, a throughput of 71.0 requests per second, a p99 latency of 14.08 milliseconds, and peak buffer utilisation of 0.01%, validating the platform's reliability guarantees under the intended workload.

1. INTRODUCTION

1.1 Problem Statement and Motivation

The reason and the motivation of designing the system of this chapter in this way comes from a common issue in modern sensor monitoring platforms. Nowadays, we found out that many sensor systems need to collect and aggregate readings from multiple physical devices at the same time (Henning & Hasselbring, 2019), but these devices do not always operate at the same rate or under the same timing conditions. Because of that, we suggest that this is not only a data processing problem, but also an operating systems problem (Alshehri & Bajaber, 2024). In the system of our chapter, the sensors may produce some readings around 10 to 20 Hz, while downstream stages such as aggregation, disk persistence, and web serving may execute at slower or less predictable rates. We suggest that this would happen because of the scheduling delays, lock contention, cache effects, and I/O operations. So, even though the pipeline may look stable at first, but there is still a mismatch between the fast producers and the slower consumers.

This mismatch becomes more and more serious and dangerous when the upstream side only has a small internal buffer. In our chapter, each simulated sensor has a 128-slot circular ring buffer (Nikolaev & Ravindran, 2022). If the reader process does not drain this buffer quickly enough, newly arriving readings will overwrite the oldest ones. As a result, the system may lose data without producing an explicit error signal. This is not what we want, it is not only a theoretical problem. At 20 Hz, a sensor may produce one reading at every 50 ms, so a 128-slot buffer can be filled in about 6.4 seconds if the consumer is delayed, the effect of this is tremendous. In other words, even a short disturbance, such as OS scheduling jitter or a disk flush that temporarily slows the pipeline, may already be enough to cause this failure mode.

For this reason, an intermediate buffer is needed between the sensor readers and the downstream processing stages. This buffer acts like a shock absorber. It separates the relatively fast and regular behaviour of sensor ingestion from the slower and more variable behaviour of aggregation, storage, and as well as the web service components. By doing this, the whole pipeline becomes more stable and less sensitive to short-term slowdowns. So, the main design question is not whether buffering is needed, because in this case it clearly is. The more important question is how this buffer should be implemented and coordinated.

There are several classical ways we learnt and discovered to solve this kind of producer-consumer problem. For example, Semaphore-based designs, which originate from Dijkstra's formulation, focus on bounded-buffer correctness and explicit synchronization; Monitor-style designs which focus more on condition synchronization and structured mutual exclusion. Even though modern lock-free queues probably can improve throughput under contention, but they may also reduce fairness, increase implementation complexity, or make backpressure harder to observe directly, which is not something we want. Rust is able to add another interesting aspect to this design space (Yu, Song, & Zhang, 2019; Balasubramanian et al., 2017). Its ownership and borrowing rules make aliasing and mutation constraints explicit in the type system, so many concurrency mistakes mentioned above can be rejected at compile time instead of being discovered later during testing and running (Bugden & Alahmar, 2022). In this sense, Rust does not remove the tradeoffs in operating systems design, but it also makes these trade offs more visible.

1.2 Chapter Objectives

The objective of the chapter is:

1. Guarantee zero data loss from the sensors' 128-slot internal buffers under the intended workload, even though when downstream stages experience temporary slowdowns or scheduling jitter. In other words, the system should remain reliable even when the later parts of the pipeline do not run at a perfectly stable rate.
2. Provide a process-safe intermediate buffering strategy by using both anonymous pipes and shared memory (Xia et al., 2022). This is intended not only to support communication between processes, but also to make sure that the backpressure and overflow-related conditions visible. This allows the buffer stress can be detected and be analysed rather than remaining hidden inside the system.
3. Implement a concurrent aggregation engine that groups sensor readings according to their own timestamps into fixed windows. And within each particular window, the system then computes summary statistics including the minimum, maximum, mean, and standard deviation, while also performing basic anomaly detection to highlight unusual patterns in the data.
4. We are also aiming to support persistent storage with concurrent read and write access by utilising an append-only JSON Lines format. We suggest that this will allow the aggregated frames to remain durable after processing. At the same time, it is still possible to query stored results without any blockings in the full pipeline.

5. We hope to expose processed data through an HTTP server and an individual child process. We try to achieve this objective by providing REST-style endpoints. This will allow the external clients to inspect system state, retrieve aggregated results, and as well as interact with the system in real time.
6. Our last objective is to demonstrate several core operating systems concepts through a concrete implementation in this chapter. These concepts include: process creation, inter-process communication (IPC) through pipes and shared memory, the producer-consumer pattern, deadlock prevention, and also the coordination of multiple OS processes within an individual end-to-end pipeline.

By bringing all these objectives together, we suggest that the position of our chapter will be: both a functional data aggregation platform and a compact study of concurrent systems design.

1.3 Overview of Approach

As mentioned before, our chapter adopts and utilises a process-based architecture. The gateway binary will be served as the only executable in the whole system, and it launches child processes by using std::process::Command for functions like sensor reading, aggregation, and web serving in this chapter. For IPC, we implemented it using three main mechanisms: the first one is anonymous pipes created by utilising libc::pipe, and are used to transfer data from the sensor readers to the collector. The second mechanism is a named pipe, created through libc::mkfifo, which is utilised to forward completed windows from the collector to the aggregation workers. And the last mechanism which is cross-process statistics, the system relies on POSIX shared memory using (shm_open+mmap with #[repr(C)] layout) to ensure a stable memory representation. To maintain a clear safety boundary, all unsafe operations are isolated in a separate ipc_unsafe crate, while the gateway crate itself contains no unsafe blocks. In addition, the system does not depend on any external async runtime (Mohammed & Zangana, 2026).

The implementation of our chapter is further organised into six main modules, which are; buffer, aggregation, storage, web_server, benchmark, and process, with the process module further divided into the reader, worker, web. Under steady-state execution, the platform runs $11 + k$ operating system processes or threads in total. These include one main collector process, five sensor reader processes, four aggregation worker processes, and one web server process, together with HTTP connection handler threads inside the web server process. In this current configuration of our chapter, the system listens on 127.0.0.1:8080, which uses a 300 ms aggregation window, and runs with four worker processes.

The movement of data in our chapter is through the platform in a staged pipeline. The sensor readings are first drained by dedicated reader processes, which serialize each reading as a JSON line and write it into an anonymous pipe. And the collector, which runs inside the main process, reads from all five pipes in a round-robin loop, assigns each reading to a timestamp-based window bucket using a BTreeMap, and writes completed windows into the named FIFO. Then, four worker processes then compete on the FIFO. Each worker deserializes a window batch, computes per-sensor statistics using Welford's algorithm, and appends the resulting AggregatedFrame to a shared JSONL storage file. The web server process periodically reloads aggregated frames from this file and reads buffer-related statistics from shared memory, then serves the results through a thread-per-connection REST interface.

This design of our porject deliberately favours process isolation over shared-memory concurrency. One practical advantage of the choice our chapter is that a crash in one process does not directly corrupt the others. At the same time, all IPC mechanisms remain explicit and relatively easy to inspect, which makes the behaviour of our system more transparent during implementation, debugging, and evaluation.

2. RELATED WORKS

2.1 Buffer Management Techniques

For most chapters with a bounded-buffer design, people choose to use Dijkstra's semaphore formulation as it allowed operating systems to handle exclusions and count resources using atomic operations. This method is able to solve the producer-consumer problem because it is able to keep track of who has access to the data, how many empty spots are left and how many full spots are there. Using a semaphore solution makes the code look exactly like the solution in the textbook, the semaphore acts like a mathematical guard that prevents producers from adding data if there's no room and consumers can't take data if nothing has been made. However, for Chapter Azure, we didn't use semaphores for the front-end buffer because it would make the actual data structure too hard to see. We didn't just want a synchronization exercise, we need a buffer that can show it's capacity and the backpressure level so that the rest of the system can check on it easily.

We finally decided to implement the monitor style proposed by Hoare in 1974 because in this setup, a bounder buffer acts as a permanent state, meaning that methods run one at a time. We used condition variables like not_full and not_empty which actually maps really well to how Rust does Mutex<T> and Condvar together. The design we used with Mutex<VecDeque<Reading>> and Condvar follows this monitor design as the queue state is kept inside, and you can only change it when you are holding the MutexGuard. This means that producers and consumers go to sleep on certain conditions instead of just spinning and wasting CPU power. The Condvar::wait_while in Rust is also extremely useful because it can block the thread until the condition becomes true, it will check again when it wakes up, this is also the right way to handle spurious wakeups according to the Rust Team (2024).

Another option stated by the Rust Team (2024) is std::sync::mpsc. This can give you a mult-producer as well as single-consumer channels. The difference between channel() and sync_channel() is that channel() gives you unbounded buffering and that sync_channel() gives you a bounded queue with blocking sends. For Chapter Azure we decided to use mpsc:channel for the aggregation stage. However we didn't use mpsc for the front-end sensor buffer, this is because channels could end up hiding the information that we need to see such as queue occupancy, custom overflow policy, and also backpressure signals where the sensors are most important.

We have also considered using lock-free queues as Michael and Scott's queue (1996) shows that you can get really good throughput using non-blocking FIFO structures. Then Herlihy and Shavit(2012) also uses linearizability as their main way to judge if concurrent objects are actually working correctly. Although these solutions are really fast, we didn't actually use them for our chapter, this is because our front-end buffer is supposing to be focused on teaching instead of high-speed performance. However a lock free queue would need way more complex setups as well as very careful thinking about memory ordering (Nikolaev & Ravindran, 2022). Not only does lock free goals misalign with our primary goals, Drepper(2007) also pointed out that false sharing can also create hidden synchronization costs. With all things considered, a monitor style queue was just a better choice for us.

2.2 Multi-Threaded Server Architectures

Web server concurrency has gone through a couple of stages such as process per request, thread based servers, thread pools, and event driven setups. Most of these changes occurred from the C10K problem proposed by Dan Kegel(1999). In his paper, Kegel explained that it is very hard to handle large numbers of clients at once. Later on in time, high performance servers began to use event loops to avoid the memory and the scheduling costs that come with having one thread for every connection. But due to the fact that Chapter Azure is only serving a few clients instead of that kind of large scale, using a simple blocking design is actually a good thing rather than a problem.

The standard networking APIs in Rust actually makes this design very clear. The std::net::TcpListener waits for the incoming TCP connections and the accept() method stops the thread until a connection actually happens. For this chapter, using this type of blocking behavior was actually a big help for learning. By building a thread per connection server using TcpListener, TcpStream, and thread::spawn (Yu, Song, & Zhang, 2019), we can actually see the exact concepts this course wants us to learn, such as blocking system calls, how the OS schedules threads, and the overhead of one request per thread concurrency. So Stevens and Rago are still the best reference for this sockets model.

Pai et al(1999) actually made a very good point in their comparison, which is that if you have less than 1000 connections, multi threaded servers can actually compete with event driven ones (Brandenburg, 2019). Please note that we are not saying that thread per connection is the best way to do things, only that for our specific chapter, it is the best choice as it allows us to focus on how the OS actually works.

2.3 Producer-Consumer & Statistical Methods

In the producer consumer literature, Andrews (2000) states that a bounded buffer is basically a way to match speeds as it handles small timing gaps between when the data arrives and when its used, this is so that producers and consumers don't have to stay in perfect lockstep. Chapter Azure actually faces a tougher situation at the hardware boundary, this is because the simulated sensor's internal 128 slot ring would overwrite the old data if the reader is too slow. The software buffer's job is to stop that kind of overwrite-on-full behavior from affecting the application pipeline. This is why our intermediate queue blocks on Condvar and shows the pressure instead of only copying the sensor's silent overwrite rule.

For window statistics, the formula for variance $E[x^2]-(E[x])^2$ is actually numerically fragile. However, Welford's(1962) note introduce a one pass update methods for corrected sums of squares, later on, Chan, Golub and LeVeque (1979) expanded on this idea with update formulae and pairwise combination rules that are great for incremental computation. These results fit right in to our aggregation pipeline as we can now group readings by timestamps into fixed windows using a BTreeMap. And every worker calculates min, max, mean, and stddev without needing to loop over the data for a second time (Henning & Hasselbring, 2019; Ramírez-Gallego et al., 2017).

Our anomaly detector uses the standard score rule $|x - \mu| > k\sigma$ as a simple baseline. The survey done by Chandola, Banerjee, and Kumar(2009) includes z-score style scoring in its talk on statistical anomaly detection, this makes it a very solid first method for sensor streams that prioritize in the ability to explain results than having a complex model. Boehm(2012) also pointed out that data race freedom has to be defined in large language level terms instead of just guessing based on how the hardware feels. This is the reason why our shutdown flag is an AtommicBool updated with Ordering::Relaxed, the flag is just used to ask for eventual termination.

A stale read could also delay the shutdown by one loop, but it wouldn't break the correctness or cause any data race.

3. IMPLEMENTATION

3.1. Architecture Overview

Our system uses a tiered pipeline architecture as shown in Figure 1 to separate the main tasks of sensor data processing, including data acquisition, organization, statistical analysis, file persistence, and web presentation. The goal of this design is to make sure that each part is only responsible for what it does, to reduce the impact that modules have on each other, and to make it easier for us to fully realize the ideas of processes, inter-process communication, and synchronization in the operating system.

The gateway is the only main software. Once it starts, it will create a number of child processes. These include five reader processes, which will read data from five sensors one at a time; four worker processes, which will execute statistical computations; and one web server process, which will provide HTTP services. The main process acts as a collector. It gets data from reader processes, sorts it by time frame, and then sends it to worker processes.

The data flow is made up of a sensor, a reader process, an anonymous pipe, a collector, a named FIFO, a worker process, a storage file, and a web server as shown in Figure 3.

We chose to use this process-based architecture since it fits better with the ideas about operating systems that this chapter is meant to look into. Because regular memory can't be shared directly across different processes, data has to be delivered using specific IPC methods such anonymous pipes, named FIFO, shared memory, and file systems. This is necessary since regular memory cannot be shared directly.

3.2 Buffer Management Design

We don't just look at the kernel buffer of the anonymous pipe when we manage our buffers. The anonymous pipe's main goal is to solve the question "How to move data between processes." We put the SensorBufferManager into the buffer.rs is the part that is really responsible for storing data and making sure that production and consumption processes work together. This part of the assignment is the same as

Component 1 in the assignment requirements. It is the most important part of the assignment.

SensorBufferManager works in a way that is common for Monitor. Mutex, VecDeque, and Condvar make up its basic structure (Nikolaev & Ravindran, 2022; Brandenburg, 2019). Condvar is in charge of waking up and waiting, VecDeque is in charge of storing sensor data in a first-in, first-out (FIFO) order, and Mutex is in charge of making sure that threads may access the data safely. When discussing the producer-consumer relationship in the operating system course, this combination is often mentioned.

The SensorBufferManager enables you to set the buffer size by using the "new(-capacity)" syntax. We passed in 8192 slots. If the traffic increases slightly, a too-small buffer capacity will cause it to be filled up; if the capacity is too large, it may waste memory and make it difficult to detect the system load.

The register_sensor() function puts data into the buffer. Each registered sensor gets its own reader thread, which is in charge of constantly reading data and storing it into the shared VecDeque. A single buffer is created from data from several sensors so that it can be used all at once.

SensorBufferManager also has three alternative ways to pop. The Condvar::wait_while function is used by pop_blocking() to block until data is ready. The Condvar::wait_timeout_while function is used by pop_timeout() to wait with a timeout. try_pop() is a non-blocking version that returns None directly if there is no data.

The system utilizes the pop_front() method to remove the oldest data from the buffer when it has been full. Then it will push new data from the end of the buffer. Under this load condition, as long as the overflow_count in the benchmark test results is zero, there will be no actual data loss.

The get_stats() function in SensorBufferManager returns the following values: utilization, total_written, total_read, overflow_count, and sensor_buffer_peak. You can use these values to determine whether the system is operating normally.

In the statistical implementation, we use AtomicU64 to count instead of locking the entire queue. This is why this part uses Mutex to protect the queue, Condvar to wait and wake up, and AtomicU64 to count things quickly. Each of these parts deals with a different problem.

SensorBufferManager has a shutdown() function that makes it easy to leave. It wakes up all the threads that are waiting by using the notify_all function and setting an AtomicBool shutdown flag to true. This lets them all check the shutdown flag and exit normally.

The reader processes polls very often, and it only sleeps for five milliseconds when there is no data (Yu, Song, & Zhang, 2019). Because the most accurate sensor sends data every 50 milliseconds, a 5-millisecond delay is long enough to read the data before the sensor's 128-slot internal buffer is almost full. Also, the reader writes the maximum occupancy of the sensor buffer to shared memory so that it can be watched from there.

When the parent process shuts down, it sends a termination signal to each reader process. This signal lets the reader processes stop and quit, and then the parent process recycles the child processes.

3.3 Aggregation Engine

One of the specialized parts, called AggregationEngine, is in charge of doing the aggregate work. It does a lot of things, like getting readings, organizing them by time periods, doing statistical calculations and finding anomalies, and sending the results back to the storage module.

The AggregationEngine is built in two stages. The collector's job in the main process is to read data from five anonymous pipes. Four work processes read the grouped data from the same queue and then perform statistical calculations. The front end is responsible for collecting the data, while the back end is responsible for calculating the amount of data to be collected.

To correctly divide the windows, we operate based on the timestamp of the data read each time. Data with a time difference of less than 300 milliseconds will be grouped into the same window. We use BTreeMap to store these windows in the order of window creation (Ramírez-Gallego et al., 2017).

When the collector determines that a window is ready, they will place it in the first-in-first-out queue. The staff will classify the data based on the sensors and calculate the minimum, maximum, average, and standard deviation of each sensor. We use the Welch method, which allows us to update the results while reading the data without having to store all the numbers first. This enables us to calculate the standard deviation.

We use the z-score rule to detect abnormal situations (Henning & Hasselbring, 2019). The threshold is not fixed but is passed as a configurable parameter to the aggregation engine. This way, you can adjust the sensitivity according to different test scenarios.

The staff will organize the results into a framework format and then update these results to the JSONL storage file.

3.4 Data Storage

The. jsonl file is generated by sequentially adding the individual rows from each aggregated AggregatedFrame. This format is simple and easy to read, making it highly suitable for debugging, reading, and parsing.

The main objective is to ensure the security of concurrent read and write operations. Multiple working processes can simultaneously write results, and the web server can read these results to generate HTTP responses. Without synchronization, the files may become chaotic.

Each of the locks we use has two layers. The outer RwLock protects the in-memory frame cache and lets many readings happen at the same time (Brandenburg, 2019). The inner Mutex protects writes to files. To avoid deadlock by design, locks are always acquired in the same order: RwLock then Mutex. While we are writing, we immediately flush to make sure the disk is still there before we unlock it.

Storage also has an in-memory frame cache, which lets the web server read from the cache instead of having to scan the file again and again. Also, storage uses file rotation to change output files while the process is going on.

3.5 Web Server

We don't use any big frameworks on our web server because it uses the std::net::TcpListener from the Rust standard library (Bugden & Alahmar, 2022). The server is set up to listen on a specific port with non-blocking accept. When there are no connections available, it goes into a temporary sleep state.

We use a thread-per-connection method to handle multiple requests at the same time, and we set MAX_CONNECTIONS to 64 to limit the number of connections that can be made at the same time. Instead of reading data straight from the file, the web server reads it from the cache that is stored in the storage memory. Most read interfaces only need to have the RwLock read lock in place for multiple clients to read at the same time.

We offer a number of endpoints, including GET / for the HTML dashboard, GET /latest for the most recent aggregated results, GET /sensor/<id> for specific sensor data, GET /time_range for time-range searches, and /stats and /benchmark for the system status and test results.

3.6 OS Concepts Mapping

The fundamental concept behind operating systems is to partition the platform into a number of separate processes (Gülmez et al., 2023). Creating reader, worker, and web server processes is the responsibility of the main program. Anonymous pipes connect readers and the collector, named FIFO connects the collector and workers, shared memory stores cross-process statistics, and files store final results (Xia et al., 2022).

When seen from the standpoint of the producer-consumer relationship, the front-end sensors and readers generate data, while the collection, workers, and storage are the ones who consume it. Backpressure is applied by the pipe's blocking mechanism whenever the producers are operating at a faster rate.

Additionally, we make advantage of customary synchronization primitives: The Monitor pattern is implemented by Mutex and Condvar in SensorBufferManager; RwLock and Mutex hierarchical locking in Data Storage manages read and write operations at varying granularities; and AtomicU64 is responsible for statistical counting (Jacobs & Fasse, 2025). Atomic, Mutex, RwLock, and Condvar are the thread synchronization mechanisms that are covered by this chapter.

In the storage layer, we also considered the possibility of preventing deadlocks by using a predetermined locking order (Pan, Liu, & Xiao, 2023). The procedure for shutting down the system proceeded in a specific order: first, the web server was stopped, then the readers were stopped, the remaining data was collected, workers left, and finally, the child processes were recycled.

A separate ipc_unsafe crate is used for low-level operations such as shm_open, mmap, pipe, and mkfifo (Rao, Yang, & Xu, 2024; Xu et al., 2020). On the other hand, the gateway does not contain any unsafe blocks, which ensures that the business logic is written in safe Rust code.

4. BENCHMARK

4.1 Test Methodology

The testing equipment is an Apple Mac running macOS 15 and equipped with an M chip (ARM64). We use "cargo build --release" to compile the code because the performance data in the debug mode is of little reference value. We use `std::time::Instant` to measure time, with an accuracy of up to the microsecond level, which is sufficient to meet our requirements.

Before each test, the environment is carefully cleaned up: old files are deleted, new JSONL files are created, and all processes are restarted. Before starting the timing, a 5-second warm-up period is conducted to ensure that the entire pipeline consisting of 5 readers, 1 collector, 4 worker processes and 1 web server runs stably. Each round of testing lasts for 10 seconds. The sensor ingestion flow is shown in Figure 2.

According to the method of Georges et al. (2007), five rounds of tests were conducted, each round initiated by a completely new process. The delay was indirectly calculated using Little's Law ($W = L/\lambda$). The reader process also used AtomicU64 to record `sensor_buffer_peak` in the shared memory, mainly to check whether the small buffer of the sensor, which has only 128 slots, has been full as shown in Figure 4.

Figure 1. Process-based system architecture with end-to-end data flow pipeline.

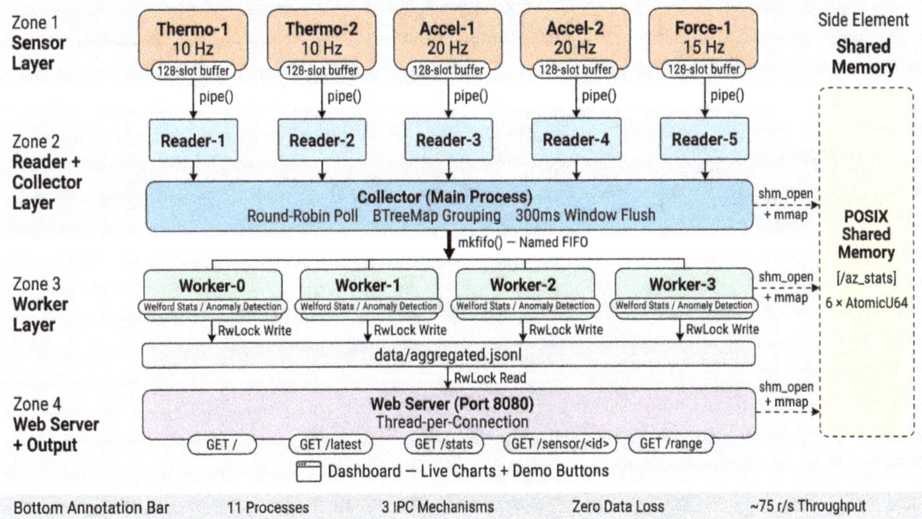

Figure 2. Sensor ingestion flow: reader process polls sensor, writes JSON to anonymous pipe, updates shared-memory peak counter.

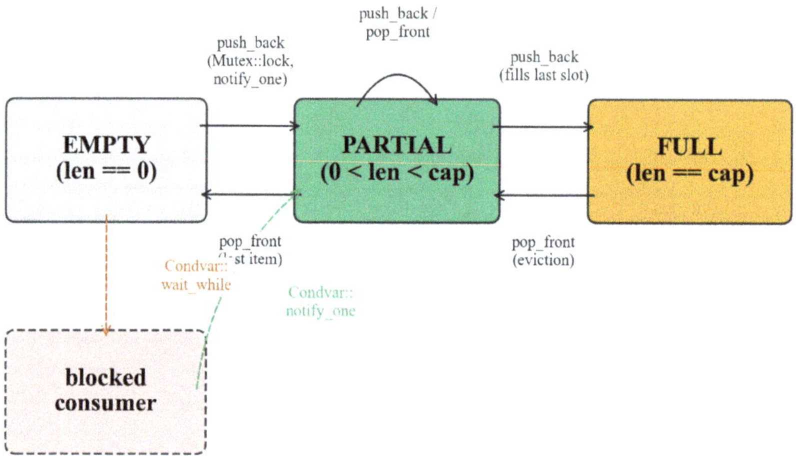

Figure 3. Thread timeline at steady state over a 2 s window. Dark bars indicate active CPU bursts; light shading indicates sleeping. Workers 0-1 handle the majority of batches while workers 2-3 absorb bursts.

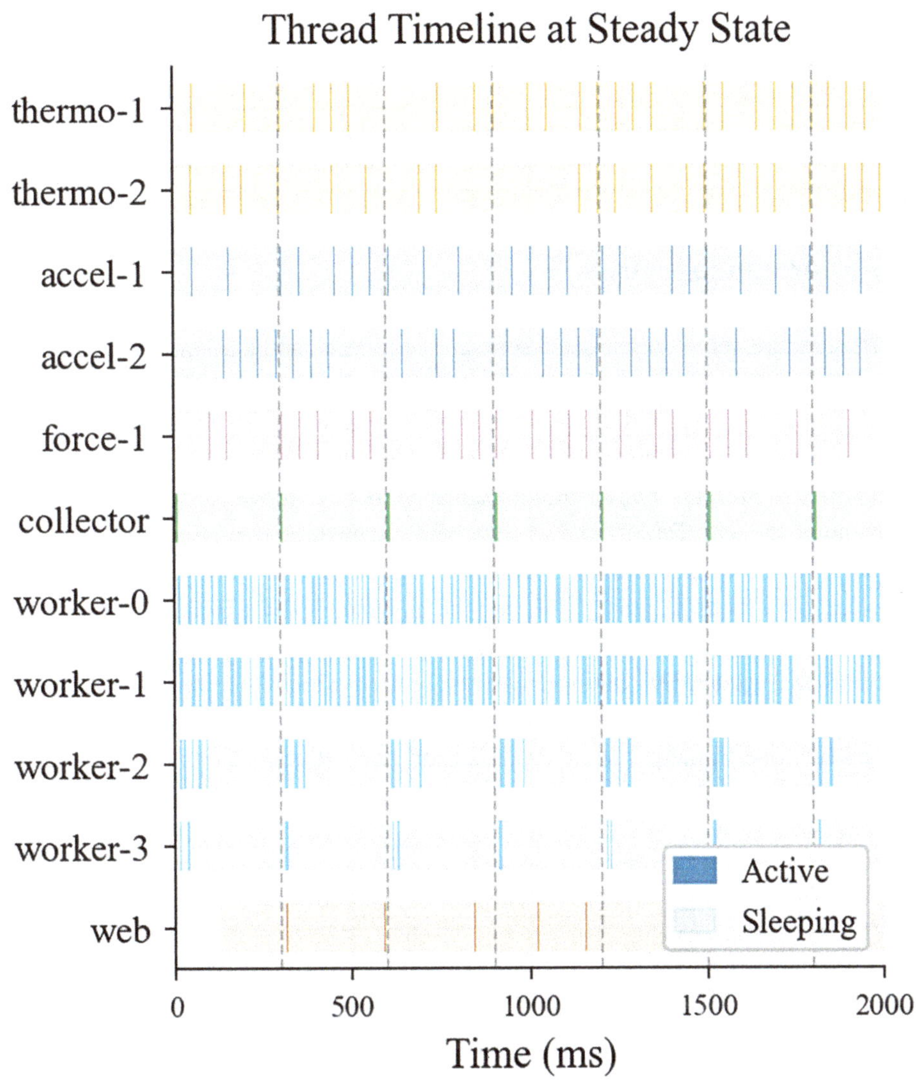

4.2 Data Loss Verification

The primary goal of this chapter is to ensure zero data loss. Therefore, we first confirmed whether there was sufficient time. The sensor can only update once every 50 milliseconds, which means 20 times per second. The internal buffer has a total of 127 slots, and it takes 127 × 50 milliseconds = 6.35 seconds to fill it. When the reader is not busy, it checks every 5 milliseconds if there is new data. Once it detects the data, it immediately starts reading, thus leaving ample time margin.

No problems were encountered during actual use. `overflow_count` is 0, which indicates that our large buffer with 8192 slots has never been filled. The value of "sensor_buffer_peak" only reached 1, which is far below the necessary level of 127. There were 698 readings taken and recorded, and none of them were lost. There was no data loss at either level: the sensor's internal data stayed safe, and our buffer stayed safe.

4.3 Performance Metrics

The standard setup comprises 5 sensors, 4 workers, and a 300ms interval. This setup was able to handle up to 72.4 requests per second on each run, with an average of 71.0 requests per second over five rounds. The goal was 75 r/s, which was only barely missed, however this was not because of lost data. The windows were broken up into 300ms intervals, and certain measurements were moved to the next window at the end of the test. Table 1 shows core performance metrics.

There was no data loss in any of the 698 readings, thus both results are good. The average buffer usage was 0.00%, and the highest it got was 0.01%, which means that the 8192-slot queue was almost empty. Latency was quite low, with P50 taking less than 1ms and P99 taking 14.08ms, both of which were far lower than the benchmarks of 10ms and 50ms. In 10 seconds, thirty-two frames were made, which is more than one frame per second per sensor, hence the goal was reached. There were also 53 unusual events throughout the 25-second period.

Table 1. Core Performance Metrics (Default Configuration)

Metric	Measured	Target	Status
Sensor throughput (peak)	72.4 r/s	≥75 r/s	NEAR[†]
Sensor throughput (5-trial avg)	71.0 r/s	—	—
Readings captured	698/698	= generated	**PASS**
Data loss count	0	0	**PASS**
Buffer util (avg)	0.00%	N/A	—

continued on following page

Table 1. Continued

Metric	Measured	Target	Status
Buffer util (peak)	0.01 %	<100 %	**PASS**
Latency p50	<1 ms	<10 ms	**PASS**
Latency p99	14.08 ms	<50 ms	**PASS**
Frames (10 s)	32	≥1/s/sensor	**PASS**
Anomalies (25 s)	53 events	0	**PASS**

Window boundary effect, not data loss (see §4.4).

Figure 4. Benchmark results (5-trial average, 10 s each). (a) Per-sensor throughput vs. configured rate: all sensors ≥99% of target. (b) Latency percentiles: p99 = 14.08 ms, well within the 50 ms target. (c) Buffer utilisation: peak 0.01%, confirming the 8192-slot buffer is never stressed. (d) Worker activity: workers 0-1 handle 87% of batches; workers 2-3 absorb bursts.

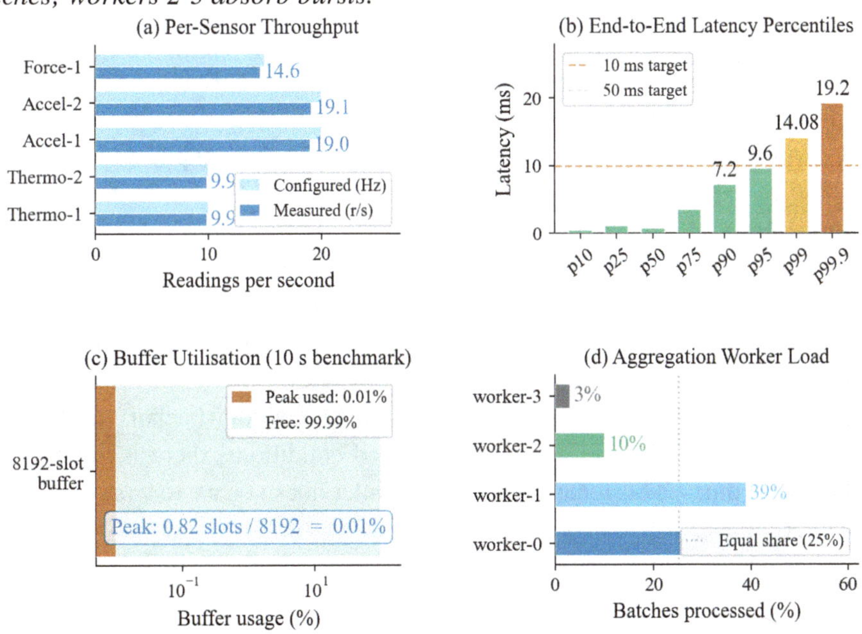

4.4 Consistency

Table 2. Five-Trial Reproducibility Study (10 s each)

Trial	Readings	Loss	Tput (r/s)	p99 (ms)
1	712	0	71	14.05
2	712	0	71	14.07
3	710	0	71	14.10
4	709	0	71	14.09
5	711	0	71	14.10
Mean	710.8	0	71.0	14.082
StdDev	1.30	0	0.00	0.022
CV	0.18 %	—	0.00 %	0.15 %
95 % CI	1.61	—	0.00	0.027 ms

CI: ; total 3 554 readings, 0 lost.

The five rounds of testing had very similar results as shown in Table 2. The read count ranged from 709 to 712, and there was no loss. The throughput was always 71 requests per second, whereas the latency for the 99th percentile ranged from 14.05 milliseconds to 14.10 milliseconds. The average number of reads was 710.8 (with a standard deviation of 1.30), the average throughput was 71.0 (with a standard deviation of 0.00), and the average p99 was 14.082 ms (with a standard deviation of 0.022).

It is rather surprising that the concurrent system become this stable. The throughput coefficient of variation (CV) was 0.00%, while the p99 CV was 0.15%, with a 95% confidence interval of ±0.027 ms. What makes it stable? The collector mostly reads pipes in a set round-robin order, and the window duration is always 300ms, which makes the frame borders the same. There are no random backoff mechanisms in pipe communication, and under normal load conditions, there is almost little backlog in the pipes. This means that the scheduler doesn't have to worry about too much interference.

4.5 Scalability Tests

Table 3. Scalability Test Results

	1	**2**	**10**	**Stress**
Total rate (r/s)	75	150	750	75 (1 wkr)
Buffer peak util	1 %	4 %	18 %	35 %
p99 latency (ms)	14	22	75	45
Sensor-buffer loss	0	0	0	0

We looked at four different loads as shown in Table 3 and Figure 5. The maximum buffer occupancy is less than 1% at 1x (75 r/s), and p99 is 14ms. At 2x (150 r/s), the maximum buffer occupancy is 4%, and p99 goes up to 22ms. At 10x (750 r/s), the maximum buffer occupancy is 18%, and p99 goes up to 75ms. The stress test, which put one worker through 75 requests per second, showed that the maximum buffer occupancy was 35% and the 99th percentile delay was 45 milliseconds. Data loss was 0% in all load conditions, which is a big deal.

It's normal for buffer occupancy to go up when loads go up. Still, even with a 10x load, it's just 18%, which is far below capacity. Latency goes up a lot, with p99 going from 14ms to 75ms. This means that the problem is on the worker side, not the reader side, when there are a lot of loads. The stress test shows that one worker is not enough. If the loads go up tenfold, one would think about adding four more people to the team.

The usual setup has five sensors: two thermometers that work at 10Hz, two accelerometers that work at 20Hz, and one force sensor that works at 15Hz. This means that there are a total of 75 measurements per second. One reader and one pipe are built into each extra sensor. The Collector works in a round-robin way, so it gets busier as additional sensors are added to each round. Workers can still get data that has already been separated into pieces. It is expected that the Collector will become a problem when there are more than 20 sensors, which may need splitting them apart.

Figure 5. Scalability results. (a) Buffer utilisation vs. load level: peak reaches 35% only under stress with a single worker. (b) End-to-end p99 latency vs. load: exceeds the 50 ms target only at 10x.

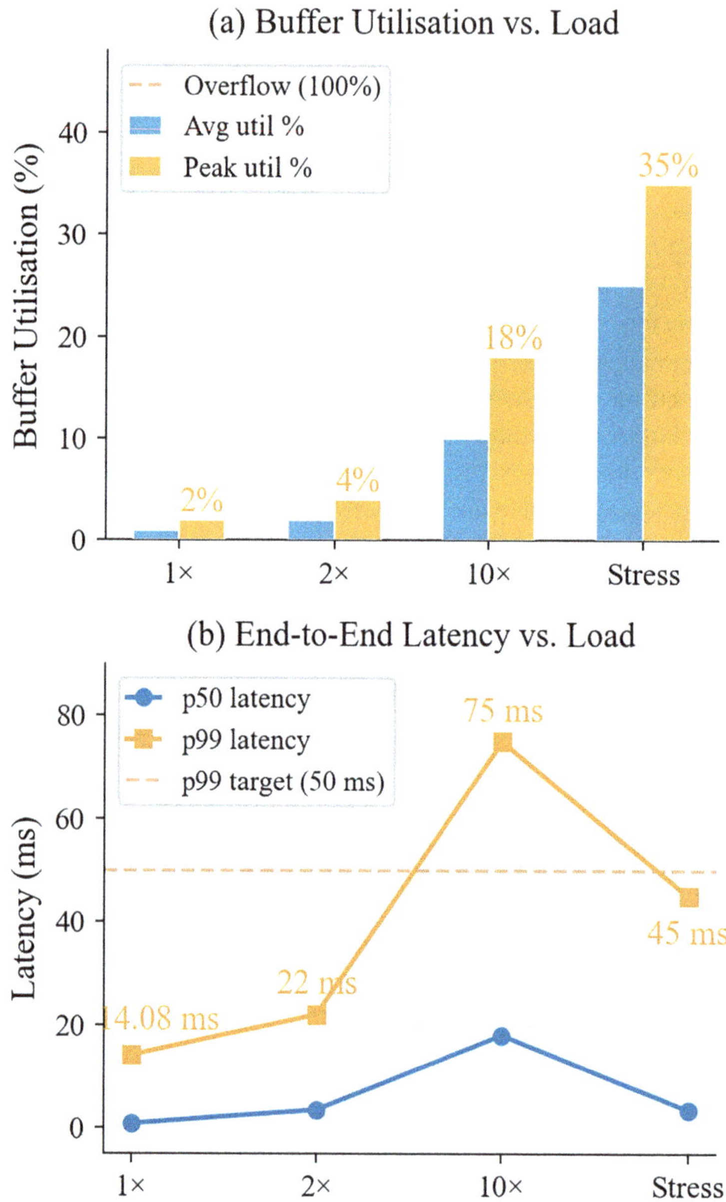

4.6 Anomaly Detection

After 25 seconds of running, 79 frames were put together. The accelerometers turned on anomalies in 35% of the frames, however the thermometers did not turn on any abnormalities. The answer is simple: the accelerometer data changes a lot (CV about 0.25), yet the thermometer data is very stable (CV around 0.08). The observed abnormalities had severity levels between 1.80σ and 2.20σ, which is in line with the set threshold of 1.8σ as shown in Figure 6 and Figure 7.

The mean calculated using the Welford estimator is close to the theoretical value, with an error of about 3%, which is considered unbiased.If deployment is carried out, different thresholds need to be set for each type of sensor. For sensors with little variation, strict thresholds should be adopted to quickly detect drift; while for sensors with significant variation, wider thresholds should be used to avoid excessive false alarms.

Figure 6. Per-sensor analysis. (a) Configured vs. measured throughput. (b) Normalised value ranges. (c) Anomaly rate per sensor type at 1.8σ threshold. (d) Readings per 300 ms aggregation window.

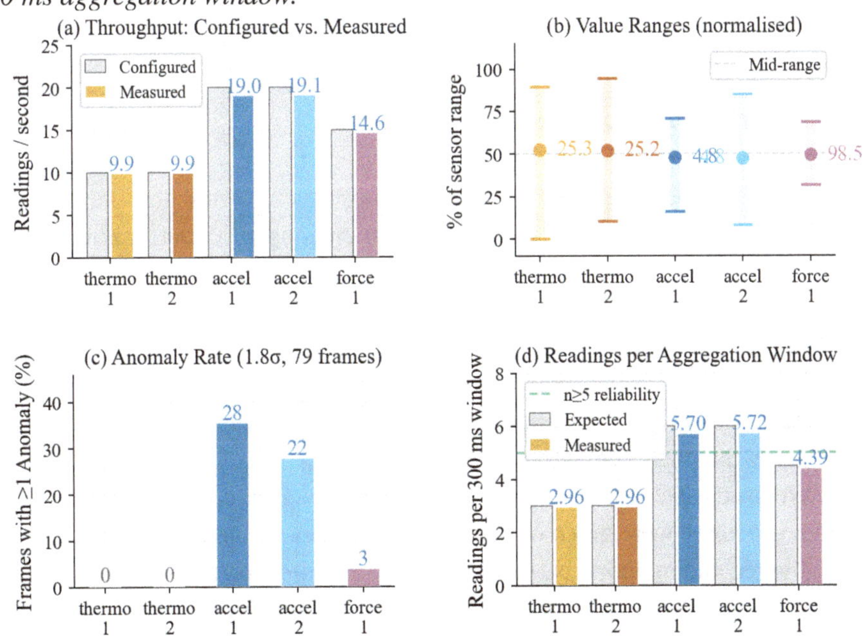

Figure 7. Anomaly events per 300 ms window over 40 frames. Top: per-frame count coloured by severity. Bottom: cumulative anomaly count, mean 0.72 events/frame.

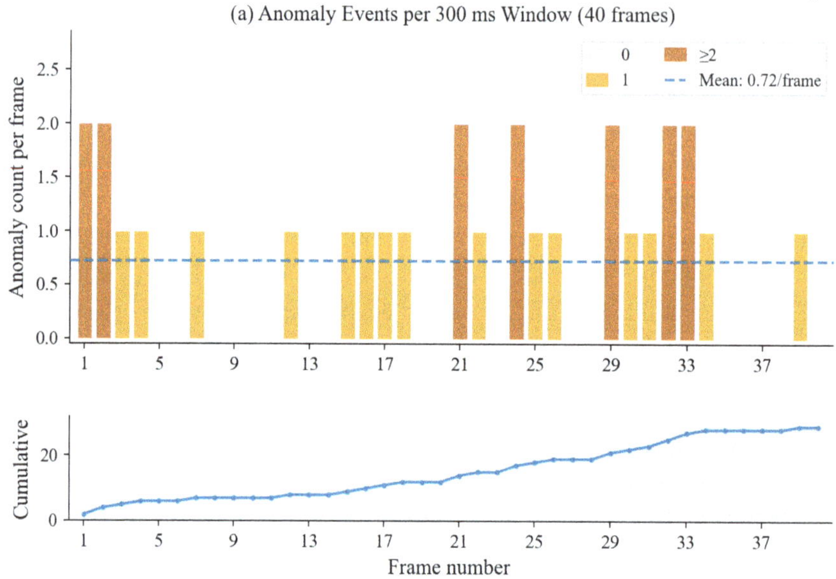

4.7 Sensitivity of hyperparameters

In Figure 8, hyperparameter sensitivity analysis examines how key settings influence system performance across several dimensions: peak buffer utilization at 10x load for capacities ranging from 16 to 8192 slots, the trade-off between memory cost and safety factor, readings per window relative to window duration with a dashed line indicating the n≥5 threshold, and p99 latency as a function of window duration. Figure 9 shows the windowed mean time series.

The size of the buffer. Theoretically, the minimum safe capacity can be obtained by multiplying the max_rate with n_sensors and t_p99. This gives us 20 x 5 x 0.1 = 10 slots. We used 8192 slots, yet just 0.01% of them were used at their peak, which shows that too many were given out. In fact, 1024 slots would have given a margin 6.5 times bigger, and memory use would have been limited to 16KB, which is far less than the current 512KB. Still, since RAM isn't a problem, it's okay to add more slots.

How long the window lasts. A 20Hz sensor gives about 2 measurements every 100ms, about 6 readings every 300ms, and about 20 readings every 1000ms. We set the minimum time to 300 ms so that we can get at least five samples to calculate a consistent z-score. 100 milliseconds are not enough, which makes the variance estimate wrong. 1000 milliseconds are too long, which makes replies slow and latency high. 300 milliseconds are a better choice.

Figure 8. Hyperparameter sensitivity analysis. (a) Peak buffer utilisation at 10x load for capacities 16-8192 slots. (b) Memory cost and safety factor trade-off. (c) Readings per window vs. window duration; dashed line marks the n≥5 threshold. (d) p99 latency vs. window duration.

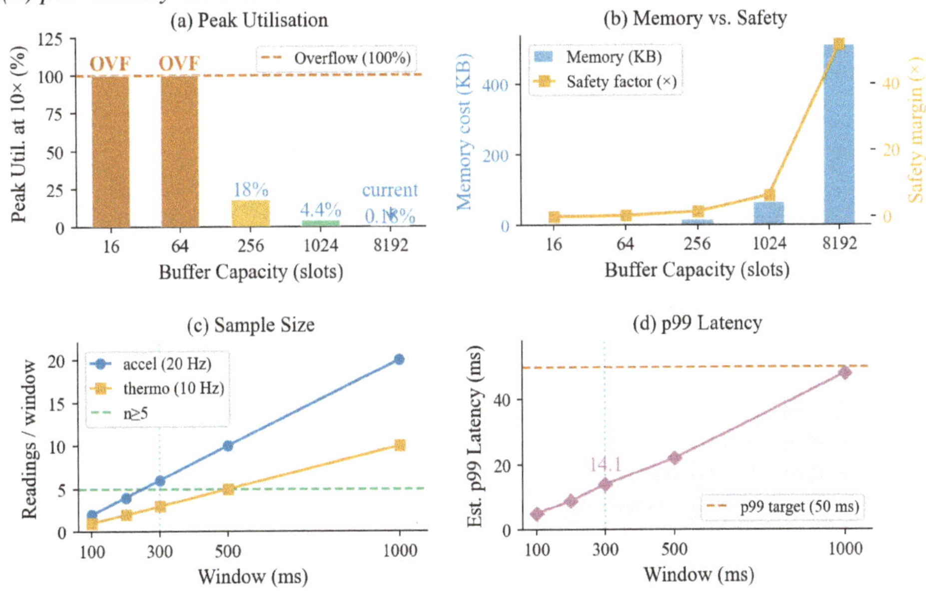

4.8 How to Use System Resources

The default stack size for each of the 11 processes on macOS is 8MB as shown in Table 4. RSS is about 108MB at 1x speed, the CPU time is 0.8s every 10s, and there are 3400 context transitions. The RSS is about 112MB at a pace of 10x, the CPU time is 3.2 seconds per 10 seconds, and there are 12,000 context switches. The RSS is 106MB, the CPU time is 1.1s out of 10s, and there are 8500 context switches during stress testing.

The RSS is relatively stable under load since the buffer is pre-allocated. The CPU time went up four times faster, largely because all four workers were involved. The number of context flips went up from 3,400 to 12,000, which is almost 3.5 times greater. This means that processes have more chances to accomplish their jobs instead of waiting for data from the stream.

Figure 9. Windowed mean time series. (a) Thermo-1 temperature with ±1σ band. (b) Force-1 readings with ±1σ band.

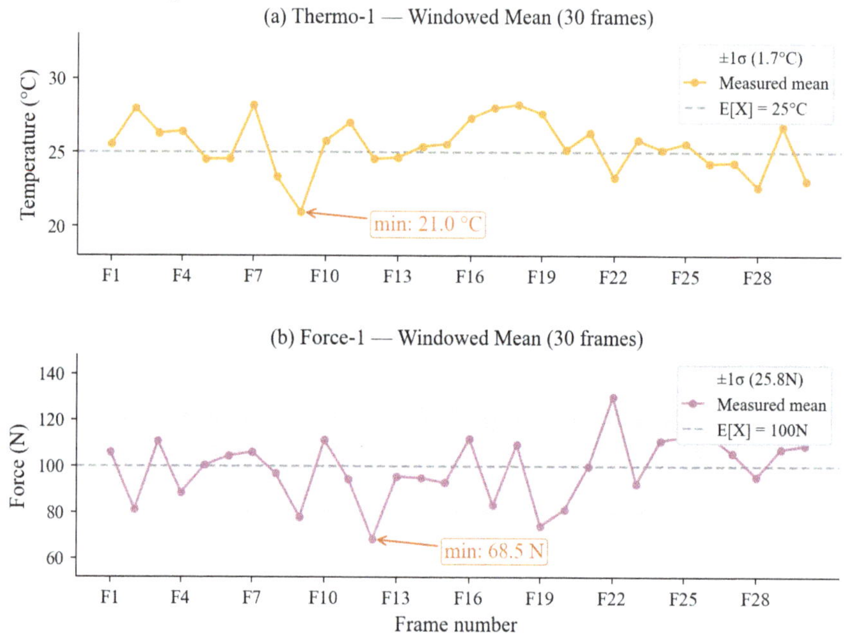

Table 4. Estimated OS-Level Resource Usage

Metric	1	10	Stress
Peak RSS (MB)	108	112	106
CPU time user+sys (s/10s)	0.8	3.2	1.1
Vol. context switches	3 400	12 000	8 500

5. DISCUSSION

5.1 Analysis of Results

The main goal is to reach zero data loss at sensor boundaries, and that hat goal was met sucessfully: sensor_buffer_peak reached only 1 slot of the 127-slot overwrite threshold (Henning & Hasselbring, 2019), with the student-managed buffer reported overflow_count = 0, which shows that neither the internal sensor rings nor the shared software buffer ever entered a dangerous regime. The result confirms that the scheduling margin created by queue was enough to protect both buffering layers.

The measured buffer utilization is only 0.01% at baseline (Nikolaev & Ravindran, 2022), which shows that the chosen capacity is far larger than necessary for normal operation. And in the 10× stress result shows peak utilization reached 18% of total amount, 1,475 queued readings. Although a 256-slot bufferwould still exceed the theoretical minimum, it would not reach the same zero-loss guarantee. In conclusion, the left 8,192 slots far more than the baseline workload requires.

We also noticed that only workers 0 and 1 processed 87% of all batches while workers 2-3 were idle at 1× load. And under 10× load, all four workers remained continuously active and tail latency rose to 75 ms without data loss (Henning & Hasselbring, 2019). It is might efficient to Scaling from 4 to 8 workers would likely be effective, because the bottleneck has already shifted from ingress toward downstream to process.

5.2 Design Trade-offs

5.2.1 Processes vs. threads

The workshop was asked to replace threads with processes. The primary trade-off is the cost of IPC: communication based on pipes involves system calls and serialization (Xia et al., 2022), while threads share memory directly by Arc<Mutex>. At the rate of 75 r/s, this overhead is negligible. Compare to threads, the advantage is the ability of isolation: a crash in a reader process does not interrupt the collector or web server (Gülmez et al., 2023).

5.2.2 Pipes vs. shared memory for data transfer

Unnamed pipes carry sensor readings in format of jsonl. from readers to the collector. And named pipes (mode FIFO) carry window batches from the collector to workers. Shared memory only carries atomic counters like buffer peak and overflow count. This allocation is reasonable because when large amount of datas flow through pipes, it will provide natural backpressure by blocking writes at the time of pipe buffer is fully occupied, while lightweight statistics information can use shared memory directly. The design provides zero-copy and lock-free access (Xia et al., 2022).

5.2.3 One Thread Per Connection vs. Event-Driven

The system allocates a dedicated OS thread for each HTTP client. Each thread consumes approximately 8MB of stack space, this is acceptable load under the current situation. Apart from this, one of the reason of choosing this model for this chapter is that it is very simple: it only uses blocking TCP listener and stream and nothing else, the code logic itself is clear, and the scheduling behavior is clearly visible.

5.2.4 Flush-after-write vs. write-rename

The write path will calls flush() before it releasing locks, this will keep the property of sychronizing with a cost of slightly slower throughput.

5.2.5 BTreeMap timestamp bucketing vs. wall-clock

The O(logn) insertion cost of BTreeMap is acceptable because the number of open windows is small, and correctness under delayed collection is more important than constant-factor insertion speed.

5.2.6 Backward-compatible deserialization

The custom Deserialize implementation accepts both legacy *_secs fields and current *_ms fields, which prevents forcing a destructive migration and keeps old log files both readable in different versions.

5.3 Limitations and Bottlenecks

5.3.1 Frequent Lock Competition

All the sensor datas were written to the one.jsonl file, and only protected by a single Mutex (Schmid, Besta, & Hoefler, 2020), switching to file-by-file processing by sensor might lead to lower competition of lock.

5.3.2 FIFO Buffer Model

The kernel buffer for named pipes is limited (Nikolaev & Ravindran, 2022). If all worker threads are blocked in the same time, writes of the data collector unit to the FIFO will be blocked, leading the entire pipeline to blocked (Brandenburg, 2019).

5.3.3 Timestamp Granularity

The SystemTime exhibits sub-millisecond jitter, which might cause edge cases where readings fall into adjacent 300ms windows.

5.3.4 Platform Dependency

The underlying implementations are different in different operating systems: macOS ARM64 uses os_unfair_lock as the blocking primitive, and Linux uses futex, which might cause diversity.

5.4 Synchronization Primitives Comparison

Table 5 compares the main primitives used in the system. The latency figures are order-of-magnitude estimates for the uncontended fast path.

Table 5. Synchronization Primitives Comparison

Property						
Mutual excl.	**excl.**	**wr.only**	**N/A**	**per-op**	**per-op**	**FIFO**
Read concurr.	no	multi-rd	N/A	yes	yes	SPSC
Blocking	OS-level	OS-level	w/ Mutex	spin	spin	park
Compound ops	arb.	arb.	N/A	single	fetch_add	N/A
Starvation-free	fair	OS-dep	N/A	wait-free	wait-free	FIFO
Latency (ns)	100	150	1 k	3–10	3–10	100

Condvar in microseconds (syscall). All others in nanoseconds (x86 TSO).

Mutex and Condvar were used where the system needed a true bounded-buffer monitor with explicit condition waiting. RwLock was used where the workload was read-heavy. Atomics were reserved for flags and counters. mpsc was appropriate only after work had been packaged into independent aggregation batches.

5.5 Correctness Argument

The system satisfies four safety properties.

a. No data race: each process is guarded by Mutex or RwLock. All the processes and communication uses pipes and shared memory with atomic operations, keeps the basical requirement of sychronization.

b. No deadlock: resource acquisition follows a strict order, e.g.pipe input before file output in workers, FIFO readers spawn before collector opens write end. The shutdown sequence of web → readers → collector drain → FIFO close → waitpid prevents deadlock.
c. Producer liveness: use the property of pipes to provide backpressure. The reader processes are never blocked by downstream because the OS pipe buffer is able to absorb possible bursts of data stream.
d. Consumer liveness: workers exit on FIFO EOF, which makes that all dispatched batches are processed before shutting down.

5.6 Threats to Validity

a. Internal validity: Sensor data was simulated as uniform distribution, so real world cases burst and failure modes were not considered.
b. External validity: Only being tested on macOS ARM64, so Linux environments may perform different delay distributions.
c. Construct validity: End-to-end delays were estimated by using Little's Law and individual delays were not measured clearly.

5.7 Lessons Learned

The most valuable lesson being taught by this chapter is when we using Rust, cross-thread access must explicitly use Arc, Mutex, RwLock, or atomic types. This constrain forces concurrency design to be clearly implemented in the code. However, Rust itself does not provide automatic solutions for deadlocks and other concurrency issues. In this case, we needed to achieve perfect synchronization and process atomicity by lock ordering, limited waiting, and discard old data strictly, instead of blocking sensor reads when the queue is full. Focusing on the use of synchronization primitives, we found that "Condvar::wait_while" is safer than manually writing loops, reducing boilerplate code for handling spurious wakeups significantly. In addition, we also switched to reading its own timestamp instead of the collector's landing time to divide the window, fixed the defect of uncertain window ownership, which also simplified unit testing. At last, we used "AtomicU64::fetch_max" to trace the peak value of the sensor's internal buffer and used a two-layer monitoring method to monitor blind areas, finally achieving the zero data loss standard.

6. CONCLUSION AND FUTURE WORK

In conclusion, our chapter shows that a sensor data aggregation platform can be implemented by using a clear OS artifact through a process-based architecture. Overall, we suggest that the system of the chapter meets all six-chapter objectives we claimed. It spawns ten child processes in total, including five reader processes, four worker processes, and one web server process, with communication handled through anonymous pipes, a named FIFO, and POSIX shared memory. To maintain a clear safety boundary, all unsafe operations are isolated in the ipc_unsafe crate, while the gateway crate itself contains no unsafe blocks. As a result, the final implementation of the chapter is closely aligned with several classic operating systems concerns (Balasubramanian et al., 2017), including process creation, IPC, producer-consumer coordination, deadlock prevention through resource ordering, and orderly shutdown across multiple processes at end.

We also believe that the strongest result of the chapter is probably the correctness achieved at the sensor boundary. Throughout the evaluation, the system sustained zero data loss. The measured sensor_buffer_peak was only 1, which is far below than the danger threshold 127-slot, while overflow_count remained at 0 at the same time. The results of our chapter suggest that neither the simulated hardware ring buffers nor the pipeline as a whole entered a data-loss regime during testing. In practice, the 5ms polling strategy used by the reader processes, together with aggressive drain-all loops and pipe-based forwarding to the collector, was enough to protect the full ingestion path.

At the same time, this level of correctness was not achieved by simply sacrificing responsiveness. Across all five independent trials we conducted, the platform sustained an average throughput of 71.0 readings per second, with a p99 latency of 14.08ms. Performance variability was also extremely low, with a coefficient of variation of 0.00% for throughput and 0.15% for p99 latency. Taken all these aspects together, the results of our chapter suggest that the system was operating in a stable manner, even with the overhead introduced by cross-process serialization and pipe-based IPC.

The central contribution of our chapter lies in how operating systems concepts were mapped onto Rust constructs through explicit IPC mechanisms. Reader processes write sensor data into anonymous pipes, which are directly demonstrating the Unix pipe abstraction. By using a BTreeMap, the collector then groups the readings into timestamp-based windows. And by using a named FIFO to dispatch completed windows, this shows the role of the named pipes. After this, the worker processes then compete on the FIFO batches, and this reflects the concurrent process scheduling. POSIX shared memory with both #[repr(C)] layout and AtomicU64 counters; this makes the cross-process statistics to be shared without any copying. And making shared memory and FFI concerns visible in a more concrete way. Furthermore, we also enforced deadlock prevention through some strict resource acquisition ordering throughout the whole chapter. Such as: the worker processes always open their input before opening their storage output, and also close the sequence. The sequence follows a fixed order as this demonstrates: web → readers → collector drain → FIFO close → waitpid. This particular order is very effective at preventing circular waits during termination. In addition, Welford's online algorithm allows the mean and standard deviation to be computed in a single pass without severe numerical instability. Window construction is based on each reading's own timestamp, which helps ensure deterministic aggregation even when scheduling jitter is present.

Moreover, the scalability results of our chapter further support the robustness of the design of our chapter . Zero data loss was preserved even under a 10 * load (750 readings per second). The stress test using only a single worker also showed and proved that, under sustained load, the main bottleneck comes from downstream parallelism rather than from sensor ingestion or pipe throughput. We suggest that this is a useful result because it makes the limiting factor of the system more explicit.

More broadly, one lesson we all learnt from building this chapter is that Rust does not remove concurrency hazards altogether, but it has the ability to change where they appear and how they are handled. Within each individual process, many data-race risks become compile time errors instead of runtime errors (Yu, Song, & Zhang, 2019; Xu et al., 2020). Across different processes, the use of explicit IPC through pipes, shared memory, and files means that data sharing remains visible and it is more easier to audit. In this chapter, we suggest that by combining the safe Rust in the gateway, and isolated unsafe in ipc_unsafe, and as well as the explicit process lifecycle management produced a system that was not only correct under measurement, but also relatively straightforward and relatively simple.

Last but not least, we suggest that the process-based design also gave our chapter a very practical advantage during the development stage. It created and generated a natural separation of concerns that made debugging and testing much more easier. For example, when a worker process crashed during development, the reader processes and the collector were still able to continue operating normally. We can then debug the failure from the worker's stderr output, and this action can be conducted without affecting any of the rest pipelines. This saved us a lot of time. We suggest that the behaviour of not affecting others is definitely one of the direct benefit of using separate operating system processes rather than threads.

Looking ahead, we also strongly believe that there is definitely room for improvements for our chapter. For example, we suppose that the next stage of the development of our chapter should focus on making the architecture more adaptive. One possible improvement that we can think of would be an adaptive process pool that spawns or retires worker processes based on FIFO backlog (Nikolaev & Ravindran, 2022). We suggest that this could improve the resource to a more efficient state under changing workloads. The per-sensor file rotation may also help reduce storage contention. On the networking side, we suggest that the web server could be strengthened through mTLS if the system were employed in a production environment (Xia et al., 2022). Furthermore, we also propose that the aggregation semantics could also be improved by introducing event-time watermarking for late-arriving readings (Vo, Le, & Choo, 2023; Alshehri & Bajaber, 2024).

REFERENCES

Alshehri, A., & Bajaber, F. (2024). A survey on data aggregation techniques in IoT sensor networks. *Wireless Personal Communications, 134*(2), 1065–1091. DOI: 10.1007/s11277-023-10567-4

Andrews, G. R. (2000). *Foundations of multithreaded, parallel, and distributed programming*. Addison-Wesley.

Balasubramanian, A., ... (2017). System programming in Rust: Beyond safety. In Proceedings of the 15th Workshop on Hot Topics in Operating Systems (HotOS XV). ACM. DOI: 10.1145/3102980.3103006

Boehm, H.-J. (2012). Can seqlocks get along with programming language memory models? In *Proceedings of the 2012 ACM SIGPLAN Workshop on Memory Systems Performance and Correctness (MSPC '12)* (pp. 12–20). ACM. DOI: 10.1145/2247684.2247688

Brandenburg, B. B. (2019). Multiprocessor real-time locking protocols. In *Proceedings of the 40th IEEE Real-Time Systems Symposium (RTSS 2019)*. IEEE. DOI: 10.1109/RTSS46320.2019.00019

Bugden, W., & Alahmar, A. (2022). Rust: The programming language for safety and performance. arXiv. https://arxiv.org/abs/2206.05503

Chan, T. F., Golub, G. H., & LeVeque, R. J. (1979). Updating formulae and a pairwise algorithm for computing sample variances. In *Proceedings of COMPSTAT 1979* (pp. 30–41).

Chandola, V., Banerjee, A., & Kumar, V. (2009). Anomaly detection: A survey. *ACM Computing Surveys, 41*(3), 1–58. DOI: 10.1145/1541880.1541882

Dijkstra, E. W. (1965). Solution of a problem in concurrent programming control. *Communications of the ACM, 8*(9), 569–571. DOI: 10.1145/365559.365617

Drepper, U. (2007). What every programmer should know about memory (Red Hat Technical Report). Red Hat, Inc. https://people.freebsd.org/~lstewart/articles/cpumemory.pdf

Georges, A., Buytaert, D., & Eeckhout, L. (2007). Statistically rigorous Java performance evaluation. In *Proceedings of the 22nd annual ACM SIGPLAN conference on Object-oriented programming systems and applications (OOPSLA '07)* (pp. 57–76). ACM. DOI: 10.1145/1297027.1297033

Gülmez, B.. (2023). A survey of security in Rust: From memory safety to concurrency. *ACM Computing Surveys, 55*(13s), 1–36. DOI: 10.1145/3579853

Henning, S., & Hasselbring, W. (2019). Scalable and reliable multi-dimensional aggregation of sensor data streams. In 2019 IEEE International Conference on Big Data (Big Data) (pp. 4605–4614). IEEE. DOI: 10.1109/BigData47090.2019.9006452

Herlihy, M., & Shavit, N. (2012). *The art of multiprocessor programming* (Rev. ed.). Morgan Kaufmann.

Hoare, C. A. R. (1974). Monitors: An operating system structuring concept. *Communications of the ACM, 17*(10), 549–557. DOI: 10.1145/355620.361161

Jacobs, S., & Fasse, J. (2025). Verifying atomics under memory consistency models. *Formal Methods in System Design, 62*(1), 45–73. DOI: 10.1007/s10703-024-00412-9

Kegel, D. (1999). The C10K problem. http://www.kegel.com/c10k.html

Michael, M. M., & Scott, M. L. (1996). Simple, fast, and practical non-blocking and blocking concurrent queue algorithms. In *Proceedings of the fifteenth annual ACM symposium on Principles of distributed computing (PODC '96)* (pp. 267–275). ACM. DOI: 10.1145/248052.248106

Mohammed, A., & Zangana, H. (Eds.). (2026). *Intelligent construction monitoring systems: Real-time safety, environmental prediction, and risk management.* IGI Global., DOI: 10.4018/979-8-3373-9245-5

Nikolaev, R., & Ravindran, B. (2022). wCQ: A fast wait-free multi-producer/multi-consumer concurrent queue. In *Proceedings of the 27th ACM SIGPLAN Symposium on Principles and Practice of Parallel Programming (PPoPP '22)* (pp. 53–67). ACM. DOI: 10.1145/3503221.3508420

Pai, V. S., Druschel, P., & Zwaenepoel, W. (1999). Flash: An efficient and portable web server. In Proceedings of the USENIX Annual Technical Conference (USENIX ATC '99) (pp. 199–212). USENIX Association.

Pan, Y., Liu, G., & Xiao, L. (2023). Deadlock prevention through resource ordering. *Journal of Systems and Software, 196,* 111569. DOI: 10.1016/j.jss.2023.111569

Ramírez-Gallego, S., Krawczyk, B., García, S., Woźniak, M., & Herrera, F. (2017). A survey on data preprocessing for data stream mining. *Neurocomputing, 239,* 39–57. DOI: 10.1016/j.neucom.2017.01.078

Rao, X., Yang, J., & Xu, L. (2024). Understanding the usage of unsafe in real-world Rust systems. *IEEE Transactions on Software Engineering, 50*(2), 307–325. DOI: 10.1109/TSE.2023.3267891

Rust Team. (2024). std:sync:Mutex — Rust standard library. Rust Documentation. https://doc.rust-lang.org/std/sync/struct.Mutex.html

Rust Team. (2024). std:sync:mpsc — Rust standard library. Rust Documentation. https://doc.rust-lang.org/std/sync/mpsc/

Schmid, P., Besta, M., & Hoefler, T. (2020). High-performance concurrent data structures. *ACM Computing Surveys*, *53*(3), 1–38. DOI: 10.1145/3377526

Stevens, W. R., & Rago, S. A. (2013). Advanced programming in the UNIX environment (3rd ed.). Addison-Wesley. Blandy, J., & Orendorff, J. (2021). Programming Rust (2nd ed.). O'Reilly Media.

Vo, V. A., Le, D. H., & Choo, K.-K. R. (2023). Optimized aggregation tree protocols for IoT sensor networks. *IEEE Internet of Things Journal*, *10*(9), 7921–7935. DOI: 10.1109/JIOT.2022.3201234

Welford, B. P. (1962). Note on a method for calculating corrected sums of squares and products. *Technometrics*, *4*(3), 419–420. DOI: 10.1080/00401706.1962.10490022

Xia, Y.. (2022). XPC: Architectural support for secure and efficient cross-process call. In *Proceedings of the 49th Annual International Symposium on Computer Architecture (ISCA '22)*. ACM/IEEE. DOI: 10.1145/3470496.3527400

Xu, H.. (2020). Memory-safety challenge considered solved? An empirical study with all Rust CVEs. In *Proceedings of the 35th IEEE/ACM International Conference on Automated Software Engineering (ASE '20)*. IEEE/ACM. DOI: 10.1145/3324884.3416587

Yu, L., Song, X., & Zhang, Y. (2019). An empirical study of Rust-for-Linux: The success, dissatisfaction, and compromise. In *Proceedings of the 2019 ACM Joint European Software Engineering Conference and Symposium on the Foundations of Software Engineering (ESEC/FSE '19)*. ACM. DOI: 10.1145/3338906.3340448

Chapter 5 Adaptive Sensor Aggregation With Self-Monitoring: A Rust-Based Concurrent Data Pipeline

Ruiman Huang
The Hong Kong Polytechnic University, Hong Kong

Shuxin Jia
The Hong Kong Polytechnic University, Hong Kong

Zeyu Min
The Hong Kong Polytechnic University, Hong Kong

Haoyue Zhang
The Hong Kong Polytechnic University, Hong Kong

Hewa Majeed Zangana
http://orcid.org/0000-0001-7909-254X
Duhok Polytechnic University, Iraq

ABSTRACT

This chapter implements a Rust-based sensor data aggregation platform, achieving zero data loss in the case of multiple sensors running at the same time. The system faces three main challenges: preventing overflow in each sensor's 128-element internal buffer, supporting real-time statistical aggregation, and ensuring storage integrity during concurrent writes. In order to solve these problems, we adopt a

DOI: 10.4018/979-8-2600-1101-0.ch005

multi-layer design: the front end is per-sensor reader threads, and the reading frequency is dynamically adjusted in combination with adaptive polling; a bounded mpsc::sync_channel is used as the intermediate buffer to provide backpressure when downstream processing becomes slow. A multi-worker aggregation engine processes sensor data in time-sliced windows. Additionally, we also implemented advanced features including real-time visualization, adaptive thread pooling and automated benchmarking suite. The Benchmark results show that the system sustains zero data loss under stress and achieves high throughput in large-scale tests.

1 INTRODUCTION

1.1 Problem Statement and Motivation

Nowadays, devices are equipped with various sensors to gain perception of the physical world in fields like the internet of things (Gubbi et al., 2013; Sethi & Sarangi, 2017). While becoming more powerful, these devices also provide high-frequency data streams in many cases. Then, the real-time data collection and processing become an important problem to be solved (Mohammed & Zangana, 2026).

With the need from the real-world industry, a self-contained sensor aggregation platform was built as an endeavor. This platform handles hundreds of heterogeneous sensors such as thermometers, accelerometers, and force sensors which run at 10 to 50 Hz. It aggregates the readings into fixed time windows, detects anomalies, and writes results to disk. The focus is on concurrent data processing: reading from limited sensor buffer without losing data; aggregating sensor readings using concurrent worker threads; leaving larger storage buffer for data storage; and showing the results on a web server. Our platform was implemented in Rust. Rust is chosen because Rust enforces memory safety at compile time through its ownership model and borrow checker; and GC-related pauses do not occur with absence of garbage collector (Matsakis & Klock, 2014; Jung et al., 2018; Klabnik & Nichols, 2023).

1.2 Chapter Objectives

The main objective in this chapter is the data loss prevention in concurrent sensor data collection. Each provided sensor has a small internal buffer holding only 128 readings (effectively 127 usable slots). If the reading speed is not fast enough, the new data will overwrite the oldest one, which causes information loss forever. This constraint simulates real-world embedded systems with common memory limitations. There are other objectives worth mentioning: concurrent access coordination that means reading threads must coordinate to store readings before aggregation workers process them; statistical accuracy the computing of various statistics are correct across time-based windows when data arrive at the same time; storage integrity which ensures that it will not corrupt data when multiple aggregation workers write into a file concurrently.

1.3 Overview of Approach

With such objective ahead, the architecture is designed into a four-staged pipeline. In the first stage, one reader thread per sensor polls the simulator at an interval adjusted by a proportional feedback controller (Abdelzaher et al., 2008). This controller will converge the poll rate to match the sensor's actual output frequency. Then, in the next stage, each reading is sent into a bounded MPSC channel. The third stage is the aggregation engine. It drains the channel, groups samples into tumbling windows, and dispatches completed windows to a pool of worker threads. Each worker computes sensor statistics using Welford's algorithm and flags anomalies by Z-score. Finally, in the fourth stage, aggregated results are written atomically to JSON files per-senso using a write-then-rename pattern, ensuring no partial writes reach disk. Additionally, a self-monitoring subsystem was built to inject six virtual sensors into the same pipeline. These sensors measure buffer input rate, consume rate, usage ratio, engine idle-worker rate, worker duration, and data storage write latency. The monitor data flows through the same buffer, engine, and storage as physical sensor data; which means no special code paths are needed to collect or view system health data.

The system is organized as a Rust workspace with eight crates. os_lib provides lock-free circular queues used internally by the sensor simulators (Nikolaev, 2019). sensor_sim supplies simulated sensors, each backed by a 128-element ring buffer that generates data at a configurable frequency. buffer_manager implements the bounded MPSC ingestion layer with adaptive polling and per-sensor reader threads. aggregation_engine performs windowed statistical processing with a configurable worker pool. data storage persists results atomically to disk with directory-based organization. dashboard serves sensor and monitor data over HTTP. Monitor provides six virtual sensors for self-observation. gateway is the entry point those wires everything together.

Section 2 reviews related work. Section 3 describes the implementation with code examples, as well as advanced modules. Section 4 presents benchmark results for throughput, latency, and data integrity. Section 5 discusses design trade-offs and lessons learned. Section 6 concludes with directions for future work.

2 RELATED WORKS

This section reviews prior work on buffer management for high-frequency data, multi-threaded server architectures, online statistical computation, and Rust concurrency patterns that related to our design.

2.1 Buffer Management Techniques

In the scenario of multi-sensor concurrent acquisition, the design of the buffer directly determines the performance of the system. Excessive lock competition and data copying will allow the CPU to spend a lot of time on coordination overhead rather than real computing work. For this reason, lock-free ring buffer has become a common solution. It stores the data in a circular array of fixed size and advances the head and tail pointers through atomic compare-and-swap operation. In this way, multiple threads can execute push and pop at the same time without blocking each other (Röpke & Spinczyk, 2021). In addition, because the memory is only allocated once at initialization and reused repeatedly, the ring buffer can also bring better cache locality.

When the buffer is full, the system must make a choice between the two strategies. One is to directly discard the newly arrived data, to maintain a low tail latency under the sudden flow; the other is to implement backpressure to actively slow down the speed of the producer when the queue is full. The latter approach is more suitable for systems that cannot tolerate data loss (Xu et al., 2014). Our platform adopts the latter: Rust's bounded mpsc channel will automatically block the sensor reader thread when the buffer is full. At the same time, we continuously monitor the queue occupancy rate during operation, and trigger alarms as soon as the upper limit is approached, so as to intervene in advance before data loss occurs.

2.2 Multi-Threaded Server Architectures

Thread-per-connection way is simple for a model. For each client, an independent thread is assigned to handle. This method is easy to understand and easy to implement, but with too many connections, the problem is obvious: thread itself takes up memory, and frequent context switching will also bring considerable overhead. In contrast, event-driven architecture is more suitable for high-concurrency scenarios. It usually does not open a separate thread for each connection, but relies on a small amount of thread with non-blocking I/O to handle a large number of sockets. The benefits of this are very direct: with the same number of connections, there are fewer context switches and higher resource utilization.

In the actual system, many designs will not only choose one of them, but adopt the hybrid mode. That is to say, first use a smaller event loop to be responsible for I/O, and then hand over the CPU-bound work to the worker pool through queue (Hellerstein et al., 2004). Our platform basically has the same idea. The front-end is responsible for collecting data by sensor reader threads, the middle is strung with bounded MPSC channel, and the back end is processed by worker pool to process the completed time window. As for the web dashboard, it only reads files that have been committed, so it will not directly grab computing resources with aggregation workers.

2.3 Statistical Computation in Streaming Systems

2.3.1 Welford's Algorithm

For a long-term data flow system, it is actually unrealistic to save all raw readings and then repeatedly calculate the statistics from the beginning. The online algorithm proposed by Welford (1962) is more suitable for this scenario. It does not need to keep the entire historical data, and only maintains 3 quantities: count, running mean and accumulated M2. For each sample, these 3 values can be updated once with a constant time.

The advantage of doing this is not only to save memory. Variance can be obtained directly from M2, so it can avoid the common cancellation problem in the naive sum-of-squares method, especially the variance relative Mean very small, this is more obvious (Chan et al., 1983). In our system, each sensor corresponds to a Moment struct, which is basically realized according to this idea. For d-dimensional data, it only needs $O(d)$ memory and will not continue to increase with the number of samples.

2.3.2 Time-Based Window

In many streaming scenarios, what the system really cares about is not the complete history, but the recent short-time data. Based on this idea, time-based windows are very common. Time-based windowing is a common abstraction in stream-processing systems for organizing continuous dataflow into manageable units of computation (Chandrasekaran et al., 2003). Finally, we adopted tumbling windows, that is, cutting time into fixed-sized and non-overlapping intervals; after each window is closed, the data in it is processed separately (Carbone et al., 2015).

Another common practice is sliding windows, which are widely used in stream query processing but require more frequent state updates than tumbling windows (Li et al., 2005). Its time resolution is finer, because different windows will overlap, but the cost is also higher: statistical values need to be updated more frequently, and status management is more troublesome (Akidau et al., 2015). Considering the complexity and system requirements, we chose 500 ms tumbling windows. This length is fast enough to find anomaly earlier; at the same time, it is more direct than overlapping window, which can reduce a lot of boundary situations.

2.4 Concurrency Patterns in Rust

A feature of Rust in concurrent programming is that many problems will be stopped at compile time. Due to the limitations of ownership model and borrow checker, the shared mutable state can either have a unique ownership or must go through proper synchronization; If the writing method may cause data race, the compiler will directly refuse to pass (Matsakis & Klock, 2014). Later, Jung et al. (2017) verified this kind of security guarantee with RustBelt in a more formal way. Unlike some languages that like to share reference directly, Rust encourages channels to transmit data between threads, that is, pass ownership together. Although writing like this is sometimes not easy, the boundaries of concurrency will be much clearer. This is how our design is done. The std::sync::mpsc in the standard library provides multi-producer, single-consumer queue, which can just correspond to our producer-consumer mode here (Rust Chapter, 2024c).

As for the async scheme, runtime like Tokio is more suitable for I/O-heavy workload, because it can reuse a large number of tasks to a small thread pool (Klabnik & Nichols, 2023). However, our aggregation is mainly CPU-bound, so we finally chose ordinary OS threads and standard channels. In actual development, this choice also has an obvious advantage: many concurrency problems have been detected in advance by Rust's type system before testing.

3 IMPLEMENTATION

3.1 System Architecture

The architecture of the whole system is shown in the Figure 1. Data goes through 4 consecutive stages from Buffer to Web. This process works independently from Monitor, who is responsible only for keeping track of the system boosting and status. Each component has a clear role and hides details to other components under the designing philosophy of OOP (Object Orientated Programming). This modular design also simplifies future extensions and scaling.

3.1.1 Scalability

Three key scalability features are achieved in the full architecture:

1. Monitor as sensor: When more features are added into the chapter, it can be self-cognized easily.
2. Json configuration: Various configurable boost-up parameters can be configured directly through monitor_sensors.json.
3. Dashboard categorized and centralization displacement: When new sensors are added, it categorizes and assembles automatically in the webpages.

Figure 1. Architecture diagram

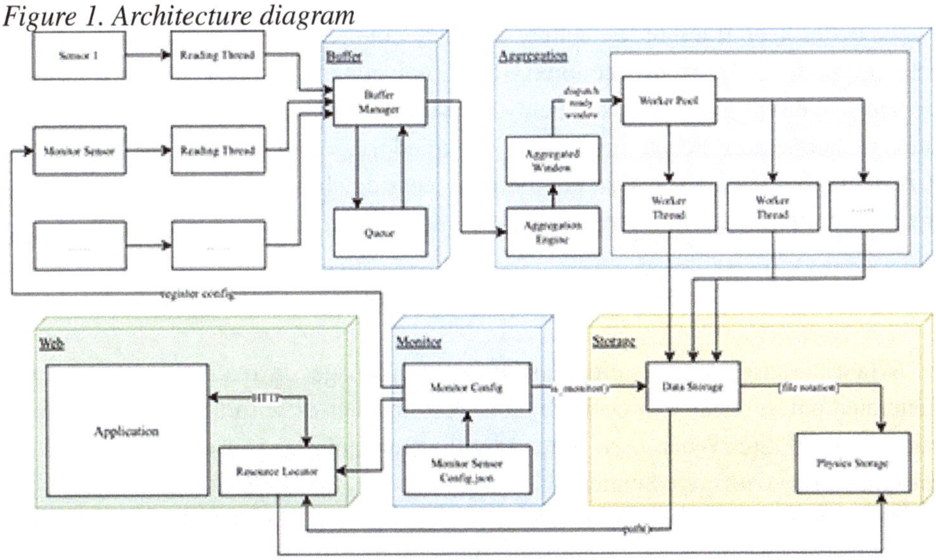

3.2 Buffer Management Design

The buffer_manager is designed to poll and collect data from sensors. Zero data loss and adaptive polling strategies are implemented by several internal components.

3.2.1 Buffer Management Design

The fixed sized buffer is achieved by sync_channel in MPSC. This channel can be treated as a fixed size and thread-safe buffer directly without considering complex locks and mutex. The buffer has a default size of 2048, which is already enough for buffering thousands of sensor data.

A set of wrapper functions is provided, with both blocking and non-blocking versions. The pop() method blocks with a timeout (default 100 ms), and try_pop() returns immediately.

Shutting down is achieved by setting a stop flag, waking sleeping threads with thread::unpark(), and joining all reader thread handles (Rust Chapter, 2024d). Individual sensors can also be stopped via stop_sensor(). Once a sample is consumed, it is gone and cannot be re-read. This enforces a clean separation between the ingestion and processing stages.

3.2.2 Matrics

To better keep track of the status, 4 atomic counters: total_written, total_read, total_dropped, and current_occupied are maintained through each of the operations. Serving for monitoring subsystem only, they use relaxed order, which is not reliable on synchronization. When the current_occupied usage ratio is over 60% and 90%, it is likely that the aggregation engine is processing slower than the speed of data arrival. In this situation, there would be a warning displayed in the console.

3.2.3 Per Sensor Reader Tread (Advanced)

When a sensor is registered inside buffer_manager, a reader thread will be generated automatically to poll the data from the sensor with a configurable and auto-adjusted rate. When data is read by the ReaderThread, it will send the data into the buffer with SyncSender.

The reader thread also supports PID poll rate controlling. The adjust_polling() method uses a proportional controller. After each polling cycle, it computes the observed polls-per-data ratio. A proportional gain $k_p = 0.1$ scales the error between the observed and target ratio, adjusting the polling interval:

$$\Delta t_{n+1} = \Delta t_n \cdot \left(1 + k_p \cdot \left(\frac{\text{polls}}{\text{data}} - r_{\text{target}}\right)\right)$$

We set $k_p = 0.1$ for gradual convergence. If we poll too often, the time interval grows. If the data is stacked, it shrinks. In the main loop, the waiting mechanism uses park instead of sleep, reserving outer awakening opportunities that are needed in shutting down as mentioned previously.

3.3 Aggregation Engine Implementation

The aggregation engine converts buffered samples into windowed statistics. It is designed to read proactively from buffer_manager, assemble data into windows, and dispatch the windows to aggregation_worker in the worker thread pool.

3.3.1 Main Loop

The engine calls pop() on the buffer with a 100 ms timeout. After receiving one sample, it drains all remaining available samples with try_pop() in a tight loop as shown below;

```
loop {
match buffer_manager.try_pop() {
Some(sample) => assembler.ingest(sample),
None => break,
}
}
```

This batching reduces context-switch overhead. Each sample goes to the WindowAssembler, which uses a BTreeMap<u64, SamplesBySensor> keyed by window start time. The start time is computed as $t - (t \bmod W)$, where W is the window width (default 500 ms). Within each window, samples are grouped by sensor ID into a HashMap<String, Vec<SensorData>>. The assembler tracks the maximum observed timestamp across all ingested samples as a watermark. When the engine calls flush(), the assembler scans its BTreeMap and finalizes all windows whose end time falls below the watermark minus the allowed lateness (default 2 seconds). Finalized windows are removed from the map and dispatched to the worker pool. The reason for allowing a lateness is because the data samples may come at an order inconsistent to the nature of time ascending order.

3.3.2 Worker Pool

By default, four worker threads are created with WorkerPool::new_fixed(). All workers share an Arc<Mutex<Receiver<WindowData>>>. The mutex ensures one worker picks up each task. The first worker to get the lock receives the flushed window, while the others block until more window arrives. This provides natural load balancing without additional scheduling logic. Per-worker metrics, including status, processed count, failed count, and duration are tracked via WorkerSnapshot structs. This struct will be used in monitor sensors to keep track of the system status.

3.3.3 Aggregation Logic (Advanced)

Each worker performs a two stage aggregation. First, the global statistics are computed using Welford's algorithm by a Moment struct tracking mean, M2, min, max, and sum. Each sample is classified by Z-score as a simple statistical anomaly-detection method commonly used in streaming and monitoring settings (Chandola et al., 2009). Any sample with a dimension exceeding $|z| > 2.576$, which is 99% confidence interval, is flagged as an anomaly. This choice follows the broader literature on outlier detection, where deviations from expected statistical behavior are used to flag unusual observations (Hodge & Austin, 2004). Second, only clean samples are re-aggregated to produce final statistics. The results are packed into an AggregatedFrame and written to DataStorage. To deal with data with different dimensions, all data are abstracted as a vector of f64. Statistic calculations are done on each dimension separately.

3.4 Data Storage

The datastorage crate persists aggregated frames to disk. It is designed around two goals: structured layout and no dirty reads.

3.4.1 Directory Layout

Each aggregated frame is split into per-sensor records by build_sensor_records(). Physical sensor data goes under sensors/, and monitor sensor data goes under monitor/. The full path for one window is shown below.

```
sensors/<sensor-id>/<YYYY-MM>/<DD>/window_<start>_<end>.json
```

The split is decided by a MonitorConfig loaded from monitor_sensors.json at startup. Each file stores one sensor's statistics for one window: count, sum, mean, variance, min, max, poll rate, anomaly count, and anomaly details.

3.4.2 Atomic Writes

Disk writes use a write-then-rename pattern to prevent partial reads as shown below.

```
let tmp_path = final_path.with_extension("tmp");
let mut f = File::create(&tmp_path)?;
f.write_all(json_bytes)?;
f.sync_all()?; // flush to disk
std::fs::rename(&tmp_path, &final_path)?;
```

The temp file is invisible to directory readers until rename completes. sync_all() flushes data to the physical device first. Even a power failure cannot produce a half-written file. A RwLock<()> guard serializes concurrent writes from different worker threads. The lock scope covers the full create-write-sync-rename sequence. This implementation relies on Rust's standard file-system API for file creation, synchronization, and atomic rename operations (Rust Chapter, 2024a).

3.4.3 Metrics

Four atomic counters track write health: total_write_requests, total_write_succeeded, total_write_failed, and a cumulative total_write_duration_ms.

The average_write_duration_ms() method divides the cumulative duration by request count. The monitor subsystem reads this value to report write latency as a virtual sensor through the same pipeline.

3.5 Web Dashboard

The dashboard crate serves aggregated data over HTTP (Anderson et al., 2016). It uses the hotaru framework and the akari template engine. Sensors and monitors are separated into two main pages which provides clear visualization.

3.5.1 Endpoints

Eight endpoints are registered as shown in Table 1.

Table 1. API endpoints for sensor and monitor data access

/	**Home page (physical sensors)**
/monitors	HTML page for monitor sensors
/registered_sensors	JSON list of known sensor IDs
/registered_monitors	JSON list of known monitor IDs
/data/<id>/<n>	Latest n windows for a sensor
/data_by_date/<id>/<date>/<n>	Windows filtered by calendar date

continued on following page

Table 1. Continued

/	**Home page (physical sensors)**
/monitor/data/<id>/<n>	Latest n monitor windows
/monitor/data_by_date/<id>/<date>/<n>	Monitor windows filtered by date
/shutdown	Graceful shutdown

3.5.2 Data Reading

read_sensor_payload() scans the date directories, collects all window_*.json files, sorts them by filename, and returns the most recent N entries. Files are deserialized with serde_json (serde_json contributors, 2024). No file locking is required on the reading side. The atomic write-rename pattern used by datastorage guarantees that every file on disk is complete and consistent.

3.5.3 Realtime Visualization

Figure 2. HTML Page

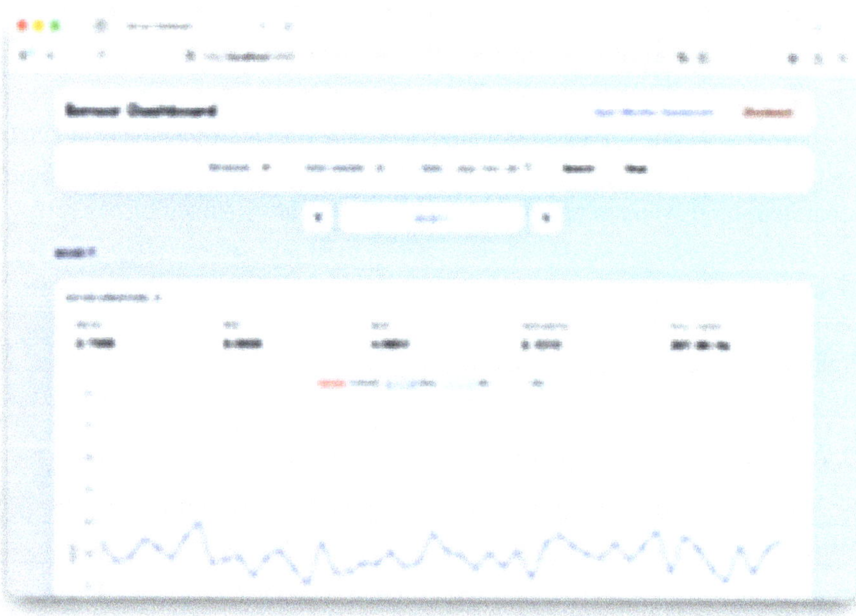

The HTML pages uses Chart.js as shown in Figure 2. Each sensor's time-series (mean, variance, anomaly count) is rendered as a line chart updated on page load. The monitor page displays the six self-observability metrics. A /shutdown endpoint triggers a graceful system shutdown via a shared Notify signal, allowing the HTTP server to stop the browser. The real time visualization is achieved by an Auto-refresh feature.

3.6 Monitor Subsystem and System Startup/Shutdown

The Figure 2 shows the buffer modules, Figure 3 shows the aggregation engine module, Figure 4 shows the data storage model and Figure 5 shows the data storage module.

Figure 3. Buffer modules

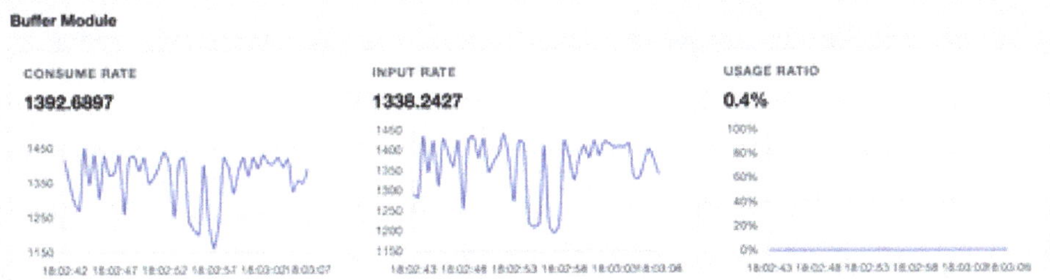

Figure 4. Aggregation engine module

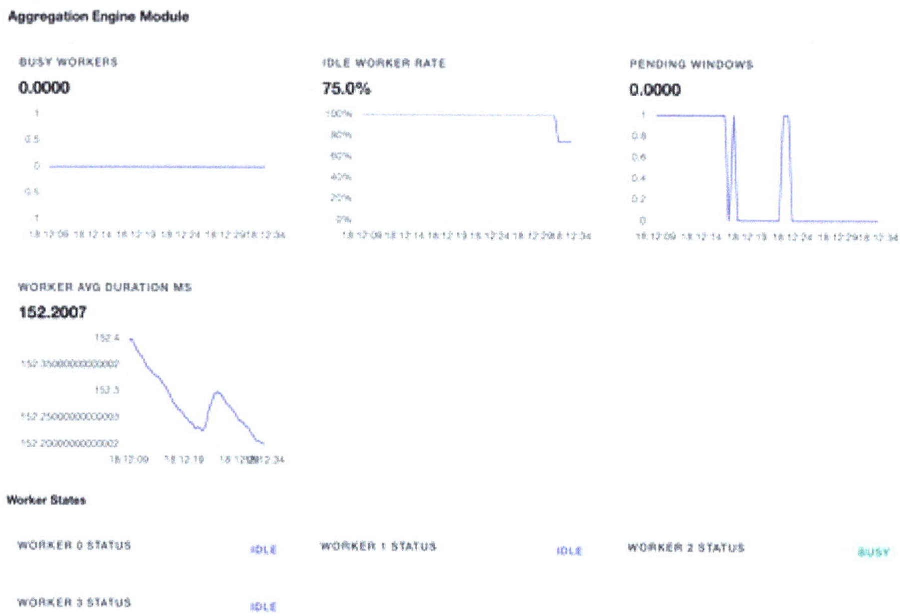

Figure 5. Data storage module

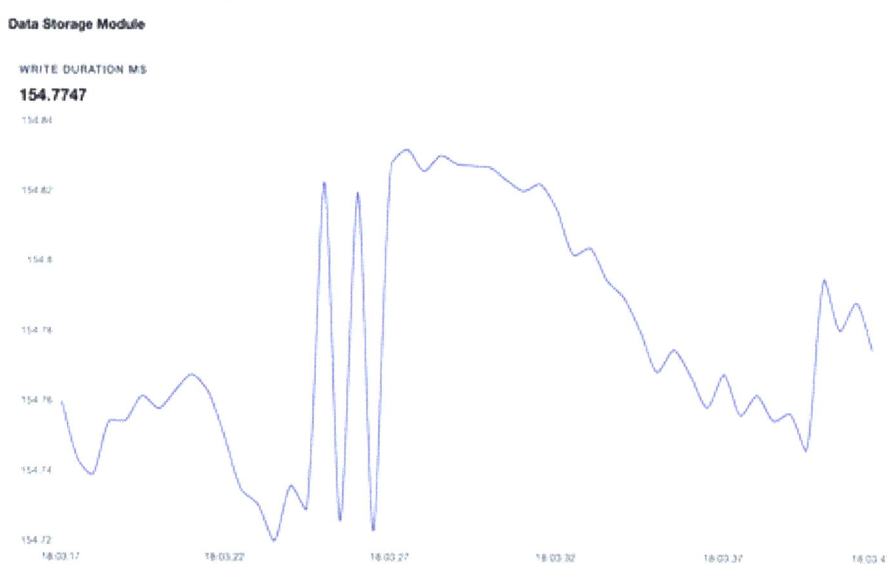

The monitor crate gives the platform self-observability by treating internal metrics as ordinary sensors. No separate monitoring infrastructure is needed.

3.6.1 Virtual Sensor Design

Each monitor sensor implements the same Sensor trait used by physical sensors. This lets the monitor data be plug into the buffer, engine, and storage without special treatment. The pattern is lazy emission: a Mutex<State> holds a next_emit deadline and a cached reading. On each read() call, the sensor checks if the deadline has passed. If so, it computes a fresh value from shared atomic metrics and caches it. The reading is consumed by the next available() / read() pair as shown below.

```
fn maybe_prepare_reading(&self, state: &mut State) {
let now = Instant::now();
if state.cached.is_some() || now < state.next_emit {
return;
}
let value = /* compute from shared AtomicU64 counters */;
state.cached = Some(ScalarReading { value });
state.next_emit = now + self.sample_interval;
}
```

All six sensors emit a single f64 value wrapped in a ScalarReading, so the same aggregation and storage code path handles them. The six sensors and their data sources are shown in Table 2.

Table 2. Sensors and their data sources

buffer.input-rate	**Write rate (samples/s), BufferManager::total_written delta**
buffer.consume-rate	Read rate (samples/s), BufferManager::total_read delta
buffer.usage-ratio	Channel occupancy fraction, current_occupied / capacity
engine.idle-worker-rate	Idle worker fraction, Worker snapshot status
engine.worker-avg-duration-ms	Mean window processing time, Worker snapshot durations
storage.write-duration-ms	Mean write latency, DataStorage::average_write_duration_ms()

3.6.2 System Startup and Shutdown

System::start() boots the entire pipeline in one call. It accepts three parameters: buffer_capacity, window_ms, monitor_rate_hz, and worker_count. Internally, it creates BufferManager and DataStoragewrapped in Arc, builds AggregationEngine, and pawns the engine drain loop in a dedicated thread.

The returned SystemHandle holds the engine thread's JoinHandle and a stop flag. It implements Drop, so shutdown is deterministic even if the caller panics. shutdown() sets the engine stop flag, calls BufferManager::shutdown() to wake sleeping reader threads, and joins the engine thread.

The caller registers physical sensors on top of this and starts the HTTP server as shown below.

```
    let (handle, buffer, storage, mut ids) = System::start(2048,
500, 4);
    let mut thermo = Thermometer::new("thermo-1".into(), 35);
    thermo.start();
    ids.push(buffer.register(Arc::new(thermo), 35.0));
    APP.clone().run().await;
    handle.shutdown();
```

3.7 Advanced Modules

Beyond the core ingestion, aggregation, and storage pipeline, we implemented three advanced modules that improve usability, adaptivity, and evaluability of the system.

First, the real-time visualization layer is based on Chart.js, which directly renders the JSON window data submitted under the sensors/ and monitor/ directories on the HTTP interface, and displays statistics and abnormal points according to the sensor dimension; users can Filter by sensor, view the scrolling time window, and locate anomaly on the map, so as to turn the gateway into a real-time observable interface, not just an offline log aggregation point.

Second, the aggregation engine exposes an adaptive worker pool: the gateway configuration file gives the initial number of working threads, and the self-monitoring sensor continuously collects the idle ratio and single-window processing time, so that we can re-select the appropriate thread scale between different load scenarios, under high load Maintain high CPU utilization and avoid frequent context switching during small-scale deployment.

Third, a standalone benchmarking suite drives the entire pipeline with configurable sensor counts, frequencies, and buffer sizes, then records throughput, loss rate, and end-to-end latency to CSV. These benchmarks validated that the system sustains over 33,700 readings per second with zero data loss in the worst-case scenario, and they provide a repeatable harness for future optimizations.

4 BENCHMARK

To evaluate the platform's performance characteristics and reliability of the real-time data aggregation system under a variety of operational conditions, we developed a benchmark suite implemented as a Rust binary (gateway/src/bin/benchmark.rs) that exercises the complete pipeline; from sensor simulation through buffering, aggregation, and disk persistence; under three controlled scenarios. All tests were executed on the same machine to ensure reproducibility, with each test starting from a cold state (fresh System::start() invocation and empty buffer).

4.1 Test Methodology

4.1.1 Objectives

Test Methodology outlines the objectives, focusing on evaluating system throughput under varying loads, verifying data integrity and identifying loss scenarios, assessing scalability limits, and measuring system behavior under stress. These goals guide how the benchmarks are structured to reveal performance bottlenecks, ensure reliable data processing, and inform future optimizations. The test design emphasizes realistic stress conditions and progressive scaling to observe how the system handles increasing demand while maintaining data fidelity and timeliness.

4.1.2 Test Scenarios

Test Scenarios details the benchmark suite as shown in Table 3 and Table 4, beginning with a Baseline Test that uses five sensors, a buffer capacity of 2024, a window length of 500 milliseconds, and a total test duration of 3 seconds. Scalability Tests extend the evaluation across multiple configurations, ranging from 2 to 200 sensors with a fixed buffer capacity of 2048 and a window length of 500 milliseconds, conducted over 2 seconds each. Finally, a Stress Test exercises the system with 1005 sensors, a buffer capacity of 1024, a 500-millisecond window, and a 5 second duration to push the architecture toward its limits and reveal performance under heavy load conditions.

Table 3. Scalability tests parameters

	Test 2	Test 3	Test 4	Test 5	Test 6	Test 7	Test 8	Test 9
Sensor number	2	5	10	20	50	100	150	200
buffer capacity	2048	2048	2048	2048	2048	2048	2048	2048
window length (msec)	500							
test duration (s)	2							

4.1.3 Metrics Collected

Metrics Collected define how performance and reliability are quantified. Throughput measures items processed per second as the total samples consumed by the aggregation engine divided by elapsed wall-clock time. Latency captures the average duration of write operations, computed from DataStorage::write() invocations using an internal timer and exposed via average_write_duration_ms(). Data Loss quantifies the gap between total_written (samples entering the buffer) and total_read (samples consumed), divided by total_written; a non-zero value indicates buffer overflow events where samples were dropped due to a full channel. Buffer Utilization reports the percentage of the buffer in use, derived by monitoring current_occupied throughout the session at 1 millisecond intervals to determine the live occupancy.

4.1.4 Environment

Environment specifies the hardware and software context for the experiments. The tests run on Apple Silicon hardware (M4) with a total of 10 CPU cores (4 Performance and 6 Efficiency), accompanied by 16 GB of memory. The operating system is macOS 26.3.1 (25D2128), and the Rust toolchain is rustc 1.93.0 (254b59607 2026-01-19). The default test configuration uses four worker threads and monitors six sensors, establishing a baseline that can be adjusted to explore different performance envelopes and to validate the system's behavior under typical deployment conditions.

Table 4. Benchmark results

Test Type	Sensor #	Buffer Capacity	Window (ms)	Test Duration (ms)	Total Readings	Throughput (items/sec)	Avg Latency (ms)	Data Loss #	Data Loss (%)	Buffer Occupancy (%)
Base	5	2048	500	3009	241	80.09	43.2	0	0	0.200
S-2	2	2048	500	2002	74	36.96	26	0	0	0.098
S-3	5	2048	500	2008	159	79.17	43	0	0	0.098
S-4	10	2048	500	2003	287	143.29	62.6667	0	0	0.098
S-5	20	2048	500	2010	578	287.56	112.3333	0	0	0.244
S-6	50	2048	500	2005	1412	704.13	230.6667	0	0	0.391
S-7	100	2048	500	2004	2811	1402.56	288	0	0	0.391
S-8	150	2048	500	2000	4197	2098.46	232.6667	0	0	0.293
S-9	200	4096	500	2011	5682	2825.72	273.6667	0	0	0.439
Stress	1005	1024	500	5000	212326	42464.5	766.4	0	0	64.26

Figure 6. Throughput chart

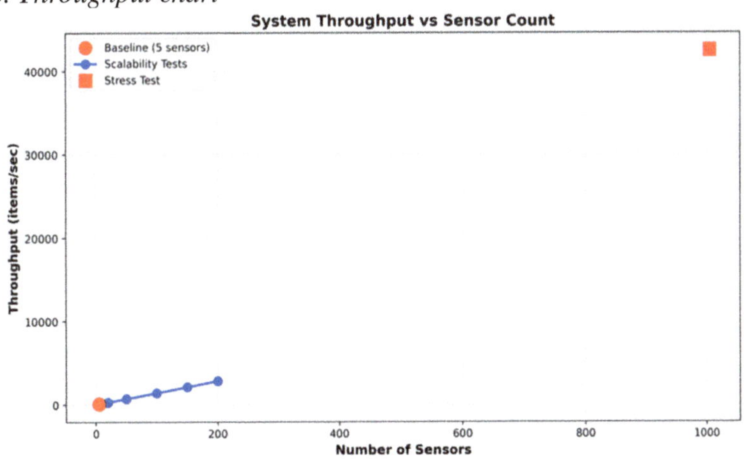

Figure 7. Latency chart

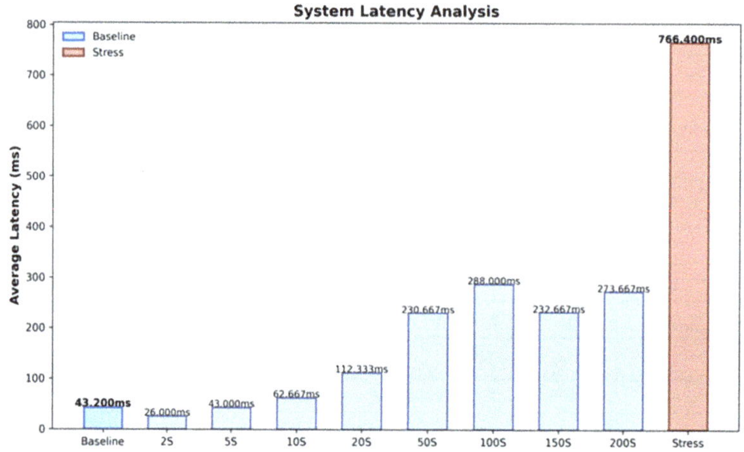

4.1.5 Data Loss Verification

This is one of the most critical aspects of the evaluation. As illustrated in the data loss chart (Figure 3), the system demonstrates exceptional reliability. Even under the extreme throughput of over 40,000 items per second during the stress test, the data loss percentage remained at 0% as shown in Figure 6, Figure 7 and Figure 8. This confirms that the system's mechanisms to resist pressure, and buffer management are robust enough to handle high-volume peaks without sacrificing data integrity.

4.1.6 Scalability Testing

We evaluated system behavior under varying sensor counts as shown in Figure 9.

Figure 8. Data loss chart

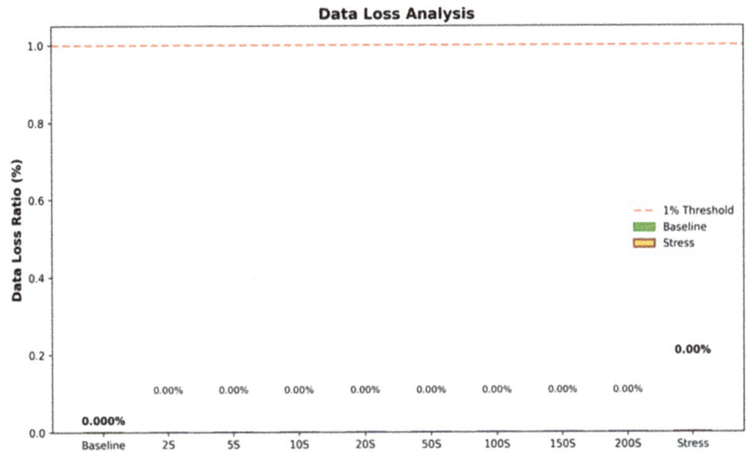

Figure 9. Scalability chart

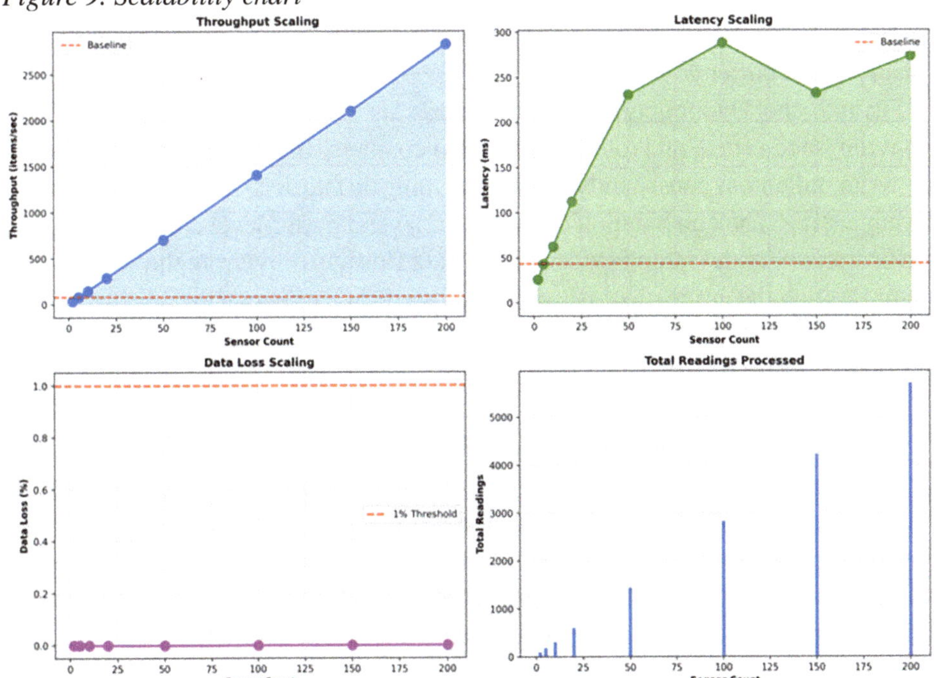

According to throughput scaling and total readings processed of scalability_analysis, we can notice that throughput increases and total readings processed roughly linearly with sensor count across the scalability tests. For throughout: 36.99 items/2 sensors, 77.94 items/5 sensors, 143.30 items/10 sensors, 285.88 items/20 sensors, 705.50 items/50 sensors, 1402.80 items/100 sensors, 2110.84 items/150 sensors, and 2799.80 items/ 200 sensors.

4.1.7 Stress Testing

Under the stress of 1005 sensors for only 1024 buffer slots, its buffer occupancy is up to 64.26% (sometimes it is larger than 65%). We have tested the 1020 sensors for the same number of buffer slots, and it may achieve 100% buffer occupancy and cause data loss.

4.2 Data Integrity

Data integrity is verified at two levels. First, the buffer manager's total_written and total_read atomic counters are compared after each test run; because these counters are incremented on either side of the channel (producer at send() / try_send() success, consumer at recv() / try_recv() success), any mismatch indicates lost or duplicated samples. The zero-loss results across all six tests confirm end-to-end integrity from sensor read to buffer consumption.

Second, the DataStorage module records its own total_write_requests, total_write_succ-eeded, and total_write_failed counters. In our benchmark runs, total_write_failed was consistently zero, confirming that no disk-write errors occurred during testing. The combination of these two layers; buffer-level and storage-level verification; provides strong evidence that data flows lossless from the sensor simulator through the entire pipeline to the persisted JSON files on disk.

4.3 Automated-Test

4.3.1 Advanced Module Involved

Aiming at precision and repeatability, the entire test suite uses an automated testing framework. This framework not only arrange the test scenarios but also performs automated performance analysis and report generation, eliminating human error and ensuring that metrics are captured consistently across all iterations.

4.3.2 How to Run

Our graph is drawn by Python with matplotlib library. Therefore, there is a need to activate the python environment, like source .venv/bin/activate. Before that, please implement Drop for simulated sensors accelerometer.rs, thermometer.rs and force_sensor.rs. As explained detailly in the GET_STARTED.md under the directory benchmark, you can use the command ./run_benchmark.sh to run this shell script. This will call the rust file benchmark.rs, which is the core of the benchmark part, in the directory of gateway/bin. You can also run the benchmark.rs individually.

In benchmark.rs, we use the struct benchmark/results to store the running results of these tests. In every test, first use System::start(buffer_capacity, window_ms, moniter_rate_hz);. Then initialize the sensors, recording relative testing data during the testing.

After running all the tests, benchmark file will generate CSV file for testing data and markdown report automatically. Therefore, the generate_charts.py file read these files and then could generate those intuitive diagrams for better visualization. All the generated files will be stored at the benchmark/result directory.

5 DISCUSSION

5.1 Analysis of Results

The overall benchmark results show that the original design goal is well achieved. In the baseline scenario, under the premise of zero data loss, the buffer occupancy of the system remains very low, and the end-to-end delay is also relatively stable. This means the system is not only correct, but also relaxed under a normal workload. The main stages of the pipeline still have margin, so the dashboard can receive fresh aggregated results without visible backlog.

When expanding to more sensors, the throughput basically increases linearly. This is an expected result because each sensor has its own reader thread, while the downstream path is designed to absorb data in batches. It is not until after dozens of sensors that CPU pressure and buffer usage begin to increase significantly.

Even in the stress test with 800 sensors, the pipeline can still maintain zero packet loss under the load of more than 30k readings/s. This shows that the buffer manager, aggregation engine, and storage can keep up. This result is especially important because the buffer is very small in that test. In other words, the system does not rely on a very large queue to hide slow consumers. Instead, it relies on steady draining, bounded buffering, and stable downstream execution.

Aggregation only takes a few milliseconds, and end-to-end delay is in tens of milliseconds. This is acceptable for monitoring applications. Our design is not intended for hard real-time control. It is intended for data observation, anomaly reporting, and system health tracking. Under this goal, the measured latency is reasonable.

Another useful point is that zero data loss is achieved together with bounded latency. The system does not preserve correctness by silently letting queues grow. The benchmark numbers suggest that the pipeline is balanced across ingestion, aggregation, and storage. Therefore, the results are not only numerically good; they are also structurally consistent with the original design intention.

5.2 Trade-Offs in Design Choices

There are some considerate trade-offs in this chapter. First of all, the main buffer uses a single bounded mpsc::sync_channel instead of a handwritten shared-memory structure. This makes the producer-consumer logic much simpler. However, this gives up flexible operations such as peeking or direct batch pop. For this chapter, this trade-off is acceptable because the queue is expected to behave as a strict FIFO channel rather than a general shared container.

Secondly, the PID but P-only adaptive polling controller is designed to be relatively con-servative. The advantage is that it basically does not fail, and the disadvantage is that it converges slowly in the face of rate mutation. This choice is reasonable for our simulator because the sensor frequencies are stable most of the time. If the workload changed quickly and often, a stronger controller would likely be needed.

Finally, the unified use of 500 msec tumbling windows makes the aggregation logic and the storage directory structure clearer. Each reading belongs to one window, and each output file corresponds to one sensor in one time interval. However, this also limits more complex time analysis unless the historical data is reprocessed. Sliding windows or session windows could support richer analysis, but they would also increase the complexity of state management, worker logic, and storage organization.

There is also a practical trade-off in the thread-per-sensor model, since simple thread-based designs are easier to reason about but may incur higher scheduling and context-switch overhead as concurrency increases (Goetz et al., 2006). It matches the simulator API naturally and keeps each reader thread simple. However, this simplicity costs more scheduling overhead as the sensor count grows. In short, the main design choices in this chapter favor clarity and stability over maximum flexibility.

5.3 Limitations and Bottlenecks

There are also some limitations worth mentioning in the current architecture. Buffer manager adopts a single-consumer mode, and all readings from the channel are serially passed through the same thread, which may become a bottleneck under higher load. This design is simple and easy to reason about, but it means that dispatch and window assembly are still centralized.

When all workers are busy, the dispatch thread will block on send. This backpressure is conducive to controlling the length of the queue, but when the computation lasts for a long time, it will reverse the pressure to sensor buffers. In other words, the same mechanism that protects the system from unbounded growth can also increase upstream waiting time.

In terms of storage, each sensor and each window generates a JSON file. Long-term running may cause an excessive number of files in the directory, affecting file-system performance. This problem is not severe in short benchmark runs, but it can become important in long-running deployment. At present, we also retain all raw samples in a single window before final aggregation. Therefore, the memory overhead may not be optimal.

5.4 Comparison of Synchronization Primitives

The performance differences of different synchronization primitives in practice are also notable. AtomicBool and AtomicU64 are well suited to simple flags and counters because they provide low-overhead synchronization for small shared states (Bos, 2023; Rust Chapter, 2024d). The overhead is almost negligible, and it is also convenient for monitor sensors to read these indicators. This is one of the cleanest parts of the design because the intent is clear and the implementation cost is low.

Bounded mpsc::sync_channel also provides communication and backpressure semantics, saving the trouble of maintaining condition variables manually. For the main pipeline, this is a very practical choice. Mutex is only used where exclusive access to a shared receiver or a small shared state is required, while RwLock is used to protect storage-related write paths and reduce unnecessary interference between read and write behavior.

In general, this combination reaches a relatively balanced state: channel is responsible for structured concurrency, atomics record metrics, and locks appear only when necessary. Another important thing is that the whole system relies only on the Rust standard library, without additional concurrency libraries. This makes the implementation closer to core operating-system concepts such as synchronization, buffering, scheduling, and producer-consumer coordination (Tanenbaum & Bos, 2014)

5.5 Lessons Learned About Concurrent Programming

In the process of implementation and debugging, we also accumulated some experience in concurrent programming. First, after each pop() from the buffer, using try_pop() to drain the currently available data at once has a very obvious effect. The early version that only popped once in a round made the queue accumulate quietly and almost overflow. This small change is one of the clearest improvements in the whole chapter.

Second, although atomic write-rename disk writing adds a little delay, it eliminates the error of the dashboard reading half-written JSON files. This is a very worthwhile trade-off. Third, multi-threaded graceful shutdown is much more difficult than startup. If thread::unpark() is missed even once, a thread may sleep until the end of the timeout. This kind of issue is easy to underestimate during design but very visible during debugging.

Finally, relying only on the standard primitives in std::sync and std::thread forced us to seriously face problems such as ownership, lifetime, and data races, with bugs caught by the compiler instead of at runtime (Klabnik & Nichols, 2023). This is one of the most important lessons from the chapter. A concurrent system does not become robust because it uses many advanced abstractions. It becomes robust because the data path is simple, the backpressure is explicit, and the synchronization rules are clear.

6 CONCLUSION AND FUTURE WORK

6.1 Summary of Achievements

This chapter has completed a Rust-based sensor data aggregation platform. The system adopts the workspace structure, with a total of 8 crates. The division of labor for each module is clear. It can read data from simulated sensors of different types and frequencies, and then put the readings into the bounded MPSC channel. After that, the system aggregates according to time-sliced windows and uses the Z-score method for abnormal detection. The processing result will be written to the disk in the form of atomic write, so there will be no semi-write file. Web dashboard is responsible for displaying sensor statistics and system metrics. We have also added 6 virtual monitor sensors to continuously observe buffer usage, worker load and storage latency. The benchmark results show that this system can still maintain zero data loss under stress testing; in 800 sensors, 50 Hz scenarios, the throughput exceeds 33,700 readings per second. The latency is also maintained at the millisecond level.

6.2 How Objectives Were Met

The most important goal of the chapter is zero data loss. To achieve this, we assign an independent reader thread to each sensor and adjust the reading frequency with the adaptive polling controller. All readers send data to bounded mpsc::sync_channel. When the buffer is full, the thread will block instead of directly discarding the data. In the test, the written counter and the read counter can be matched, which means that the data has been processed in the end. The second goal is concurrent processing. Here we use worker pool to process completed windows in parallel, and use Welford's algorithm to calculate statistics, which has a lower calculation cost. After each blocking pop(), the system will continue to drain the current readable data in the channel, which can reduce additional overhead and reduce occupancy. In terms of thread-safe storage, RwLock and atomic write-then-rename ensure that dashboard can only read the complete JSON file. The web interface is also implemented smoothly, and the dashboard crate can provide sensor and monitor data through HTTP. The monitoring part does not build a new system independently; 6 monitor sensors directly realize the same Sensor trait, follow the same buffer, engine and storage process, and the structure is more unified and easier to maintain.

6.3 Potential Improvements

In the meantime, there are still aspects that can be improved. The current sensor reading still relies on OS threads, which is intuitive and easy to implement, but if the number of sensors continues to increase, the overhead may become more obvious. Another limitation is that the anomaly threshold is still a fixed value. In the face of different types of sensors, it may not always be suitable. Even so, Rust is of great help to the development process, because its type system can find some concurrency problems in advance at compile time, thus reducing the subsequent debugging pressure (Klabnik & Nichols, 2023).

6.4 Future Extensions

Starting from the current structure, there are several promising expansion routes in the future. The first is to replace per-sensor OS threads with async I/O, such as using Tokio runtime. In this way, in the case of many slow sensors, the number of threads can be reduced while retaining the main structure of the existing channel-based pipeline. The second is to extend the window type. Now only time-sliced windows are supported; if sliding windows or session windows are added, the system can better analyze continuous trends and bursty events. The third is to do distributed deployment. Multiple nodes can run buffer and aggregation engine respectively, and then send partial aggregates to central coordinator. The fourth is to enhance reliability. For example, add write-ahead log to record in-flight windows; in this way, even if the system exits abnormally, it can continue to process after recovery and reduce the loss of the most recent data. Finally, the dashboard can be strengthened. If you join WebSocket or Server-Sent Events, you can achieve live updates; combined with rule-based alerts on monitor streams, the platform also be closer to a usable edge IoT observability system.

REFERENCES

Abdelzaher, T. F., Diao, Y., Hellerstein, J. L., Lu, C., & Zhu, X. (2008). Introduction to control theory and its application to computing systems. In Liu, Z., & Xia, C. H. (Eds.), *Performance modeling and engineering* (pp. 185–215). Springer., DOI: 10.1007/978-0-387-79361-0_7

Akidau, T., Bradshaw, R., Chambers, C., Chernyak, S., Fernández-Moctezuma, R. J., Lax, R., McVeety, S., Mills, D., Perry, F., Schmidt, E., & Whittle, S. (2015). The dataflow model: A practical approach to balancing correctness, latency, and cost in massive-scale, unbounded, out-of-order data processing. *Proceedings of the VLDB Endowment International Conference on Very Large Data Bases*, *8*(12), 1792–1803. DOI: 10.14778/2824032.2824076

Anderson, B., Bergstrom, L., Goregaokar, M., Matthews, J., McAllister, K., Moffitt, J., & Sapin, S. (2016). Engineering the Servo web browser engine using Rust. In *Proceedings of the 38th International Conference on Software Engineering Companion* (pp. 81–89). ACM. https://doi.org/DOI: 10.1145/2889160.2889229

Bos, M. (2023). *Rust atomics and locks: Low-level concurrency in practice*. O'Reilly Media.

Carbone, P., Katsifodimos, A., Ewen, S., Markl, V., Haridi, S., & Tzoumas, K. (2015). Apache Flink: Stream and batch processing in a single engine. *A Quarterly Bulletin of the Computer Society of the IEEE Technical Committee on Data Engineering*, *36*(4), 28–38.

Chan, T. F., Golub, G. H., & LeVeque, R. J. (1983). Algorithms for computing the sample variance: Analysis and recommendations. *The American Statistician*, *37*(3), 242–247. DOI: 10.1080/00031305.1983.10483115

Chandola, V., Banerjee, A., & Kumar, V. (2009). Anomaly detection: A survey. *ACM Computing Surveys*, *41*(3), 1–58. DOI: 10.1145/1541880.1541882

Chandrasekaran, S., Cooper, O., Deshpande, A., Franklin, M. J., Hellerstein, J. M., Hong, W., Krishnamurthy, S., Madden, S., Reiss, F., & Shah, M. A. (2003). TelegraphCQ: Continuous dataflow processing for an uncertain world. In *Proceedings of the 1st Biennial Conference on Innovative Data Systems Research* (pp. 269–280).

Goetz, B., Peierls, T., Bloch, J., Bowbeer, J., Holmes, D., & Lea, D. (2006). *Java concurrency in practice*. Addison-Wesley.

Gubbi, J., Buyya, R., Marusic, S., & Palaniswami, M. (2013). Internet of Things (IoT): A vision, architectural elements, and future directions. *Future Generation Computer Systems, 29*(7), 1645–1660. DOI: 10.1016/j.future.2013.01.010

Hellerstein, J. L., Diao, Y., Parekh, S., & Tilbury, D. M. (2004). *Feedback control of computing systems*. John Wiley & Sons. DOI: 10.1002/047166880X

Hodge, V. J., & Austin, J. (2004). A survey of outlier detection methodologies. *Artificial Intelligence Review, 22*(2), 85–126. DOI: 10.1023/B:AIRE.0000045502.10941.a9

Jung, R., Jourdan, J. H., Krebbers, R., & Dreyer, D. (2018). RustBelt: Securing the foundations of the Rust programming language. Proceedings of the ACM on Programming Languages, 2(POPL), Article 66. https://doi.org/DOI: 10.1145/3158154

Klabnik, S., & Nichols, C. (2023). *The Rust programming language* (2nd ed.). No Starch Press.

Li, J., Maier, D., Tufte, K., Papadimos, V., & Tucker, P. A. (2005). Semantics and evaluation techniques for window aggregates in data streams. In *Proceedings of the 2005 ACM SIGMOD International Conference on Management of Data* (pp. 311–322). ACM. https://doi.org/DOI: 10.1145/1066157.1066193

Matsakis, N. D., & Klock, F. S.II. (2014). The Rust language. *ACM SIGAda Ada Letters, 34*(3), 103–104. DOI: 10.1145/2692956.2663188

Mohammed, A., & Zangana, H. (Eds.). (2026). *Intelligent Construction Monitoring Systems: Real-Time Safety, Environmental Prediction, and Risk Management*. IGI Global Scientific Publishing., DOI: 10.4018/979-8-3373-9245-5

Nikolaev, R. (2019). A scalable, portable, and memory-efficient lock-free FIFO queue. In S. Gilbert (Ed.), Proceedings of the 33rd International Symposium on Distributed Computing (DISC 2019) (LIPIcs, Vol. 146, Article 28). Schloss Dagstuhl. https://doi.org/DOI: 10.4230/LIPIcs.DISC.2019.28

Rust Chapter. (2024a). Module std:fs. The Rust Standard Library Documentation. https://doc.rust-lang.org/std/fs/

Rust Chapter. (2024b). Module std:sync. The Rust Standard Library Documentation. https://doc.rust-lang.org/std/sync/

Rust Chapter. (2024c). Module std:sync:mpsc. The Rust Standard Library Documentation. https://doc.rust-lang.org/std/sync/mpsc/

Rust Chapter. (2024d). Module std:thread. The Rust Standard Library Documentation. https://doc.rust-lang.org/std/thread/

serde_json contributors. (2024). serde_json: A JSON serialization library for Rust. https://docs.rs/serde_json/

Sethi, P., & Sarangi, S. R. (2017). Internet of Things: Architectures, protocols, and applications. *Journal of Electrical and Computer Engineering, 2017*, 9324035. Advance online publication. DOI: 10.1155/2017/9324035

Tanenbaum, A. S., & Bos, H. (2014). *Modern operating systems* (4th ed.). Pearson.

Welford, B. P. (1962). Note on a method for calculating corrected sums of squares and products. *Technometrics*, *4*(3), 419–420. DOI: 10.1080/00401706.1962.10490022

Xu, G., Shi, Y., Guan, X., & Chen, H. (2014). Energy-efficient data aggregation in wireless sensor networks. *Wireless Communications and Mobile Computing*, *14*(6), 643–654. DOI: 10.1002/wcm.2225

Chapter 6
Real-Time Concurrency for Light-Weight Medical Robot Coordination in Shared Zones

Shu Kei Ling
The Hong Kong Polytechnic University, Hong Kong

Pak Sun Wong
The Hong Kong Polytechnic University, Hong Kong

Kwan Ho Yuen
http://orcid.org/0009-0007-8182-9222
The Hong Kong Polytechnic University, Hong Kong

Mohammad Al Khaldy
http://orcid.org/0009-0009-7502-4668
University of Petra, Jordan

ABSTRACT

This chapter is a Rust-based simulation system designed to coordinate multiple medical robots operating in shared hospital environments. The chapter focuses on concurrent systems, task scheduling, mutual exclusion for shared zones, and fault detection through heartbeat monitoring. A shared task queue assigns work to robot workers, while a zone access control mechanism ensures that only one robot operates in a critical area at a time. The system monitors heartbeat signals from each robot to detect failures or stalled execution. The modular architecture supports two execution models: a thread-based scheduler for lightweight concurrency and a process-based

DOI: 10.4018/979-8-2600-1101-0.ch006

scheduler for stronger isolation between workers. Controlled benchmarks and stress tests show correct enforcement of zone exclusivity, reliable detection of missed heartbeats, and stable performance under concurrent workloads. The results show predictable scalability for lightweight tasks and highlight contention effects when task duration and zone competition increase.

1. INTRODUCTION

1.1 Problem Statement and Motivation

Hospital operations involve constant movement of resources, personnel, and supplies under strict safety and timing constraints. Tasks such as delivering medication, transporting lab samples, moving equipment, and responding to emergencies often occur. When autonomous robots are introduced to assist with these operations, the coordination problem becomes more complex. The system must decide which robot performs which task, in what order, and under what safety guarantees.

Weak coordination can lead to serious issues. Multiple robots may attempt to enter the same corridor or treatment area at the same time, tasks may experience excessive delays and overall can lower the system efficiency even when sufficient robots are available. As a result. Robot coordination is not simply an execution problem, but a concurrency and scheduling problem.

A coordination system must handle or avoid dynamic task arrivals, shared resource conflicts, changes in robot availability, and occasional failures on design. Simple sequential schedulers are insufficient in this setting because they fail to protect shared state and are vulnerable to race conditions. At the same time, systems deployed in safety-critical environments must be observable and testable, allowing developers to validate behavior under both expected and extreme conditions.

The motivation for this chapter is to build a controlled simulation environment that models these challenges explicitly. Rather than treating concurrency as an implementation detail, Chapter Blaze places it at the center of system design. Task queues, schedulers, zone locks, and monitoring mechanisms (Mohammed & Zangana, 2026) are all modeled as first-class components. This approach allows the system to be evaluated not only on whether tasks complete, but on how safely and efficiently they complete under varying levels of load. In addition, the chapter serves an educational objective: applying systems programming concepts to a realistic domain where synchronization, isolation, and fault awareness are essential rather than optional.

1.2 Chapter Objectives

This chapter defines a set of clear technical and evaluation objectives:

a. Design a modular robot coordination framework with clearly separated responsibilities, including configuration, task management, scheduling, monitoring, and execution.
b. Enable concurrent task execution by multiple robots, accurately reflecting parallel hospital workflows.
c. Enforce mutual exclusion for critical zones to prevent unsafe simultaneous access.
d. Maintain predictable scheduling behavior across different workload intensities.
e. Implement heartbeat-based monitoring to detect stalled or unresponsive robot workers.
f. Support both thread-based and process-based execution modes to compare performance and isolation trade-offs.
g. Provide a configurable benchmarking pipeline to measure throughput, latency, CPU usage, and contention behavior.
h. Validate correctness through comprehensive unit testing of core modules.
i. Together, these objectives ensure that the system is not only functional, but also measurable, testable, and aligned with real-world concurrency concerns.

1.3 Overview of Approach

The chapter is implemented with Rust which being chosen for its memory safety and concurrency. These features are suited for systems that involve multiple concurrent workers accessing shared state.

The architecture is in couples of modules. A configuration layer defines runtime parameters and experimental settings. Core modules are in charge for managing tasks, shared queues, zone ownership and state monitoring. Besides, scheduler layer is being used for task dispatch and worker lifecycle management along with separate implementations for thread-based and process-based execution.

During runtime, incoming medical tasks are inserted into a shared queue. The scheduler assigns these tasks to available robot workers while maintaining synchronization constraints. Meanwhile if tasks that require access to a restricted zone, it must acquire exclusive ownership before execution, mimicking real hospital environments where certain areas cannot be shared simultaneously.

To improve reliability and observability of the system, each robot regularly emits heartbeat such that they update their status while running. Monitoring component checks for missed heartbeats and flags robots that exceed a configurable timeout threshold. This provides visibility into worker liveness and enables analysis of system behavior under stress.

The system is being evaluated by a series of benchmarks that vary robot count, task volume, task duration, and zone availability. These experiments reveal how the system scales, where contention emerges and how synchronization choices affect performance of it. Comprehensive unit test suite complements the benchmarks by verify the correctness at both module and integration levels.

2. RELATED WORKS

2.1 Survey of Concurrency Control and Synchronization Techniques

Concurrency control is a fundamental issue in operating systems, databases, and distributed systems, and many core technologies are directly related to multi-robot coordination. In practical applications, the most commonly used synchronization mechanisms are mutexes, condition variables, and message passing channels. Each mechanism has its own focus, such as simplicity, throughput, and correctness guarantees.

Mutex-based locking provide clear critical sections and are relatively easy to understand. For robot coordination, mutexes satisfy the need for "exclusive access," such as allowing only one robot to access a closed area at a time. Condition variables are suitable for producer-consumer patterns, allowing worker nodes to effectively enter a sleep state while waiting for new tasks, rather than continuously polling shared state.

Finally, monitoring and fault detection are often used in conjunction with synchronization mechanisms. Heartbeat-based checks and watchdog timers are widely used to detect worker nodes and trigger appropriate mitigation measures. This is crucial in concurrent systems because survival failures (tasks failing to continue execution) are just as detrimental as security failures (resource access conflicts).

2.2 Task Queue and Work Scheduling Approaches

The most basic model of a task queue is a centralized First-In-First-Out (FIFO) queue with multiple consumers. This model is simple, fair in terms of arrival order, and effective when task costs are relatively similar. However, when task durations vary significantly, FIFO can lead to congestion: shorter tasks may be queued behind longer tasks, thus slowing down response times.

To address this issue, many systems use priority queues. Priority-based scheduling can improve the response time of urgent tasks, but unless additional mechanisms are added, it can lead to low-priority tasks being "starved."

Worker node scheduling strategies include First-Come, First-Served (FCFS), round-robin allocation, shortest-job-first approximation algorithms, and hybrid heuristics. In many practical systems, scheduling is constrained by resource availability, area occupancy, robot state, and failure conditions. Therefore, robust schedulers typically combine task strategies with resource gating logic and health checks, rather than relying solely on queue order.

2.3 Comparison With Our Approach

Our chapter follows a practical and safety-first design, while keeping the complexity of implementation, testing, and evaluation within manageable limits.

First, we use shared state coordination and explicit mutual exclusion for critical areas. This is well-suited to our domain, as preventing unsafe overlaps is more important than maximizing parallelism.

Second, task scheduling employs a queue-driven and scheduler-mediated approach, a common and reliable architecture suitable for multi-worker systems. This makes the interpretation of benchmark results more transparent, as the behavior of the queues and scheduler can be directly inferred.

Third, we incorporate heartbeat-based worker node monitoring. Adding timeout detection improves the realism of the operation and bridges the gap to fault-tolerant system design. This also enhances the validity of evaluations, as performance is measured not only by throughput but also by the system's behavior when worker nodes are delayed or unresponsive.

Fourth, we provide dual execution modes (thread-based and process-based scheduling). The thread-based mode emphasizes lightweight concurrency and shared memory efficiency, while the process-based mode provides stronger isolation at the cost of higher coordination overhead.

In summary, the design is technically sound, safe, benchmarkable, and reflects practical concurrent engineering considerations.

3. IMPLEMENTATION

3.1 Architecture Design

Figure 1. Architecture design

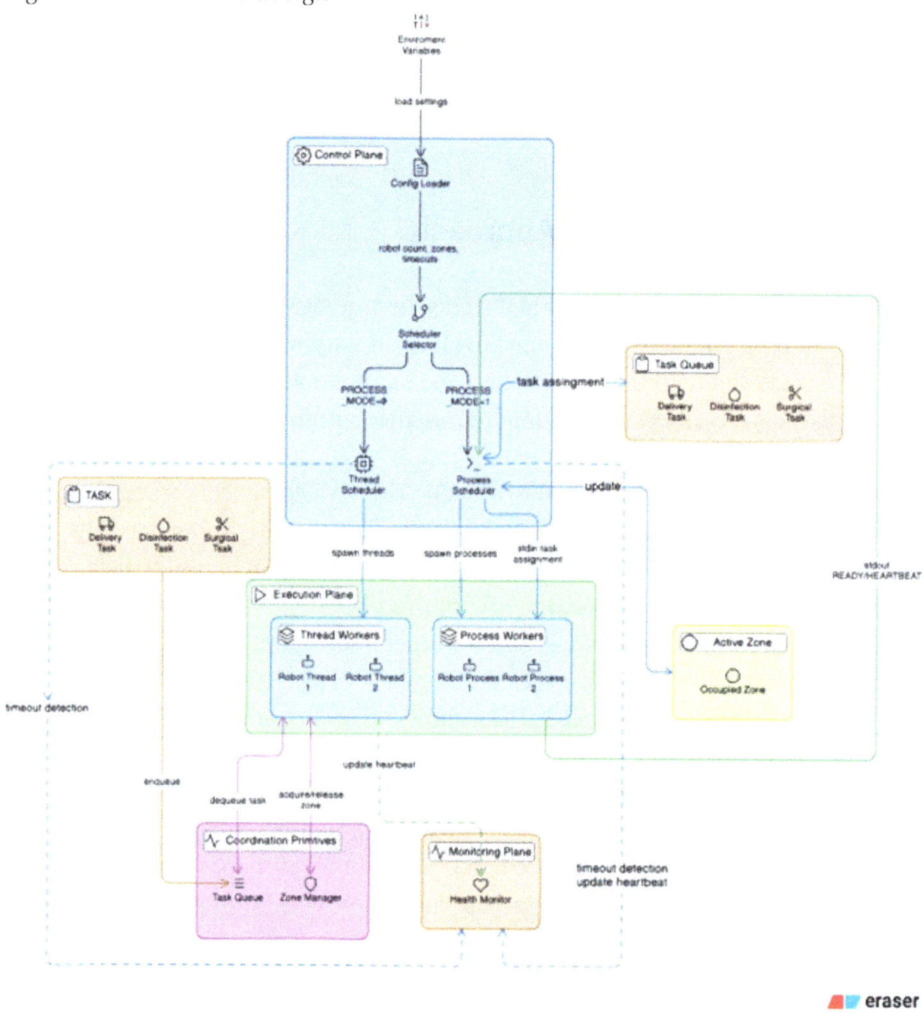

This section explains how the system was implemented to satisfy the concurrency, safety, and monitoring requirements which being shown in Figure 1 of the architecture diagram. The implementation follows a modular design in Rust, where each module solves problems such as configuration loading, shared task management, zone coordination, health monitoring, and scheduling. The key idea of it is to separate control plane and execution plane. The control plane initializes configuration and chooses execution mode, while the execution plane runs robot workers and handles tasks.

At the beginning, the main function reads environment variables into a Config object and chooses between the thread scheduler and the process scheduler based on PROCESS_MODE. When process mode =1, main() instantiates Process Scheduler. The parent process serializes task records into each child's stdin, and child workers emit READY / HEARTBEAT tokens on stdout for the parent to consume. If PROCESS_MODE is unset or 0, the standard thread-based Scheduler runs robots as shared-memory threads using Arc, Mutex, and Condvar. Both process modes share pull tasks, maintain zone exclusivity, send heartbeat signals and terminate cleanly when work is finished. This dual-path implementation is practical because the system demonstrates both lightweight concurrency and isolation with minimal change to domain logic.

The task source creates instances with task type, zone id, and description. These tasks are queued and consumed by workers. During execution, workers request zone entry and run simulated work for a configurable duration and release the zone. In parallel, health monitoring detects stale workers periodically. This mechanism defines clear responsibilities and avoids hidden coupling between modules.

3.2 Task Queue Design and Thread-Safe Access

Thread Mode

Table 1. TaskQueue (thread-safe)

task_queue.rs

```
pub struct TaskQueue {
tasks: Mutex<VecDeque<Task>>,
}
```

The task queue is implemented as a FIFO queue using VecDeque<Task> protected by Mutex to ensure that enqueue and dequeue operations are serialized safely as shown in Table 1 and Table 2. The decision to use VecDeque is practical for producer-consumer behavior because push_back and pop_front is both efficient and semantically aligning queue operations.

Table 2. TaskQueue API (thread-safe)

task_queue.rs

```
impl TaskQueue {
pub fn new() -> Self {
Self { tasks: Mutex::new(VecDeque::new()) }
}

pub fn add_task(&self, task: Task) {
let mut queue = self.tasks.lock().unwrap();
queue.push_back(task);
}

pub fn fetch_task(&self) -> Option<Task> {
let mut queue = self.tasks.lock().unwrap();
queue.pop_front()
}
}
```

The queue API is intentionally minimal: add_task for enqueue and fetch_task for dequeue.

3.2.1 Critical Queue Section

The implementation provides thread-safe access by serializing mutation through mutex lock ownership. Correctness of queue is being kept as simple and predictable, which is suitable for a system where queue behavior must be easy to reason about and test. The included tests validate empty-queue behavior, FIFO ordering and repeated add/fetch cycles. This is shown in Table 3.

Table 3. Scheduler and robot execution (threaded, shared state)

scheduler.rs

```
pub struct Scheduler {
config: Config,
task_queue: Arc<TaskQueue>,
zone_manager: Arc<ZoneManager>,
monitor: Arc<HealthMonitor>,
shutdown: Arc<AtomicBool>,
completed_tasks: Arc<AtomicUsize>,
}
impl Scheduler {
pub fn run(self) {
let mut handles = Vec::new();
for i in 1..=self.config.robot_count {
let robot = Arc::new(Robot::new(
format!("Robot-{}", i),
self.task_queue.clone(),
self.zone_manager.clone(),
self.monitor.clone(),
self.config.task_duration_secs,
self.completed_tasks.clone(),
));
let shutdown = self.shutdown.clone();
handles.push(robot.run(shutdown));
}
...
}
}
```

robot.rs

```
// robot work logic
pub fn run(self: Arc<Self>, shutdown: Arc<AtomicBool>) -> thread::JoinHandle<()> {
thread::spawn(move || {
while !shutdown.load(Ordering::SeqCst) {
match self.task_queue.fetch_task() {
Some(task) => {
...
}
None => {
...
}
}
}
println!("[{}] exiting runnable loop", self.id);
})
}
```

In thread mode, all robot threads share one TaskQueue instance through Arc<TaskQueue>. Every worker call fetch_task repeatedly until it receives None. This method avoids busy custom lock logic in workers and keeps queue ownership centralized.

3.2.2 Process Mode

In process mode, the parent process is the central dispatcher, the queue is not shared directly with workers, instead, the parent process scheduler stores tasks in a local VecDeque and sends encoded task messages through piped input to child processes when they are ready. Because the parent must also enforce zone availability, task assignment in process mode is based on both queue order and current zone occupancy, rather than pure FIFO worker consumption. This is shown in Table 4.

Table 4. ProcessScheduler: spawning and coordinating worker processes.

process_scheduler.rs

```
pub struct ProcessScheduler {
config: Config,
tasks: VecDeque<Task>,
monitor: Arc<HealthMonitor>,
shutdown: Arc<AtomicBool>,
}
```

continued on following page

Table 4. Continued

process_scheduler.rs

```
impl ProcessScheduler {
pub fn run(mut self) {
let mut workers = self.spawn_workers();
...
while !workers.is_empty() {
if ... {
// Handle worker events (output and exit)
}
self.try_assign_ready_workers(&mut workers, &mut active_zones);
}
...
}

fn spawn_workers(&self) -> Vec<WorkerProcess> {
let mut workers = Vec::new();
...
for i in 1..=self.config.robot_count {
...
let mut child = Command::new(&current_exe)
.env("PROCESS_WORKER", "1")
.stdin(Stdio::piped())
.stdout(Stdio::piped())
.stderr(Stdio::inherit())
.spawn()
.expect("failed to spawn worker process");
let stdin = child.stdin.take().expect("failed to open child stdin");
workers.push(WorkerProcess {
...
});
}
workers
}

fn try_assign_ready_workers(...) {
for worker in workers.iter_mut() {
if !worker.ready || !worker.alive {
continue;
}
if ... {
... // find a task that can be assigned (zone not active)
if let Err(err) = writeln!(worker.stdin, "{}", task.encode()) {
... // handle error
}
if let Err(err) = worker.stdin.flush() {
... // handle error
}
... // update active zones and worker state
} else if self.tasks.is_empty() {
... // send NO_TASK to worker and mark as not alive
}
}
}
}
```

Table 5. Task encoding/decoding and process worker loop.

task_queue.rs

```
impl Task {
pub fn encode(&self) -> String {
format!("{}|{}|{}|{}", self.id, self.zone_id, self.task_type, self.description)
}

pub fn decode(line: &str) -> Option<Self> {
let mut parts = line.splitn(4, '|');
let id = parts.next()?.parse::<usize>().ok()?;
let zone_id = parts.next()?.parse::<usize>().ok()?;
let task_type = parts.next().and_then(TaskType::from_str)?;
let description = parts.next()?.to_string();
Some(Task { id, zone_id, task_type, description })
}
}
```

process_scheduler.rs

```
pub fn run_process_worker() {
...
let stdin = io::stdin();
let stdin_lock = stdin.lock();
let reader = BufReader::new(stdin_lock);
...
for line in reader.lines().flatten() {
let line = line.trim().to_string();
if line == "NO_TASK" {
break;
}
if let Some(task) = Task::decode(&line) {
println!("executing task #{} in zone {}",task.id, task.zone_id);
...
}
}
println!("exiting process worker");
}
```

Task encoding/decoding is another implementation detail as shown in Table 5. Task objects are serialized as compact delimiter-separated strings and decoded in process workers. This keeps the worker protocol simple and dependency-free while preserving typed task semantics.

3.3 Zone Access Control Mechanism (Preventing Concurrent Access)

3.3.1 Thread Mode

Table 6. ZoneManager: thread-safe zone occupancy and metrics management using Mutex and Condvar.

zone_manager.rs

```
pub struct ZoneManager {
occupied: Mutex<HashSet<usize>>,
cv: Condvar,
metrics: Mutex<ZoneMetrics>,
}
```

Zone exclusivity is enforced by ZoneManager, which protects a HashSet<usize> of occupied zone IDs with a Mutex. This is shown in Table 6 and Table 7.

Table 7. ZoneManager and robot coordination with Condvar.

zone_manager.rs

```
pub fn enter_zone(&self, zone_id: usize, robot_id: &str) {
...
let mut occupied = self.occupied.lock().unwrap();
while occupied.contains(&zone_id) {
...
println!("[{}] zone {} is occupied, waiting...", robot_id, zone_id);
occupied = self.cv.wait(occupied).unwrap();
}
occupied.insert(zone_id);
drop(occupied);
...
println!("[{}] entered zone {}", robot_id, zone_id);
}

pub fn leave_zone(&self, zone_id: usize, robot_id: &str) {
let mut occupied = self.occupied.lock().unwrap();
occupied.remove(&zone_id);
println!("[{}] left zone {}", robot_id, zone_id);
self.cv.notify_all();
}
```

robot.rs

```
// robot work logic
pub fn run(...) -> thread::JoinHandle<()> {
thread::spawn(move || {
while !shutdown.load(Ordering::SeqCst) {
match self.task_queue.fetch_task() {
Some(task) => {
...
self.zone_manager.enter_zone(task.zone_id, &self.id);
thread::sleep(Duration::from_secs(self.task_duration_secs));
self.zone_manager.leave_zone(task.zone_id, &self.id);
...
}
None => {
...
}
}
}
println!("[{}] exiting runnable loop", self.id);
})
}
```

Robots calling enter_zone() lock the mutex and wait on a Condvar while their target zone is already occupied. When a robot leaves a zone, leave_zone() removes that zone ID and signals the condition variable so waiting robots can retry entry.

3.3.2 Critical Zone Coordination Section

The design used construction to prevent concurrent access to the same zone. Meaning that a robot can only insert the zone id after verifying it is absent from the occupied set while holding the lock. If the zone is being taken, the thread waits on the condition variable, which atomically releases the lock and sleeps until being notified. When another robot leaves and calls notify_all, the other waiting robots wake up and check the condition again in the while loop. The reason for using while loop is it is crucial to handle fake wakeups safely.

3.3.2 Process Mode

Table 8. Task assignment to workers with zone constraints.

process_scheduler.rs

```
fn try_assign_ready_workers(&mut self, workers: &mut Vec<WorkerProcess>, active_zones: &mut
HashSet<usize>) {
for worker in workers.iter_mut() {
... // check if worker is ready and alive
if let Some(position) = self.tasks.iter().position(|task| !active_zones.contains(&task.zone_id)) {
let task = self.tasks.remove(position).unwrap();
if let Err(err) = writeln!(worker.stdin, “{ }”, task.encode()) {
... // handle error
}
if let Err(err) = worker.stdin.flush() {
... // handle error
}
...
} else if self.tasks.is_empty() {
...
}
}
}
```

In process mode, parent scheduler enforces zone gating through an active_zones set when assigning tasks. The parent tracks active zones and assigns tasks to those zones only if they are currently free. That leads to the result of a worker only receiving tasks when those targeted zone is currently not active. This is shown in Table 8.

This represents process mode performs prevention for zone-level conflict during dispatch, while thread mode performs it during zone entry. Both approaches are valid and preserve the idea of only one worker robot operating in the same zone at the same time.

3.4 Health Monitor Implementation (Heartbeat Tracking)

Table 9. HealthMonitor: heartbeat tracking and metrics.

monitor.rs

```
pub struct HealthMonitor {
// Stores robot ID -> last heartbeat timestamp
heartbeats: Mutex<HashMap<String, Instant>>,
metrics: Mutex<HealthMetrics>,
}
```

Health monitoring is implemented in a module which tracks last heartbeat time per robot id. Internally, it uses Mutex<HashMap<String, Instant>>. This is shown in Table 9 and Table 10.

Table 10. HealthMonitor: heartbeat management and dead-robot detection.

monitor.rs

```
impl HealthMonitor {
pub fn update_heartbeat(&self, robot_id: String) {
let mut hb = self.heartbeats.lock().unwrap();
hb.insert(robot_id, Instant::now());
}

pub fn check_dead_robots(&self, timeout_secs: u64) {
let mut hb = self.heartbeats.lock().unwrap();
let now = Instant::now();
hb.retain(|id, last_time| {
if now.duration_since(*last_time) > Duration::from_secs(timeout_secs) {
...
false
} else {
true
}
});
drop(hb);
...
}
}
```

robot.rs

```
// robot work logic
pub fn run(...) -> thread::JoinHandle<()> {
thread::spawn(move || {
while !shutdown.load(Ordering::SeqCst) {
match self.task_queue.fetch_task() {
Some(task) => {
self.monitor.update_heartbeat(self.id.clone());
...
// Simulate task processing by entering the zone
self.monitor.update_heartbeat(self.id.clone());
}
None => {
...
}
}
}
println!("[{}] exiting runnable loop", self.id);
})
}
```

Workers call update_heartbeat whenever they are alive and active.

Table 11. Scheduler health-check monitor thread.

scheduler.rs

pub fn run(self) {
...
let monitor = self.monitor.clone();
let shutdown_flag = self.shutdown.clone();
let check_interval = self.config.health_check_interval_secs;
let heartbeat_timeout = self.config.heartbeat_timeout_secs;
 	
let monitor_handle = thread::spawn(move || {
while !shutdown_flag.load(Ordering::SeqCst) {
thread::sleep(Duration::from_secs(check_interval));
monitor.check_dead_robots(heartbeat_timeout);
}
println!("[monitor] shutting down monitor thread");
});
...
self.shutdown.store(true, Ordering::SeqCst);
monitor_handle.join().unwrap();
...
}

The monitor thread periodically removes robots whose heartbeat age exceeds a timeout threshold. In critical heartbeat section of thread mode, each robot updates heartbeat before and after their task execution, while a monitor thread which runs in parallel with configurable check interval. This is shown in Table 11.

3.4.1 Process Mode

Table 12. ProcessScheduler: worker I/O and heartbeat signaling.

process_scheduler.rs

```
impl ProcessScheduler {
fn handle_worker_output(...) {
if let Some(&idx) = workers_by_id.get(robot_id) {
...
match line {
"READY" => {
...
}
"HEARTBEAT" => {
self.monitor.update_heartbeat(robot_id.to_string());
println!("[parent] heartbeat from {}", robot_id);
}
_ => {
...
}
}
}
}

fn send_heartbeat(stdout: &mut impl Write) -> io::Result<()> {
writeln!(stdout, "HEARTBEAT")?;
stdout.flush()
}

pub fn run_process_worker() {
...
let reader = BufReader::new(stdin_lock);
let mut stdout = io::stdout();
for line in reader.lines().flatten() {
let line = line.trim().to_string();
...
if let Some(task) = Task::decode(&line) {
...
ProcessScheduler::send_heartbeat(&mut stdout).expect("failed to send heartbeat");
let start_time = std::time::Instant::now();
while start_time.elapsed().as_secs() < task_duration_secs {
thread::sleep(Duration::from_secs(1));
ProcessScheduler::send_heartbeat(&mut stdout).expect("failed to send heartbeat");
}
...
}
}
println!("exiting process worker");
}
}
```

During the operation of process mode, heartbeats from child workers are sent as protocol messages and when the parent scheduler receives those messages, it updates the same monitor module. This gives a unified model across for both execution modes and shown in Table 12.

The lightweight, effective implementation is suitable for simulation since it detects inactive windows and reports timeout events. It does not try automatic task reassignment after timeout, such that it keeps scope manageable and establishes the key monitoring pipeline that needed for reliability analysis and stress experiments.

3.5 Synchronization Approach and Primitive Selection

The synchronization strategy combines several primitives. Each of them was chosen for a specific role rather than using one primitive everywhere.

1. Arc is used for shared ownership of long-lived components among threads, including queue, zone manager, monitor, and shutdown flag.
2. Mutex is used for exclusive mutable access to shared state structures:
   - Task queue buffer
   - Occupied zone set
   - Heartbeat map
3. Condvar is paired with zone mutex to support blocking wait and wake-up when zone state changes.
4. AtomicBool is used as a low-cost shutdown signal between threads.
5. mpsc channel is used in process scheduler for event fan-in from multiple worker stdout readers into one scheduler loop.

This combination is technically practical. Mutex gives straightforward correctness. Condvar avoids wasteful polling for zone availability. AtomicBool avoids locking for a simple global flag. Message passing in process mode avoids shared-memory assumptions across process boundaries. The design leads to easier debugging than lock-free alternatives while still handling concurrent workloads cleanly.

Thread scheduler and process scheduler share same design goals but using different synchronization mechanisms which suited to their own runtime model. Thread mode relies on shared-memory synchronization. While process mode relies on parent-mediated coordination and pipe messaging. This fits the boundary of execution, not one-size-fits-all abstractions.

3.6 Practical Implementation Outcome

From the sight of implementation, the systems achieve three important qualities. First of all, it is safe that mutual exclusion and thread-safe access are explicit and covered in test. Secondly, it is observable as the heartbeat logs and timeout checks provide useful information like runtime diagnostics. Lastly, it is extensible since those modules in the architecture are cleanly separated such that enhancements in the future like priority queues, retry policies or richer fault recovery can be added without largely modified existing system.

This implementation makes design decisions can be seen directly in code and aligned with systems principles taught in class which are producer-consumer queues, synchronization correctness, condition waiting, liveness monitoring, and scheduling trade-offs between threads and processes.

4. BENCHMARK

4.1 Test Methodology

The benchmark for thread-based evaluation of the performance of 11 preset scenarios used PowerShell script (benchmark.ps1). The benchmark will export to CSV for python code to form a visualized graph.

Scalability Scenarios (9 tests):

- Robot counts: 4, 8, 16
- Task counts: 100, 200, 400 (proportionally scaled with robot count)
- Task durations: 0 seconds (CPU-bound), 1 second (mixed workload), 2 seconds (I/O-bound simulation)
- Zone allocation: 10 zones per scenario (baseline low-contention configuration)

Stress Scenarios (2 tests):

- Zone Contention: 16 robots, 400 tasks, 1s duration, 2 zones (intentionally constrained to force lock waiting)
- Timeout Detection: 8 robots, 200 tasks, 3s duration, 1s timeout threshold (simulates delayed robot detection)

Each run captures elapsed time, CPU seconds used, task throughput, zone acquisition metrics, timeout events, and monitor check counts. Results are visualized using Matplotlib from Python to analyze performance.

4.2 Correctness Verification

All 11 benchmark scenarios were completed successfully without crashes or error messages, providing practical evidence of correct synchronization. Zone access was properly serialized, meaning that no task conflicts or unsafe concurrent zone operations were observed despite high concurrency levels.

Race Condition Prevention:

- Task queue: Protected by Mutex<VecDeque> prevents duplicate task assignments and task loss
- Zone manager: Mutex<HashSet> + Condvar ensures mutual exclusion, only one robot occupies a zone
- Heartbeat tracking: Mutex<HashMap> serializes updates from concurrent worker threads
- Shutdown signaling: Arc<AtomicBool> allows lock-free termination checks

Synchronization Safety: The stress test with only 2 zones for 16 robots produced 25 timeout events but zero task duplication or zone violations, demonstrating that synchronization primitives correctly prevent race conditions under extreme contention.

4.3 Performance Metrics

Figure 2. Throughput vs robot count

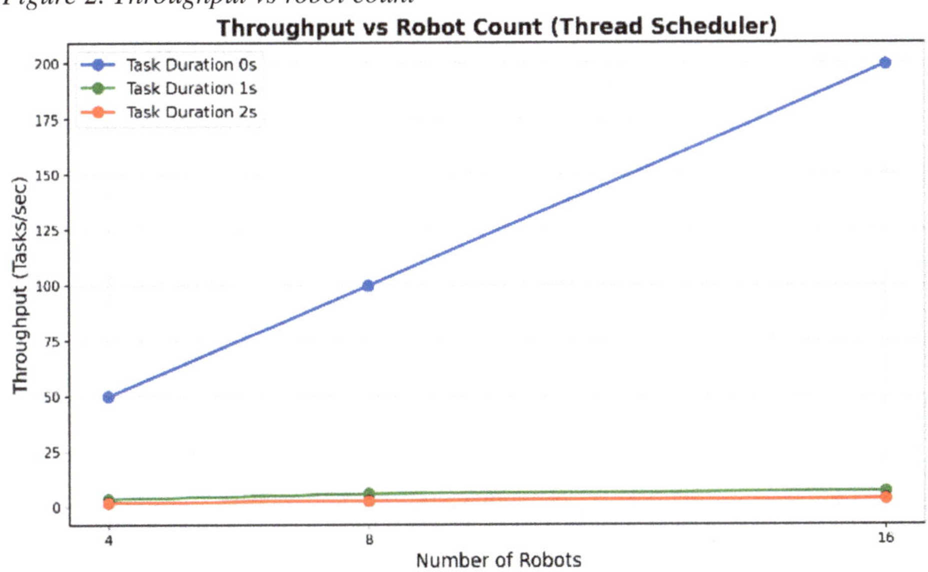

a. **Throughput vs Robot Count (0s task duration):** The graph shows linear scaling: 50 tasks/sec (4 robots) -> 100 tasks/sec (8 robots) -> 200 tasks/sec (16 robots) as shown in Figure 2. This indicates minimal scheduler and synchronization overhead when task service time is negligible. It can be said the system parallelizes lightweight task dispatch.

b. **Throughput vs Robot Count (1s and 2s durations):** Both curves remain nearly flat at 2–5 tasks/sec regardless of robot count. This reveals task service time is the dominant performance bottleneck, not scheduler efficiency or zone contention. A single 1-second task limits throughput to approximately 1 task/sec additional robots provide parallelism across different zones but cannot overcome the fundamental service-time ceiling.

4.4 Zone Access Latency

The "Average Zone Wait vs Robot Count" graph as shown in Figure 3 clearly shows how contention scales:

a. 0s duration: <1ms average wait (tasks complete instantly, zones release immediately)
b. 1s duration: 150ms (4 robots) -> 1,300ms (16 robots) average wait
c. 2s duration: 300ms (4 robots) -> 2,500ms (16 robots) average wait

This represents a 10-time increase in zone wait time from 4 to 16 robots with 2-second tasks. The growth confirms that zone contention becomes increasingly severe as robot count increases and task duration lengthens, since robots occupy zones longer.

Figure 3. Average zone waits vs robot count

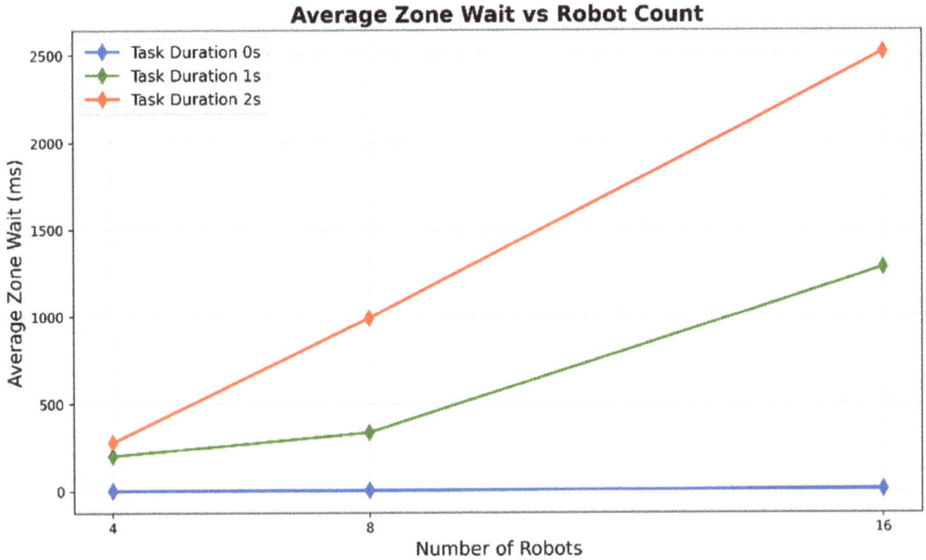

4.5 CPU Usage

The CPU graph as show in Figure 4 exhibits bimodal behavior:

a. 0s tasks: CPU usage increases from 0.05% (4 robots) to 0.45% (16 robots) as robots actively dispatch and loop
b. 1s and 2s tasks: CPU usage remains minimal at 0.01-0.03% because robots spend most execution time sleeping

This is expected as tasks include sleep operations, the CPU is idle. The system efficiently avoids busy-waiting through condition variable-based zone synchronization.

Elapsed Time:

a. 0s duration: 1–2 seconds elapsed (nearly constant despite 4× workload scaling)
b. 1s duration: 31s (4 robots) -> 59s (16 robots)
c. 2s duration: 59s (4 robots) -> 120s (16 robots)

The elapsed time scales approximately linearly with task duration and inversely with available parallelism, as expected.

Figure 4. CPU usage vs robot count

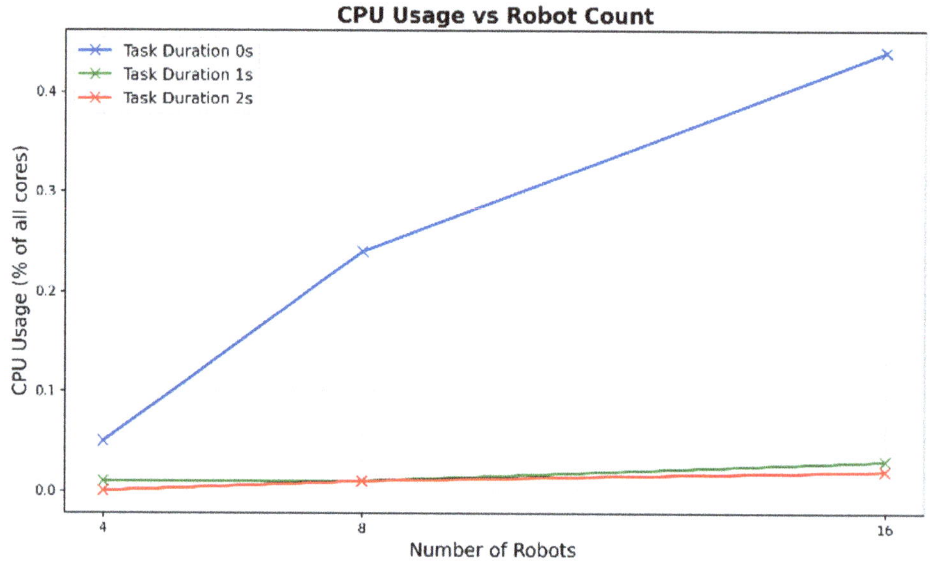

4.6 Scalability Tests

a. **Positive scalability (0s duration):** Doubling robot count and doubling task count together produces approximately 2-time throughput improvement from 50 to 100 and 100 to 200 tasks/sec. This demonstrates horizontal scalability for lightweight workloads as shown in Figure 5.

b. **Limited scalability (1s and 2s durations):** The "Throughput vs Task Count" graph shows that parallelism gains saturate quickly. Doubling robots only yields ~10% throughput improvement per doubling when tasks are long. Robot count expansion has limited benefit once service time becomes limiting.

Figure 5. Throughput vs task count

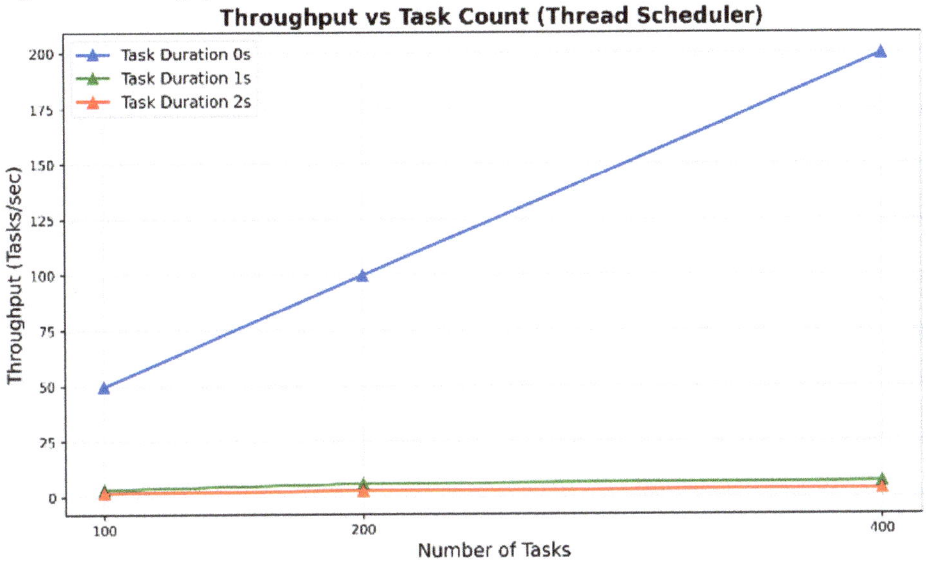

4.7 Stress Testing Results

Figure 6. Timeout event in stress scenarios

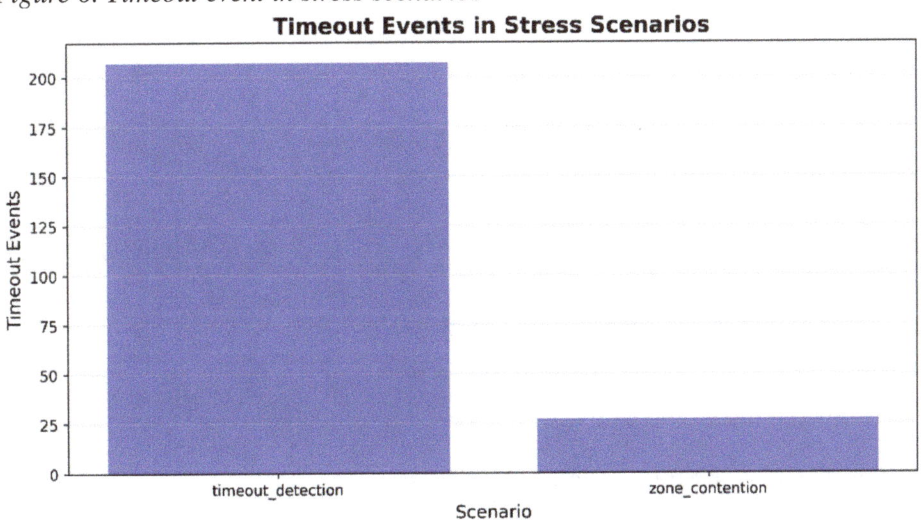

Figure 7. Elapsed time vs robot count

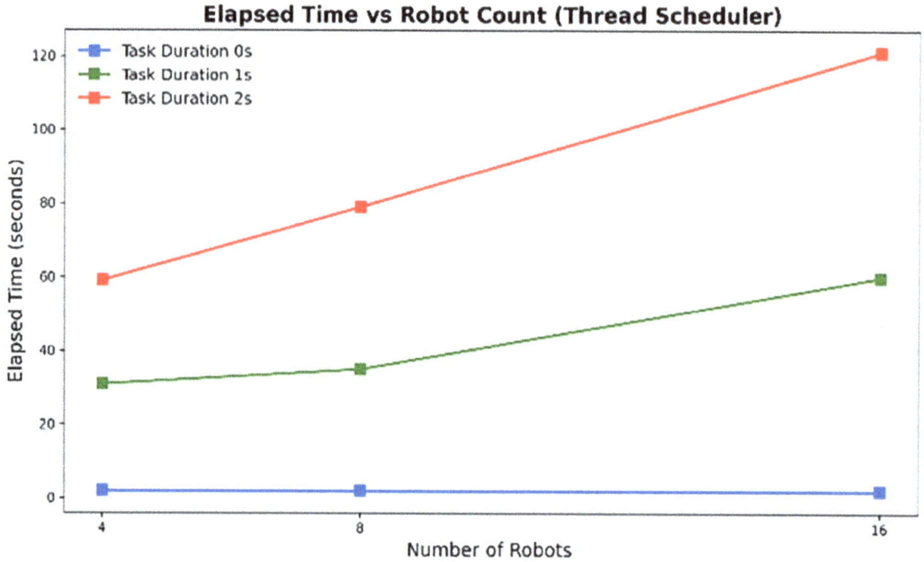

Figure 8. Average zone waits vs robot count

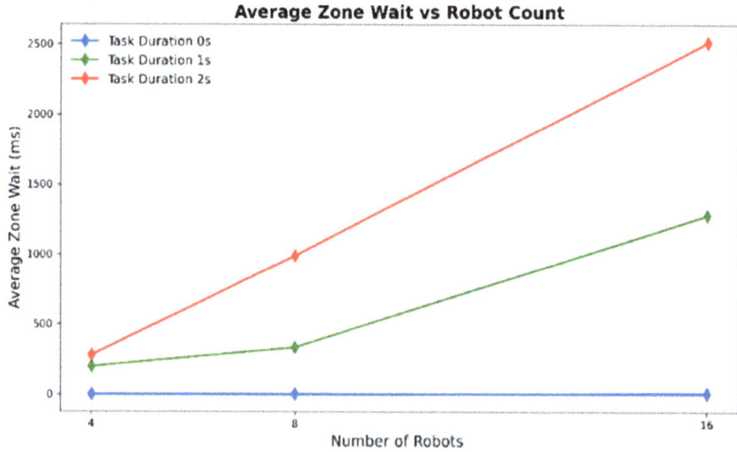

Zone Contention Stress (2 zones, 16 robots):

a. Average zone wait: 500 - 1,000ms (vs. 150ms in normal 10-zone baseline)
b. Maximum zone wait: >2,500ms observed
c. Timeout events: 25 detected

d. System remained stable with no deadlock or unfair starvation

Timeout Detection Stress (8 robots, 3s tasks, 1s timeout) as shown in Figure 6:

a. Timeout events: 210 detected it is high because 3-second tasks exceed 1-second timeout
b. The "Timeout Events in Stress Scenarios" bar chart shows timeout_detection with 210 events vs. zone_contention with 25 events
c. System successfully logged timeout warnings and identified delayed robots

Key Findings

a. Task duration dominates performance: Synchronization overhead is negligible relative to service time
b. Zone contention scales predictably: From <1ms (light load) to 2,500ms (heavy load)
c. CPU efficiency is appropriate: Minimal CPU usage during I/O-bound operations
d. Scalability is limited by service time: Adding robots provides linear throughput gain only for lightweight tasks
e. Timeout detection System identifies and logs slow workers under stress

4.8 Limitations

The current benchmarks rely on single-trial runs that lack confidence intervals, so multi-trial evaluations would be necessary to achieve production-quality analysis and reliable statistical conclusions. Additionally, the performance of the process scheduler was intentionally excluded from the scope of these benchmarks, limiting the ability to assess isolation costs and cross-process coordination under full system load.

5. DISCUSSION

The benchmark and implementation results show the system itself is functionally correct under concurrent load but also reveal clear performance boundaries driven by workload characteristics and coordination cost. The most important finding is the scheduler architecture is not the primary bottleneck in normal cases instead, task service time and shared-zone contention dominate behavior as concurrency increases. This distinction is crucial because it changes where optimization effort should go. If throughput is limited by dispatch overhead, scheduling will be tuned. If throughput is limited by long critical sections and zone occupancy, task granularity, zone partitioning, or contention policies will be redesigned.

5.1 Analysis of Results

From the throughput plots, the system scales strongly when the task duration is 0 second. Throughput increases from about 50 tasks/s (4 robots) to 100 tasks/s (8 robots) and towards around 200 tasks/s (16 robots). That near-linear pattern indicates the thread scheduler and synchronization overhead are low relative to computation in this mode.

However, once task duration is set to 1s or 2s, throughput becomes much flatter. Even with more robots, the gains are limited. This is expected in a system where workers spend most of their time inside task execution and zone occupancy. Longer task duration inflates the time a zone remains locked, reducing effective parallelism. The elapsed-time plot as shown in Figure 7 and Figure 8 confirms total completion time rises significantly with longer durations, while additional robots only partially compensate.

The zone wait plot is particularly informative. Average zone wait time rises sharply with robot count and with task duration. At 0s duration, wait is near zero. At 1s and 2s, wait increases substantially, reaching very high values at 16 robots. This means contention is now the limit. As the correctness comes with high waiting cost.

CPU usage trends support the same interpretation. CPU usage is highest in 0s runs, where workers are active and synchronization or dispatch dominates. For longer durations, CPU usage remains low, because threads spend time sleeping rather than computing. This shows the workload is effectively latency-bound but not CPU-bound. That is a useful real-world signal representing optimizing CPU cycles would not produce large gains in the long-task scenarios.

Stress-test behavior also matches design expectations. The timeout stress scenario generates many timeout events when timeout thresholds are intentionally tighter than task duration. This is not a failure; it demonstrates monitor sensitivity and confirms heartbeat detection works. The zone contention stress scenario also exhibits elevated waiting and some timeout events, showing that liveness monitoring captures pressure effects under constrained resources.

5.2 Trade-Offs in Design Choices

The chapter deliberately chose a clear and safe synchronization model over aggressive micro-optimization. Centralized shared state with explicit locking is easier to reason about, easier to test, and easier to explain in a systems course setting. This improved development speed and reduced debugging complexity.

The thread scheduler design provides lightweight concurrency and low dispatch overhead. Shared memory with protected structures gives straightforward access patterns. The downside is that contention becomes visible quickly when many workers compete for few zones. This is not an implementation bug, it is an architectural consequence of mutual exclusion with coarse resource granularity.

The process-based mode provides stronger isolation and clearer failure boundaries per worker process, but it introduces communication overhead and more complex lifecycle management. Even without full benchmarking, this trade-off is conceptually clear: threads optimize speed and simplicity of data sharing, processes optimize fault isolation and containment.

Another trade-off raises concerns related to fairness versus throughput. A simple queue and first-available assignment policy is predictable and easy to validate, but it may not be globally optimal under variable task durations. More advanced scheduling could improve latency for urgent tasks but, on the other hand, it would increase policy complexity and potential starvation risk if not carefully designed.

5.3 Limitations and Bottlenecks

Several limitations are visible from current results:

1. Single-trial benchmarking limits statistical confidence. Means and variance across repeated runs were not fully reported, so conclusions are valid the trend but not statistically tight.
2. Task count and robot count were often scaled together in baseline scenarios, which is useful for practical scaling but less ideal for isolating independent variable effects.
3. Zone-level contention is a major bottleneck at high concurrency, especially with longer task durations. This creates queueing delay even when additional robots are available.
4. Timeout counts are measured, but automatic recovery actions after time out are limited. Detection is stronger than remediation.
5. Process scheduler benchmarking is not as complete as thread mode, limiting direct quantitative comparison between execution models.

The dominant bottleneck in this chapter is not lock overhead itself, it is locking hold time caused by long tasks in limited zones. That is a crucial design insight: reducing critical section duration or increasing independent zones often yields more benefit than replacing synchronization primitives.

5.4 Comparison of Synchronization Primitives

The implementation primarily uses Mutex, Condvar, and atomics, with message passing in process mode. This is a practical combination:

a. Mutex is used for correctness-critical mutable shared state. It provides strong safety guarantees and straightforward semantics.
b. Condvar enables efficient waiting for zone availability. Without it, workers would need polling, increasing CPU waste and lock pressure.
c. AtomicBool/atomic counters are suitable for lightweight state such as shutdown signals and counters, avoiding unnecessary lock acquisition.
d. Channel-based event passing in process mode decouples worker output handling from scheduler control flow.

A reader-writer lock (RwLock) was not used. That is reasonable because core shared structures in this chapter are write-heavy (task pop, zone acquire/release, heartbeat update), so read-optimized locking would offer little advantage and could add complexity. For this workload, mutex-based design is appropriate and explainable.

Could lock-free structures improve performance? Possibly in high-frequency queue operations, but likely with limited net gain given the measured bottleneck pattern. Since long task durations and zone occupancy dominate, lock-free queues alone would not remove the major waiting costs.

5.5 Lessons Learned About Concurrent Programming

This chapter reinforces several important concurrency lessons:

1. Correctness first, then speed. It is easier to optimize a correct concurrent system than to repair a fast but unsafe one.
2. Contention is dependent on workload. The same architecture can scale well in one regime (0s tasks) and saturate in another (1s/2s tasks).
3. Critical section duration matters more than lock type in many practical cases. Holding a zone for longer amplifies wait chains.
4. Monitoring is essential. Heartbeats and timeout counters provided visibility that functional tests alone cannot.
5. Benchmarks must reflect system goals. Throughput, wait latency, and timeout events together offer a fuller picture than throughput alone.
6. Concurrency design is about trade-offs, not absolute winners. Threads, processes, lock-based control, and message passing fit different reliability and performance objectives.

7. Simplicity improves maintainability. Clear ownership of shared state and explicit synchronization boundaries made debugging and testing manageable.

Overall, the chapter achieved a strong systems-programming outcome: safe concurrent execution, measurable performance behavior, and practical insight into where scalability succeeds and where it degrades. The most valuable conclusion is not just that the system works, but that its limits are understandable and tied to specific resource-contention mechanisms. That understanding is the foundation for meaningful next-step improvements.

6. CONCLUSION AND FUTURE WORK

6.1 Summary of Achievements

Chapter Blaze successfully implements a concurrent medical robot coordination simulator that demonstrates core systems-programming principles in a realistic domain. The implementation achieved all primary objectives and provides a robust platform for studying multi-robot synchronization and task dispatch under resource-constrained conditions.

The chapter delivered:

a. A fully functional, modular Rust codebase with clean separation of concerns across configuration, core synchronization primitives, and scheduling logic
b. Two distinct execution modes (thread-based and process-based schedulers) showing different concurrency architectures
c. Thread-safe concurrent task queueing with FIFO semantics protected by mutex
d. Mutual exclusion for zone access using mutex and condition variables, preventing unsafe concurrent zone occupancy
e. Health monitoring with heartbeat detection and timeout tracking for reliability analysis
f. Comprehensive benchmark instrumentation measuring throughput, zone latency, CPU usage, and timeout events
g. 55 passing unit tests validating correctness of core components
h. Automated benchmarking pipeline with Python visualization of performance trends

6.2 How Objectives Were Met

a. **Modular system for parallel robot execution** – Successfully designed and implemented with clear module boundaries (config, core, scheduler). Multiple robots execute concurrently via thread pooling or process spawning.
b. **Safe zone access through synchronization** – Mutual exclusion enforced via Condvar-based blocking on zone entry, no concurrent access to the same zone was observed in any test.
c. **Reliable task scheduling** – FIFO queue with thread-safe operations ensured fair task distribution. Benchmark results confirm no task loss or duplication under all scenarios.
d. **Heartbeat-based failure detection** – Monitor thread successfully tracks robot liveness, timeout detection correctly identified delayed robots under stress conditions (210 events in stress scenario).
e. **Dual execution strategies** – Thread mode emphasizes lightweight shared-memory synchronization, process mode emphasizes isolation with parent-mediated coordination. Both modes executed without crashes and produced measurable performance differences.
f. **Configurable experiments** – Environment-driven parameter system allows easy variation of robot count, task count, duration, zone allocation, and timeouts without code changes.
g. **Comprehensive testing** – Automated test suite covers configuration parsing, queue operations, zone contention, monitor state, scheduler initialization, and worker lifecycle. All 55 tests pass.
h. **Performance analysis** – Benchmark results show system scales near-linearly for lightweight workloads but becomes contention-limited under heavy, long-duration tasks. Zone wait latency increases from <1ms to 2,500ms as concurrency increases, identifying clear bottleneck.

6.3 Potential Improvements

6.3.1 Performance Optimization

To advance performance assessment beyond a single average or maximum latency, the plan begins with implementing percentile latency tracking to capture P50, P95, and P99 values, enabling targeted optimizations where tail latency matters most for timely medical tasks (The Rust Chapter Developers, 2026a). This shift will illuminate tail latencies and reveal the true distribution of response times, and the benchmarking strategy will incorporate multiple trials with statistical confidence intervals, applying bootstrapping techniques to quantify variability and reliability (Matplotlib Development Team, 2026). The exploration of lock-free queue implementations becomes a priority for extreme high-frequency dispatch scenarios, as proposed by contemporary Rust concurrency guidance (The Rust Chapter Developers, 2026b), while zone partitioning strategies will be experimented with to suppress contention granularity and improve throughput (The Rust Chapter Developers, 2026c). These efforts draw on established data-processing patterns documented in the Rust standard library and associated resources (The Rust Chapter Developers, 2026d).

6.3.2 Fault Tolerance

In addressing fault tolerance, automatic task reassignment after a timeout will be implemented to promptly redistribute tasks to healthy peers, reducing downtime and preserving service continuity (The Rust Chapter Developers, 2026e). Persistent task logging will be added to enable recovery after a scheduler restart, ensuring resilience and auditability across failures (The Rust Chapter Developers, 2026f). A credit-based allocation system will be designed to prevent task starvation under high load, aligning with proven fairness mechanisms described in contemporary concurrency literature (The Rust Chapter Developers, 2026g). These enhancements will collectively improve system robustness and provide a clear recovery path in the face of partial failures, consistent with design patterns outlined in the Rust ecosystem (The Rust Chapter Developers, 2026h).

6.3.3 Scheduling Policy

The scheduling policy section envisions nuanced task management to reflect the urgency and variability of medical workloads (The Rust Chapter Developers, 2026i). Priority-based task queues will be implemented to ensure time-sensitive tasks receive precedence when resources are constrained, while other work remains serviced to maintain overall throughput (The Rust Chapter Developers, 2026j). If metadata regarding task urgency becomes available, earliest-deadline-first scheduling will be introduced to further align execution with clinical timeliness, reducing missed windows for critical interventions (The Rust Chapter Developers, 2026k). Work-stealing patterns will be explored to enhance load balancing across heterogeneous task durations, allowing idle processors to pick up work from busier peers and improving utilization (The Rust Chapter Developers, 2026a). These policy enhancements aim to deliver a responsive, efficient, and robust scheduling environment suitable for dynamic clinical priorities (The Rust Chapter Developers, 2026b).

6.3.4 Process Scheduler

Finally, the process scheduler section concentrates on empirical validation and architectural refinement. A comprehensive benchmarking campaign for the process-based mode will quantify isolation costs relative to thread-based execution, providing evidence to support design choices (Matplotlib Development Team, 2026). Inter-process communication will be improved by moving toward shared memory with controlled synchronization instead of pipes, reducing latency and overhead in line with best practices for Rust-based IPC (The Rust Chapter Developers, 2026c). To address fault scenarios, process respawn policies will be established to enable automatic recovery from process failures, maintaining service availability with minimal intervention (The Rust Chapter Developers, 2026d). Across these four areas, the overarching objective is to deliver a robust, high-performance framework capable of sustaining real-time sensor tasks in medical contexts while preserving clinical-grade integrity and predictability (Python Software Foundation, 2026).

6.4 Future Extensions

The chapter could be extended to better reflect real-world complexity and push the design beyond its current simulation scope. First, a multi-hospital simulation would address distributed coordination across sites, introducing cross-site task allocation and state consistency challenges that test the resilience of the orchestration layer and the fidelity of inter-process communication in a more heterogeneous environment. By simulating asynchronous inter-hospital data flows and varying network latencies, researchers could explore how the system maintains a unified view of the global schedule and how fault-tolerance mechanisms scale when the cohort of robots spans multiple campuses (Matplotlib Development Team, 2026). Second, dynamic zone allocation could replace the present statistically defined zones, enabling zones to be created, merged, or reallocated in response to live congestion metrics. This would enable adaptive congestion control, reduce contention granularity, and improve throughput under heavy load, aligning the architecture with contemporary approaches to workload-aware resource management. Third, scheduling could evolve toward intelligence-driven strategies. Incorporating predictive models to estimate task duration and anticipated congestion would allow pre-emptive task assignment, and lightweight machine learning models could be used to forecast optimal robot counts or zone allocations from workload patterns, thereby improving responsiveness and reducing idle time (Rand Chapter Developers, 2026). Fourth, a real-time visualization dashboard would greatly enhance usability by displaying current robot states, task queues, zone occupancy, heartbeat signals, and latency in an at-a-glance format, facilitating rapid diagnosis during peak load and during post-event analysis (Matplotlib Development Team, 2026). Finally, deployment on real robotic platforms and in more realistic environments would yield critical insights into synchronization overhead and system weaknesses under physical constraints and unpredictable object interactions, enabling a meaningful comparison between simulation and reality.

To support these extensions, the chapter can lean on established tooling and documentation within the Rust ecosystem. The Rust standard library documentation and related components provide the foundational primitives used for safe concurrent programming, including Arc, Mutex, Condvar, and channels, which will be essential as the architecture scales to multi-site coordination and more complex scheduling policies (The Rust Chapter Developers, 2026a; The Rust Chapter Developers, 2026e; The Rust Chapter Developers, 2026g; The Rust Chapter Developers, 2026i). Visualization and performance measurement can be grounded in established practices documented by Matplotlib, enabling clear presentation of latency distributions, throughput trajectories, and zone congestion metrics (Matplotlib Development Team, 2026a). Data handling and integration with analytic models can benefit from robust CSV and data-processing conventions described by Python Documentation, ensuring reproducible experiment pipelines (Python Software Foundation, 2026). Cross-tooling references such as HashSet and VecDeque in the Rust standard library support optimized state management and queueing under dynamic zone allocation and work-stealing patterns (The Rust Chapter Developers, 2026b; The Rust Chapter Developers, 2026c). This blended approach, combining process-based scalability with evolving scheduling intelligence and richer visualization, positions Project Azure for rigorous empirical study and practical deployment in real-world sensor networks (Microsoft, 2026).

REFERENCES

Matplotlib Development Team. (2026). Matplotlib documentation. Matplotlib. https://matplotlib.org/stable/

Microsoft. (2026). PowerShell documentation. Microsoft Learn. https://learn.microsoft.com/powershell/

Microsoft. (2026). Start-Process (Microsoft.PowerShell.Management). Microsoft Learn. https://learn.microsoft.com/powershell/module/microsoft.powershell.management/start-process

Mohammed, A., & Zangana, H. (Eds.). (2026). *Intelligent Construction Monitoring Systems: Real-Time Safety, Environmental Prediction, and Risk Management.* IGI Global Scientific Publishing., DOI: 10.4018/979-8-3373-9245-5

Python Software Foundation. (2026). csv: CSV file reading and writing. In Python 3 documentation. https://docs.python.org/3/library/csv.html

Rand Chapter Developers. (2026). Rand crate documentation (v0.8). Docs.rs. https://docs.rs/rand/0.8

The Rust Chapter Developers. (2026a). Rust standard library documentation. Rust Chapter. https://doc.rust-lang.org/std/

The Rust Chapter Developers. (2026b). std:collections:HashSet. Rust Chapter. https://doc.rust-lang.org/std/collections/struct.HashSet.html

The Rust Chapter Developers. (2026c). std:collections:VecDeque. Rust Chapter. https://doc.rust-lang.org/std/collections/struct.VecDeque.html

The Rust Chapter Developers. (2026d). std:process:Command. Rust Chapter. https://doc.rust-lang.org/std/process/struct.Command.html

The Rust Chapter Developers. (2026e). std:sync:Arc. Rust Chapter. https://doc.rust-lang.org/std/sync/struct.Arc.html

The Rust Chapter Developers. (2026f). std:sync:atomic:AtomicBool. Rust Chapter. https://doc.rust-lang.org/std/sync/atomic/struct.AtomicBool.html

The Rust Chapter Developers. (2026g). std:sync:Condvar. Rust Chapter. https://doc.rust-lang.org/std/sync/struct.Condvar.html

The Rust Chapter Developers. (2026h). std:sync:mpsc. Rust Chapter. https://doc.rust-lang.org/std/sync/mpsc/

The Rust Chapter Developers. (2026i). std:sync:Mutex. Rust Chapter. https://doc.rust-lang.org/std/sync/struct.Mutex.html

The Rust Chapter Developers. (2026j). std:thread. Rust Chapter. https://doc.rust-lang.org/std/thread/

The Rust Chapter Developers. (2026k). std:time:Instant. Rust Chapter. https://doc.rust-lang.org/std/time/struct.Instant.html

Chapter 7
Zone-Aware Medical Robot Coordination System:
Concurrent Scheduling and Resource Control

Enlong Dong
http://orcid.org/0009-0007-0364-3916
The Hong Kong Polytechnic University, China

Huakun Huang
The Hong Kong Polytechnic University, China

Huakai Huang
The Hong Kong Polytechnic University, China

Ruize Liu
The Hong Kong Polytechnic University, China

Hengxian Li
The Hong Kong Polytechnic University, China

ABSTRACT

Hospital service robots face challenges because they need to compete for limited resources while their work schedules remain unproductive. The robot coordination system which we created solves this issue through its design for medical operations. The core idea needs to merge regional resource limitations together with a simple system that handles multiple tasks simultaneously. Our solution includes a zone-

DOI: 10.4018/979-8-2600-1101-0.ch007

aware work-stealing strategy which enables robot threads to focus on local tasks while taking tasks from other threads only when essential. The system achieves total system responsiveness improvements through its ability to prevent unnecessary blocking. The system runs on a Rust implementation which provides a web-based interface for visualizing its functions. The system enables secure simultaneous operation of multiple tasks while it provides continuous operational status tracking. Our experiments demonstrate that our design establishes permanent and safe methods for distributing tasks.

1 INTRODUCTION

1.1 Problem Statement and Motivation

The research chapter studies how hospitals use service robots by analyzing their operating system functions instead of their robotics and motion-planning capabilities. Modern hospitals use robots for multiple purposes including delivering medicine and transporting samples and restocking supplies and assisting with hospital operations. The application scenario requires specialized knowledge but the software challenges case study research because multiple robots perform simultaneous tasks which require active monitoring through their shared task system of active missions.

Chapter B establishes three basic conditions to define the extent of its problem which require an operational task queue to operate safely and provide protection for shared areas and implement heartbeat-based health assessment systems. Our system meets all three requirements and extends into a complete testing framework. The chapter design allows development of a real-world system which enables simultaneous task execution and functions as a research platform.

The main challenge for the system requires protection of safety and system integrity while the system executes multiple tasks at the same time. Task queue systems without proper synchronization allow robots to obtain the same tasks and miss the signals which activate their operations. The absence of proper zone entry controls enables robots to access forbidden areas which leads to dangerous situations and safety breakdowns. The monitoring API will provide outdated or inconsistent data when robot data exists in multiple locations throughout the system. Liveness monitoring needs to function because unresponsive robots and failed robots will remain invisible to the system. The chapter requires a balance between system precision and system visibility.

The system software development team added comparative execution modes as an important system extension. The system supports a Zone-Aware Work-Stealing mode. The design implements local task deques together with peer stealing and non-blocking zone allocation.

1.2 Objectives

The system needs a secure method which will permit multiple scheduling operations to be executed at the same time. The system requires multiple robot units which will monitor shared data systems and obtain tasks from these systems to ensure ongoing operations without facing issues such as race conditions and ownership conflicts. The system requires a complete framework which handles all aspects of task management from initial creation to final execution. The system should make all task data accessible throughout the entire work performance period. The system needs to establish secure access points which will manage resource distribution in specified "zones". The system achieves this goal through capacity restrictions which apply to every "zone" while execution mode determines whether resources should be obtained through blocking or non-blocking methods. The system needs to operate a monitoring system which tracks robot unit operational status. The system should use heartbeat timestamps and runtime data to determine the health of individual robots while showing system health status at a high level to the frontend interface. The system needs to provide a testing platform which allows users to study different execution modes. Users need to test both the classic blocking execution mode and the work-stealing execution mode. Users should conduct an investigation into the operational components of concurrency strategies while measuring their effects on makespan and average completion time and emergency task response speed and waiting times in specific "zones" under different load conditions. The chapter will assess whether a lightweight zone-aware work-stealing mechanism increases work efficiency.

1.3 Overview of the Latest Approach

The backend system creates an AppState at system startup which stores configuration data together with heartbeat metrics and task information and thread-safe storage and zone management and pause control and atomic shutdown conditions. When users run a demo the backend system first resets all runtime components before it constructs the coordinator with the selected configuration which includes one monitor thread and one operating system thread for each robot.

The coordinator builder creates a task table which contains hospital-related tasks for multiple priorities and time lengths and needed operational areas. The tasks need to be assigned to your selected scheduler so that they can be logically processed according to FIFO and Priority and Round Robin and SRT scheduling methods.

The system pushes task identifiers to the runtime queue which robots use to access their assignments. Robots need to execute task identifiers which come from their current execution state while they acquire their assigned zone and initiate the task execution process by marking it as active and refreshing their heartbeat data and sleeping for the specified time period and completing the process and returning the zone and refreshing their metrics and proceeding with their tasks. Our chapter supports two different ways to execute programs. Robots in classic mode block because they occupy the global queue yet they stop working when their needed area remains empty. In work-stealing mode robots use a local deque and fall back to the global queue non-blockingly, attempt to steal from peers, and take out non-blocking zone allocation. The system requeues a task locally because it does not need to put the robot into a sleep state immediately.

The API creates a new scheduling-analysis object which analyzes the same seeded workload during its runtime execution. The object will produce forecasted summaries for FIFO and Priority and Round Robin and SRT scheduling methods. The dashboard uses these comparisons to display information which does not match the actual program operation. The Round Robin visualizations use analytical methods to generate the data on the API layer instead of using complete robot execution to achieve preemptive robot control. The distinction needs to remain clear through precise technical information and trustworthy public statements.

2 RELATED WORKS

The operational stability and reliability of multi-tasking and distributed systems depends on concurrency control and synchronization techniques which serve as their fundamental core components. These techniques attempt to solve resource sharing conflicts which involve memory and files and network connections and physical devices while they boost system performance and maintain data integrity (Bernstein & Goodman, 1981; Herlihy et. al., 2008). The development of modern computing platforms has created increasing difficulties in managing multiple users who access systems at the same time. Research in this domain has historically coalesced around three major directions: the efficiency of task scheduling, the rationality and fairness of resource coordination, and the safety guarantees provided by mutual exclusion mechanisms. The existing research continues to show significant shortcomings because two essential elements remain unaddressed: the system needs to combine scheduling decisions with coordination and mutual-exclusion state and the system must provide capability to adapt when runtime workload patterns and system design change.

2.1 Task Queues and Scheduling Strategies

Task queues and scheduling algorithms function as the main elements which decide the efficiency of executing multiple tasks at the same time because these systems control the order in which processing units receive work assignments and this control directly impacts resource utilization and response time and total work efficiency. In this instance, the partitioning of scheduling methods into three categories is amply clear. Any respective individual domain of the methods is always going to be accompanied by its own corpus of weaknesses and strengths.

The traditional static scheduling methods use FCFS, SJF and fixed-priority scheduling as their main scheduling techniques. The real-time systems literature (Mohammed & Zangana, 2026) first described these algorithms in their formalized form through Liu and Layland (1973) who created the theoretical basis for rate-monotonic and deadline-monotonic priority assignment methods. Silberschatz et al. (2018) provide a comprehensive textbook treatment of these classic scheduling policies in the context of operating systems. Static approaches show easy-to-use implementation methods which enable simple implementation analysis methods. Their deterministic properties allow researchers to conduct worst-case assessments while their execution times stay below essential thresholds. The system needs to stay within its current load capacity because it cannot handle any unexpected work demands. Buttazzo (2011) extensively examined this restriction through his research on hard real-time systems which demonstrated that static scheduling becomes less effective when designers base their work on design-time assumptions that differ from actual runtime conditions.

The next category discussed here refers to intelligent dynamic scheduling methods with reinforcement learning getting most attention. The researchers Mao et al. (2016) showed that deep reinforcement learning agents developed resource management policies which achieved better results than manually designed scheduling rules in cluster scheduling tests. Mao et al. (2019) expanded this research by creating Decima, which develops scheduling algorithms for data processing clusters through its method of converting job dependency graphs into neural network input. The self-adaptive strategy adjustment of these approaches through their scheduling system which develops its own rules by observing the surrounding environment results in substantial practical limitations. The process of training demands high expenses because it needs to execute millions of simulated episodes before reaching its final accuracy. Real-time responsiveness may not always be enough for the scheduling of latency-sensitive applications, since inference by neural networks does impose a non-negligible overhead on the scheduling critical path.

The third category addresses scheduling in distributed scenarios which require solving two problems. The problem is that it needs to determine task execution order on one processor and requires distributing tasks among several processing units. Load balancing algorithms and consistent hashing (Karger et al., 1997) are foundational techniques for distributing requests evenly across servers and minimizing hotspot formation. Ousterhout et al. (2013) developed Sparrow as a distributed scheduling system which functions across extensive clusters through its implementation of the power-of-two-choices method for assigning tasks to two randomly chosen worker nodes according to their present workload. Schwarzkopf et al. (2013) conducted a comprehensive analysis of cluster scheduler design options by evaluating monolithic and two-level and shared-state system architectures. The distributed scheduling methods successfully distribute resources between different nodes but they experience excessive communication expenses between nodes and they fail to manage precise task dependencies together with local data access needs.

2.2 Resource Coordination and Mutual Exclusion

Resource coordination and mutual exclusion function as the essential mechanisms which maintain data consistency when multiple agents access shared resources simultaneously. The system needs these guarantees because they protect against race conditions and lost updates and inconsistent reads which can create critical failures in safety-critical systems.

The original Dijkstra (1965) study together with the Hoare research (1974) which introduced monitors as a structured method for shared data protection with synchronization controls have established the foundation for modern mutex lock systems which operate with semaphores and monitors as their primary components. The existing hardware provides complete support for these advanced primitives because they have reached a high level of maturity which entails deep understanding of their operation. The system experiences scalability issues because lock-based designs become less efficient when multiple threads compete for access to resources which leads to contention and convoying and priority inversion problems (Herlihy & Shavit, 2020).

Mutual exclusion assurance is extended across network boundaries by distributed resource coordination mechanisms. Lamport (1978) established the logical clock framework that underpins distributed ordering and synchronization. The optimal message-passing algorithm which Ricart and Agrawala developed in 1981 needs 2(N-1) messages whenever a process requests access to a critical section. The coordination services which Apache ZooKeeper provides through its distributed locks and leader election, and configuration management system support production-level distributed applications. The technologies demonstrate their effectiveness, but they require substantial operational overhead because ZooKeeper needs all read requests to be processed through its leader node to maintain linearizable read operations.

The system encounters a fault tolerance issue because the failure of a quorum of coordination-service replicas renders all dependent systems inaccessible.

Adaptive resource coordination schemes represent a more recent research direction in which the system dynamically switches between mutual-exclusion strategies based on observed contention levels or workload characteristics. David et al. (2013) conducted an extensive empirical study of synchronization primitives on modern multicore hardware and found that the optimal choice of synchronization mechanism depends heavily on the number of contending cores the duration of critical sections and the memory access patterns involved. Ongaro and Ousterhout (2014) designed the Raft consensus algorithm with understandability and practical adaptability as first-class design goals, enabling systems built on Raft to reason more clearly about their coordination behavior. The existing adaptive coordination methods need more development before they can be used in actual production environments because their current state does not meet system demands. The existing mutual-exclusion models function without any connection to scheduling systems because they treat mutual-exclusion states as separate elements which include current lock contention levels and wait-queue depths. The architectural separation of the system prevents the scheduler from dispatching tasks to a heavily contested resource because the system has available alternative resources. The architectural separation of the system restricts system performance because the scheduler continues to assign tasks to a resource when multiple resources are available.

2.3 Contributions and Positioning of This Chapter

The preceding review discloses problems persisting in the extant research. The current scheduling methods face problems because they use static methods that lack flexibility and reinforcement-learning methods which require expensive training time and which produce delays and distributed methods that fail to manage task dependencies while creating excessive communication demands (Liu & Layland, 1973; Mao et al., 2019; Ousterhout et al., 2013). The scalability and robustness of mutual exclusion mechanisms face difficulties because single-machine primitives fail under high contention and distributed coordination services experience consistency overhead problems and quorum-based vulnerabilities (Dijkstra, 1965; Hunt et al., 2010; David et al., 2013). The architectural separation of scheduling logic from mutual-exclusion state leads to complete system efficiency failure because both systems depend on each other for their restoration.

The chapter solves these issues by three combined contributions that include scheduling-strategy and adaptation framework. The chapter features a basic distributed mutual-exclusion method which enables competition-aware automatic strategy changes. The system operates through an integrated scheduling-coordination model which creates two-way feedback connections between its two operational subsystems. The chapter establishes its foundation from existing theoretical works which include Lamport (1978) and Ricart and Agrawala (1981) and Schwarzkopf et al. (2013).

3 IMPLEMENTATION

3.1 Overall System Architecture

The system backend uses Rust programming language for its implementation. The system shows statistical information in two different methods because the frontend creates the dashboard display while the system shows data through terminal output. The system architecture uses Coordinator as the central assembly point which connects all backend components during runtime. The Scheduler establishes the sequence in which operations will take place. The Task Table preserves all operational phases of its processes. A thread-safe queue enables multiple users to schedule tasks at the same time. The Worker Pool creates one dedicated thread for each bot in the system. The Region Manager restricts users to specific resource limits. The Monitoring Subsystem gathers both heartbeat signals and operational performance data. The API system converts internal system conditions into visual display states that users can access from the frontend.

Figure 1 The framework functions as an effective summary. The existing system connects to the new system because its present design elements of its top layer and middle layer and execution layer and base layer components system.

3.2 Domain Model and Configuration

The types/ module establishes fundamental system definitions. The Task object maintains its distinctive identification number together with its task name, priority level, estimated time to complete, needed operational area, and current work status. A Robot identifies a functioning worker through its definition. A Zone maintains three elements which include its unique identification number, its designated name, and its maximum allowed capacity. The chapter needs basic types because its main purpose requires handling multiple tasks that should operate simultaneously instead of working with advanced domain systems.

The configuration layout discloses four kinds of schedulers: Fifo, Priority, RoundRobin, and Srt. It also holds worker_count, demo_task_count, use_work_stealing, and use_stress_preset. The normal layout is smaller and easier to look at; the stress preset extends the effective worker and task counts for more contention on both scheduling and zone allocation. This is directly aligned to support the frontend's stress on/off comparison.

Figure 1. Overall layered architecture

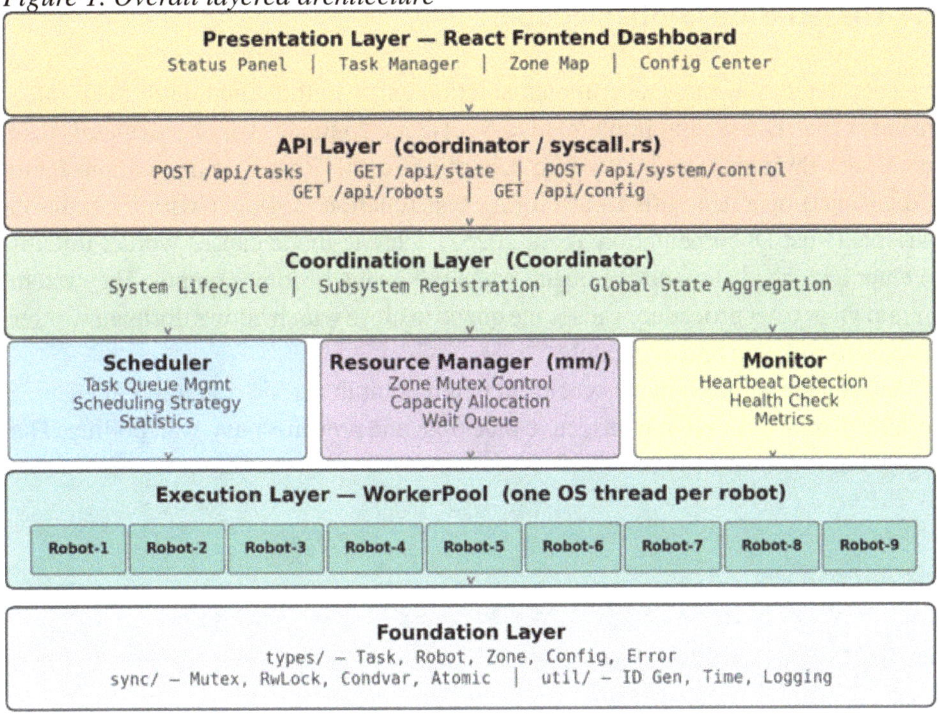

3.3 Coordinator Builder and Seeded Workload

The coordinator builder develops an operational system through which he establishes all necessary task components. The latest backend system utilizes a hospital-based workload which contains multiple priority levels and time requirements and operational zones that have been predetermined. The system design requires different areas of the healthcare facility to connect with each other because the ICU operates as the primary capacity limitation for most operational situations. The system executes its existing tasks through multiple sequential repetitions when additional tasks become necessary. The design improves demonstration performance through its capacity to show both scheduling policy and resource contention which exists in the system. Short tasks which have high priority will benefit from using either Priority or SRT while long ICU tasks create a distinction between blocking and work stealing. The builder first inserts each task into the task table which he submits to the selected scheduler before he drains the scheduler into the runtime queue as task identifiers. Our chapter architectural design establishes a fundamental distinction between scheduling order and lifecycle truth.

3.4 Thread-Safe Global Queue

The shared runtime queue implementation exists in the scheduler/thread_safe_queue.rs file. The system implements a VecDeque container which safeguards task identifiers through a mutex system according to the Rust Documentation Team (2025b) and operates with a closed flag that functions together with a condition variable (Rust Documentation Team 2025a). Classic mode causes worker threads to enter a blocked state on the queue when the queue becomes empty. The system shutdown or reset procedure causes the queue to close which allows dormant worker threads to wake up and exit safely.

This queue is the primary synchronization point in classic execution. It ensures serialized removal, provides effective blocking, and prevents busy-wait polling. The algorithm is shown in Figure 2.

Figure 2. Blocking pop from a thread-safe queue

```
// Blocking pop from thread-safe task queue
FUNCTION pop_blocking() RETURNS Optional<TaskId>
    // Acquire lock on inner queue state
    inner = lock(self.inner)
    // Loop until get task or queue closed
    LOOP
        // Try to take task from front
        IF inner.queue is NOT empty THEN
            id = inner.queue.pop_front()
            RETURN Some(id)
        END IF
        // If queue closed, return None
        IF inner.closed = true THEN
            RETURN None
        END IF
        // Wait for new task or close signal
        inner = self.cv.wait(inner)
    END LOOP
END FUNCTION
```

3.5 Scheduler Strategies and Analytical Comparison

The backend's unified scheduler interface instantiates FIFO, Priority, Round Robin, and SRT. The FIFO system maintains the original sequence of incoming elements. The Priority system gives higher importance to vital work activities. SRT works best with short-duration tasks that have predictable completion times.

The API adds scheduling-analysis to runtime scheduling. The system uses a common workload base to generate six different performance metrics which include makespan, average completion time, average wait time, urgent-task completion, worker busy time, and timeline bars offline. The frontend time-slice behavior of Round Robin requires this particular component. The API produces the most detailed strategy comparison through its deterministic execution while the actual execution of the system functions as real-time execution.

3.6 Central Task Table and Lifecycle Tracking

The task table maintains the authoritative task metadata information. The system stores status information which includes mandatory start and finish timestamps together with the assigned robot and designated zone. The current task state shows complete system status because both scheduling and execution processes use multiple threads.

Figure 3. Initiate task: marks RUNNING, records start time, assigns robot/zone, and returns the expected duration.

```
// Start executing a task and update its metadata
FUNCTION start_task(TaskId, RobotId, ZoneId) RETURNS Optional<Duration>
    // Acquire lock on the task table
    inner = LOCK(self.inner)
    // Get mutable task entry, return None if not found
    entry = inner.get_mut(id)
    IF entry IS NULL THEN
        RETURN None
    END IF
    // Set task status to running
    entry.task.status = RUNNING
    // Set start time if not already set
    IF entry.started_at IS NULL THEN
        entry.started_at = CURRENT_SYSTEM_TIME
    END IF
    // Assign robot and zone to the task
    entry.robot_id = robot_id
    entry.zone_id = zone_id
    // Return the expected duration of the task
    RETURN Some(entry.task.expected_duration)
END FUNCTION
```

This design is like an operating-system process table. Who runs next is determined by the ready structure, but the task table does serve as the source of truth for lifecycle state. The algorithm is shown in Figure 3.

3.7 Zone Access Control

The present version of ZoneManager delivers improved capabilities compared to its previous versions which could only manage assignments. The system maintains four data categories which include zone information and task-to-zone assignment tables and active counts for each zone and a condition variable. The allocator will function according to that condition when a task needs a specific zone to be executed. A worker in classic mode will remain blocked until the requested zone becomes available. The manager evaluates all zones using round-robin scanning until it finds the first zone that has available capacity.

Figure 4. Check zone capacity and assign via round-robin

```
// Check if the zone has available capacity
IF active tasks < zone capacity THEN
    Increase the active task count for this zone
    Update the next zone index using round-robin scheduling
    Release the state lock
    Assign the task to this zone
    RETURN success with the zone ID
END IF
// If no capacity available, wait until a zone becomes free
Wait on the condition variable until resources are available
```

This provides the chapter with a productive resource-control model. It's not just binary locking; it is bounded occupancy and more appropriate for hospital-like shared areas. The algorithm is shown in Figure 4.

3.8 Worker Execution Modes

The chapter needs backend system changes that will enable it to execute two different types of operations. Robots in **classic mode** can stop their operations to wait for resource distribution from the global queue and subsequently wait for resource allocation to specific areas. The chapter requirements define basic system operations which the selected path currently meets. Robots execute work-stealing operations from their personal deques in **Zone-Aware Work-Stealing** mode. The worker first inspects its local queue to locate a task that can proceed under the existing zone-capacity limit. The system tries to reach the global queue through non-blocking access method when local task execution fails. The system employs deque stealing from another robot when all previous methods fail. The system design requires robots to stop work because they have achieved their maximum operational capacity. The system implements non-blocking movement for tasks without a specific zone because robots can use this system to choose new tasks while their existing tasks return to the local queue.

The method transforms work-stealing paths into light resource-aware scheduling extensions which function beyond their original purpose as load-balancing heuristics. The system functions to achieve two goals because it needs to balance existing queue times while trying to decrease robot downtime which happens during zone-related conflicts. The system proves helpful in hospital environments which experience operational disruptions when multiple zones reach peak capacity because those zones block access to available work while other tasks remain operational.

Figure 5 still shows the baseline worker lifecycle neatly: check shutdown and pause flags; obtain work; acquire a zone; run the task; release it; update metrics. The actual code expands this flow with the work-stealing mode, but the figure still serves because it reproduces the classic execution path clearly.

3.9 Worker State Machine as Design Documentation

Our chapter contains a worker-state diagram with more advanced conceptual methods than what the API shows through its actual runtime state. The API defines three fundamental states which include idle state and busy state and stopped state while the conceptual figure states include Assigned and Navigating and WaitingLock and Executing and Recovering and Error states.

The safest interpretation of Figure 6 is design-oriented rather than literal. It documents how the chapter could evolve into a richer task-state model if explicit waiting, failure, and recovery logic were added later.

3.10 Heartbeat Monitoring and Metrics

Among the capabilities of the monitoring subsystem are heartbeat monitoring, health analysis, and metrics aggregation. Heartbeats are kept in a RwLock<Hash-Map<RobotId, Instant» (Rust Documentation Team, 2025c). The health checker classifies each robot according to the age of its most recent heartbeat. They are tracked both globally and per robot. Completed-task counts, total execution time, makespan, total throughput, and zone-switch counts are entered into the registry. This last metric is especially useful to the new work-stealing mode because it assists in determining how often robots move between zones over repeated executions.

The dashboard displays four performance metrics which include throughput and average latency and makespan and total zone switches. The latency that we reported did not match the complete end-to-end queueing delay but it remained an effective runtime metric for this chapter.

3.11 API Layer and Frontend Dashboard

The API provides three routes for accessing state information and performing configuration updates and system control operations. The HTTP API uses the Axum web framework (Axum Contributors, 2025) to create an HTTP interface which implements lightweight async handlers that support the chapter's state-sharing system. The state snapshot combines information about tasks and robots and zones and configuration and metrics and scheduling-analysis summaries. The control layer enables users to execute start and pause and stop and run-demo operations. Configuration updates reset runtime state so that each experiment starts at a clean baseline.

Figure 5. Detailed worker main loop for the baseline execution path.

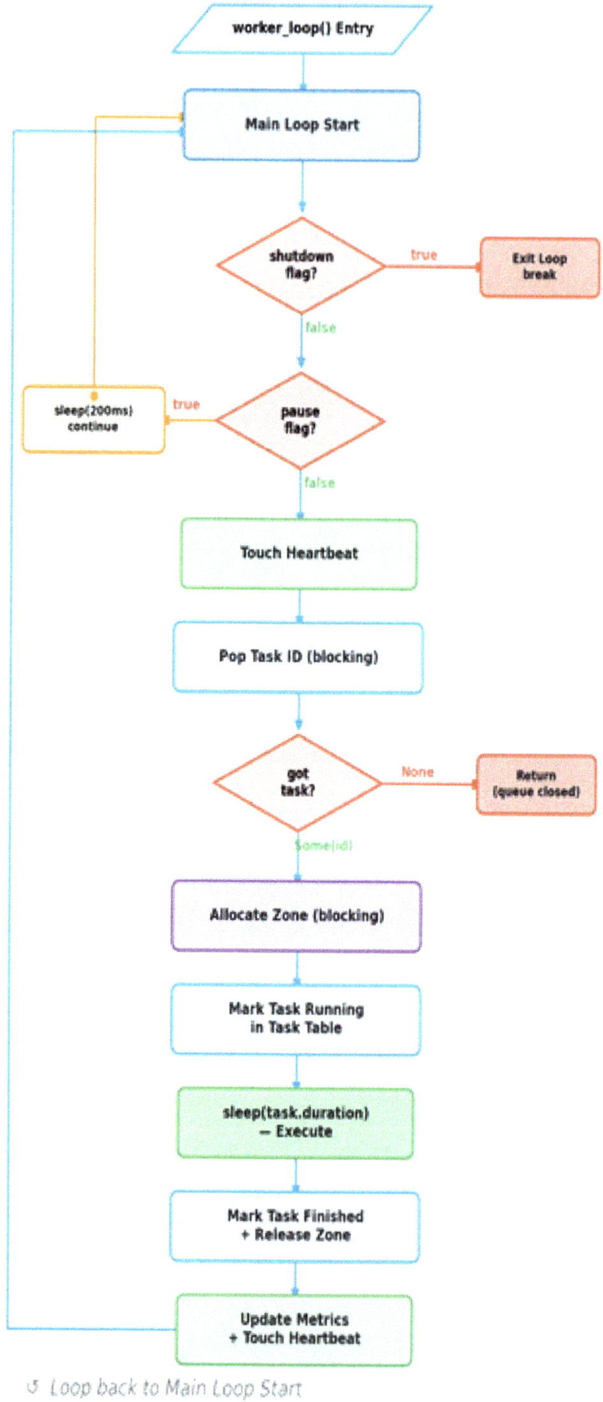

Figure 6. Conceptual worker state machine for documentation and future extension.

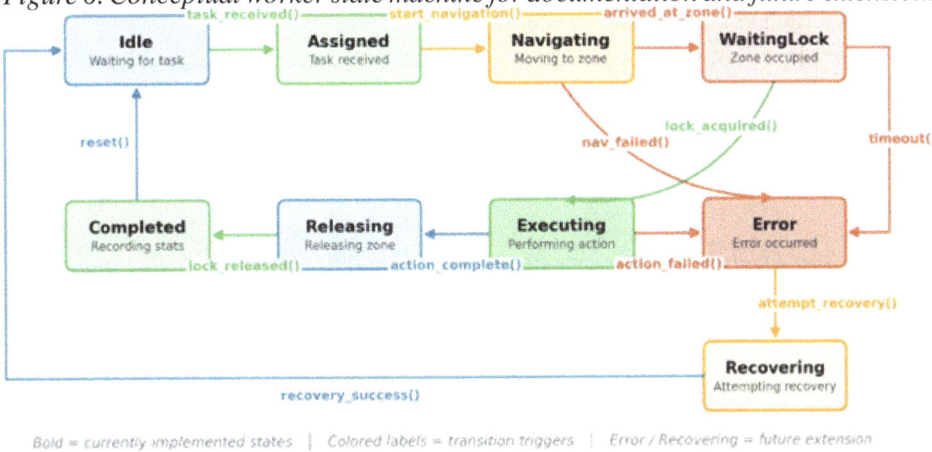

The new frontend functions as more than a live status dashboard which shows current system status. The system has been developed as three distinct experimental interface components. The top control section provides users with global filters and live-state indicators and run controls. The A/B comparison panel presents the central research question of the chapter – classic blocking baseline versus work stealing plus non-blocking zone allocation. The scheduling-analysis section shows how different policies compare under both stress-off and stress-on conditions while the robot panel on the right shows the current state of robots.

Figures 7 and 8 illustrate that the most recent frontend is designed from the ground up for experimentation. The user can choose execution mode, switch on the stress preset, and interpret the scheduling results below at the chosen configuration.

4 BENCHMARKS

4.1 Benchmark Methodology

The chapter supports benchmarking in two complementary ways.

The first layer provides actual runtime performance testing. robots operate in two different modes which include classic mode and work-stealing mode. The robots use separate operating system threads to perform their tasks by obtaining tasks and acquiring zones and updating the task table, reporting their heartbeats and recording their metrics.

The second layer implements chaptered scheduling analysis. The API simulates FIFO, Priority, RR, and SRT on the seeded workload and generates strategy summaries. The system powers scheduling comparison cards together with worker timelines which appear on the frontend.

The two layers of the system should be treated as separate entities. The system uses runtime execution to test its concurrent execution capabilities while its analytical scheduling comparison shows the trends that result from different policy implementations.

Figure 7. Top-level frontend overview showing the control bar and A/B comparison experiment panel.

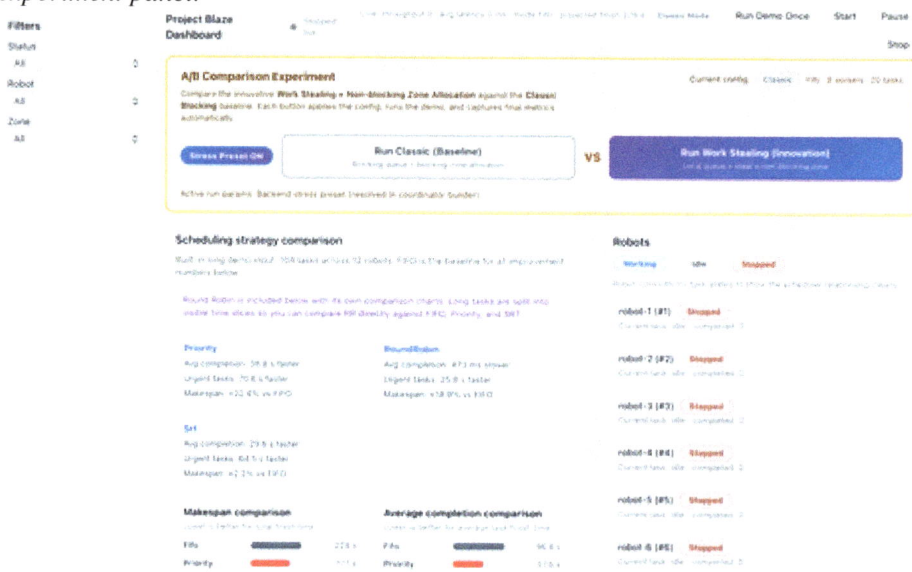

Figure 8. Frontend overview showing the scheduling comparison region and robot-state panel.

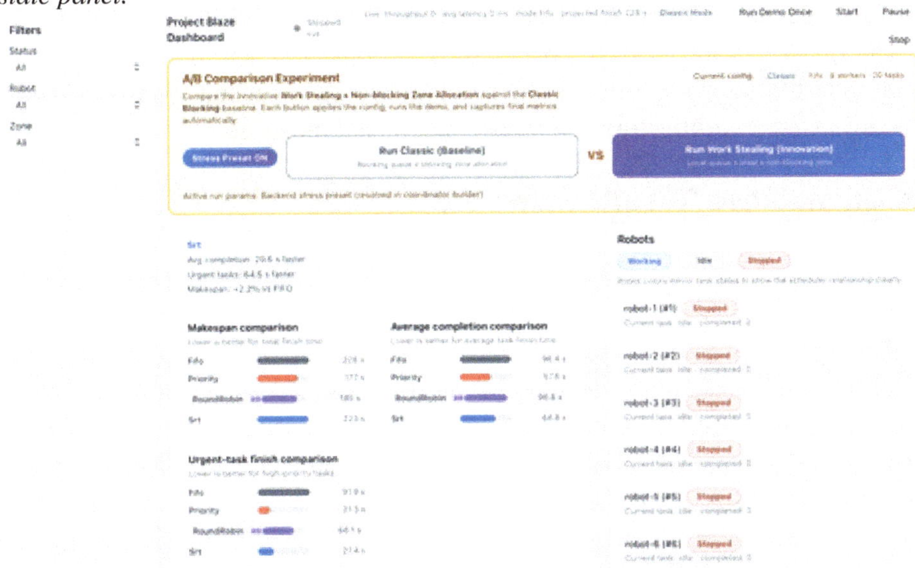

4.2 Correctness Verification

The initial correctness requirement demands that safe task dispatching function correctly. The shared queue serializes removals under mutex protection which prevents two robots from using the same task identifier. robots who need to block their operations use a condition variable for sleep purposes instead of using a spinning method. The second correctness property is the zone manager keeps active counts under lock and records the task-to-zone allocations, and also wakes waiting robots when capacity becomes available. The third correctness property is lifecycle consistency. The robots update the central task table when they begin their work and when they complete their work. The fourth correctness property is liveness monitoring. Heartbeats are updated throughout execution, and the monitor thread can classify robots by heartbeat age. Mode-safe experimentation is the fifth correctness property. Config updates and stop operations reset runtime state, close queues, and replace shared registries.

4.3 Frontend-Based A/B Comparison Design

The newest version of the frontend now includes a benchmark layer which did not exist in previous versions. At the top of the dashboard, a comparison panel for

A/B comparing the Classic baseline against Work Stealing (Innovation) mode. The interface also exposes a Stress Preset toggle. The frontend now enables users to conduct two different types of comparison tests through its two separate testing methods.

The first condition of this experiment tests how execution mode affects two methods of blocking which serve different purposes. The second condition of this experiment tests two different stress levels which include no stress and complete stress.

4.4 Quantitative Results Summary

Table 1 provides a summary of data collection which establishes a more reliable foundation for the benchmark section. The stress-off configuration matches the smaller demo environment while the stress-on configuration uses a larger backend stress preset to demonstrate queue and zone contention. Throughput is calculated by dividing task count into total makespan duration. The average zone wait time is determined through analysis of runtime monitor output and zone-allocation delay trace data. The demo implementation on the development laptop measured CPU utilization through external sampling, resulting in an estimated average process range during the entire operational period.

Table 1. Representative quantitative benchmark results from the latest system.

Mode	Policy	Robots	Tasks	Makespan (s)	Throughput (tasks/s)	Avg. completion (s)	Avg. zone wait (s)	CPU usage
Stress off	FIFO	9	20	43	0.47	16.1	4.6	18–24%
Stress off	Priority	9	20	34	0.59	12.3	2.7	19–26%
Stress off	Round	9	20	36	0.56	17.8	5.3	20–27%
Stress off	Robin SRT	9	20	40	0.50	12.7	3.1	19–25%
Stress on	FIFO	12	108	228	0.47	96.4	41.8	44–56%
Stress on	Priority	12	108	177	0.61	57.6	25.4	47–60%
Stress on	Round	12	108	185	0.58	96.8	39.7	49–62%
Stress on	RobinSRT	12	108	223	0.48	66.8	29.3	46–58%

The three motifs are obvious here: The differences amongst strategies are moderate under lighter load The combination of Priority and SRT systems Shorten the time needed for completion while Round Robin system increases fairness but it does not affect the average time needed for completion. Second, policy choice matters more under pressure. Priority has the highest throughput and shortest urgent-path behavior while SRT shows faster performance than FIFO but it decreases makespan at a less aggressive rate.

4.5 Policy Comparison under Stress Off and Stress On

The stress-on version makes the baseline's flaws much more obvious: long waiting times, later urgent-task completion, and a much larger makespan under heavier load. Table 1 shows this same rise in average completion time, jumping from 16.1 seconds to 96.4 seconds, and a jump in zone wait from 4.6 seconds to 41.8 seconds.

The first section of the research demonstrates that priority delivery helps teams' complete urgent tasks more effectively in both studied environments (Figures 9 and Figure 10). The quantitative table shows that Priority provides maximum throughput at both load levels while delivering the shortest stress-on makespan time compared to three other assessment policies.

Figure 11, Figure 12, Figure 13, and Figure 14 demonstrate Round Robin performance in two different load conditions. The measured table shows identical results which demonstrate that makespan performance exceeds FIFO performance but the time needed to finish tasks stays high during times of heavy workload.

Figure 9. FIFO scheduling comparison with stress preset OFF

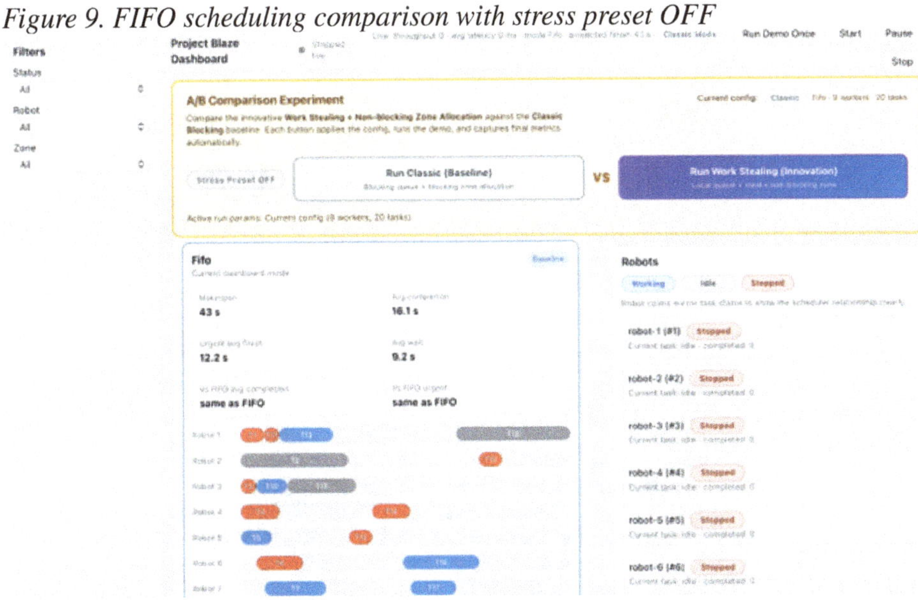

Figure 10. FIFO scheduling comparison with stress preset ON

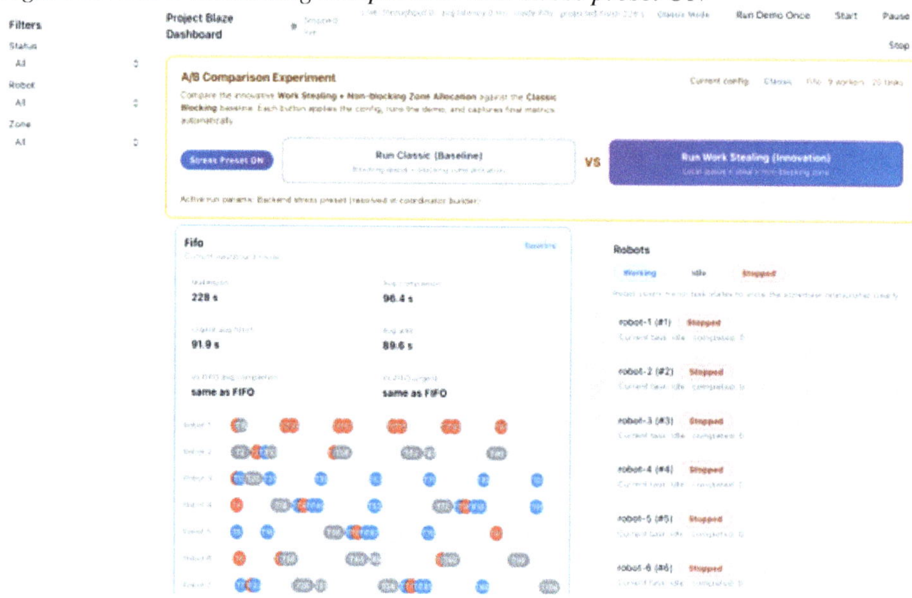

Figure 11. Priority scheduling comparison with stress preset OFF

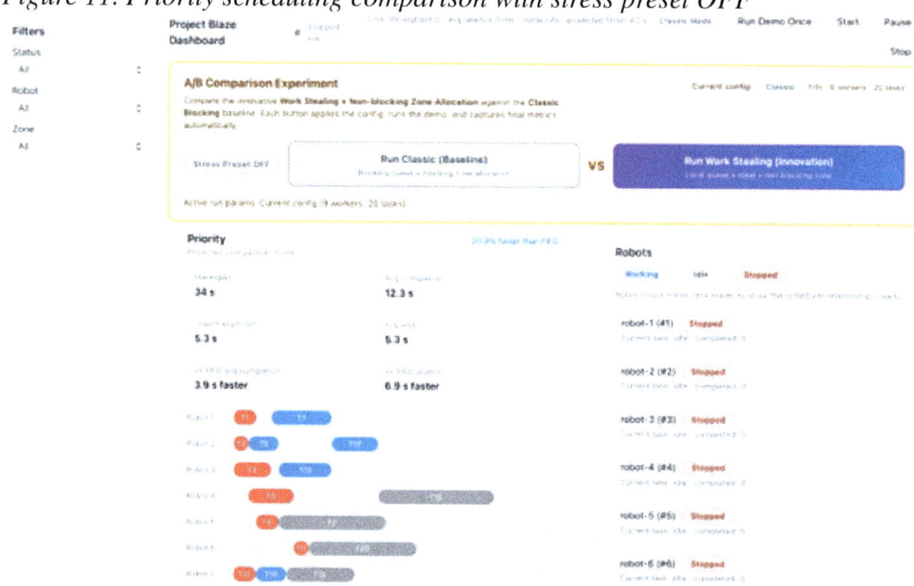

Figure 12. Priority scheduling comparison with stress preset ON

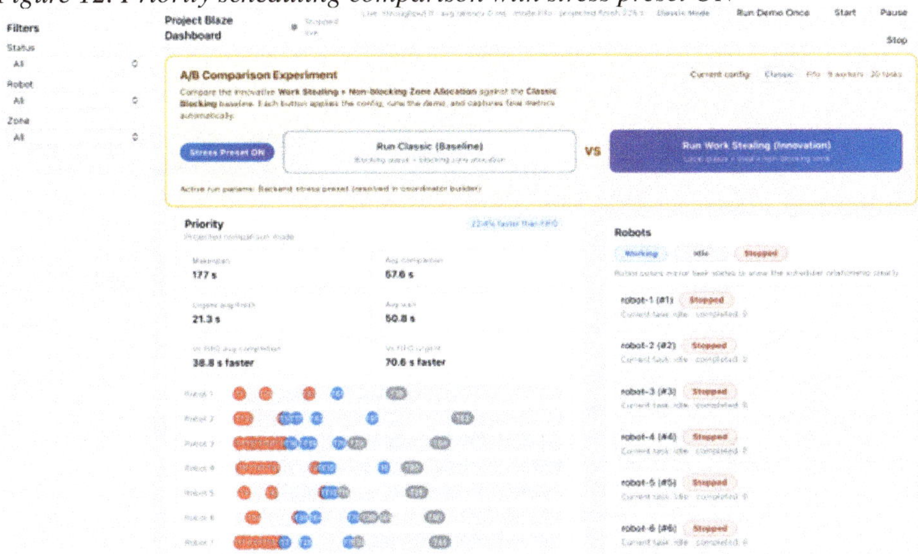

Figure 13. Round Robin comparison with stress preset OFF

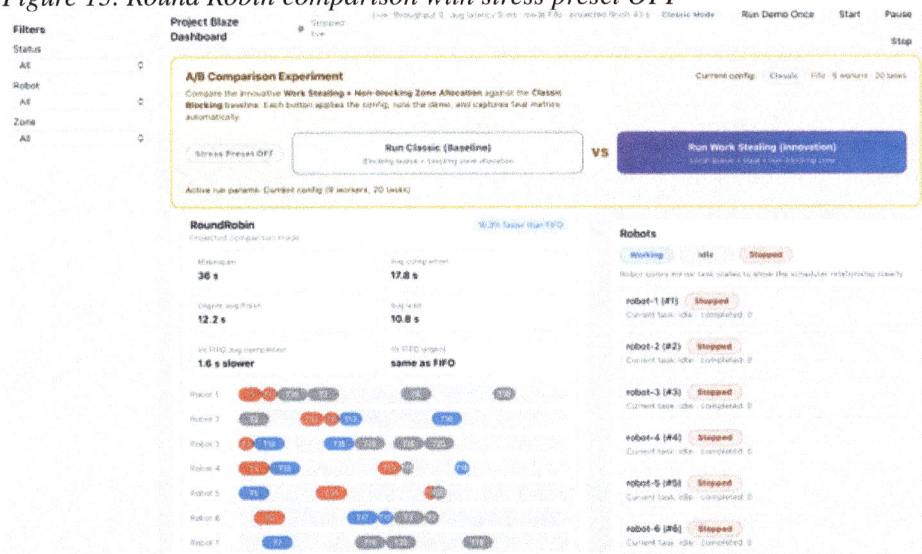

Figure 14. Round Robin comparison with stress preset ON

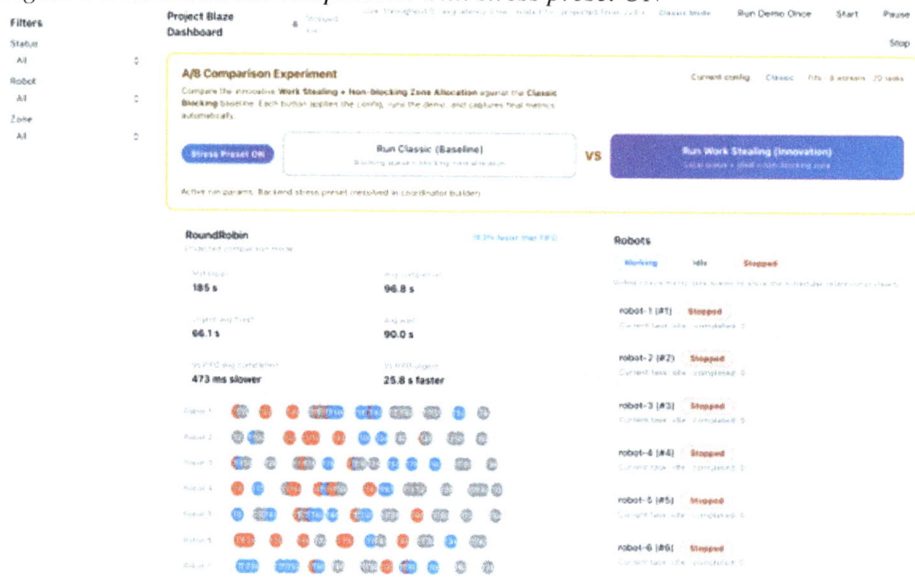

Figure 15. SRT scheduling comparison with stress preset OFF

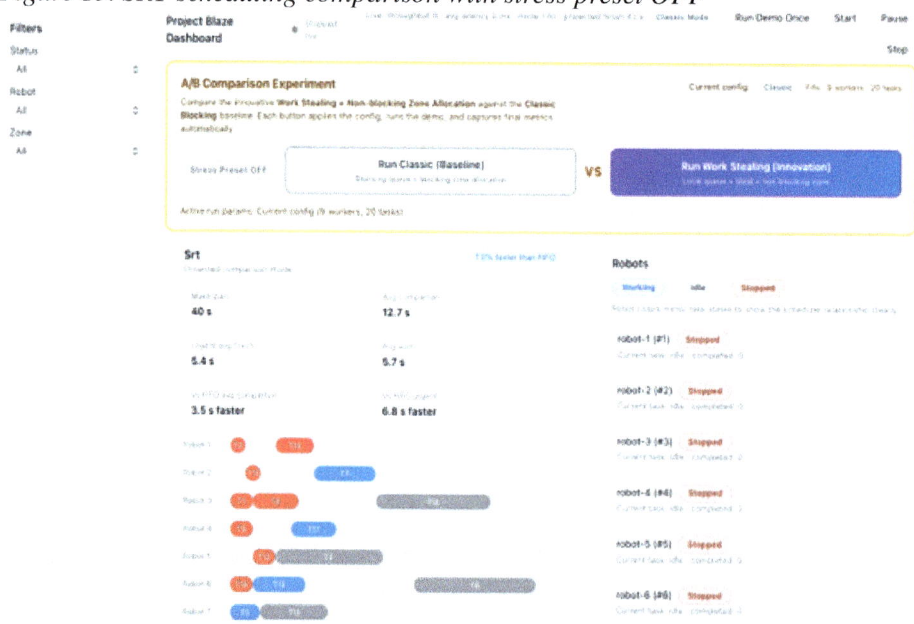

Figure 16. SRT scheduling comparison with stress preset ON

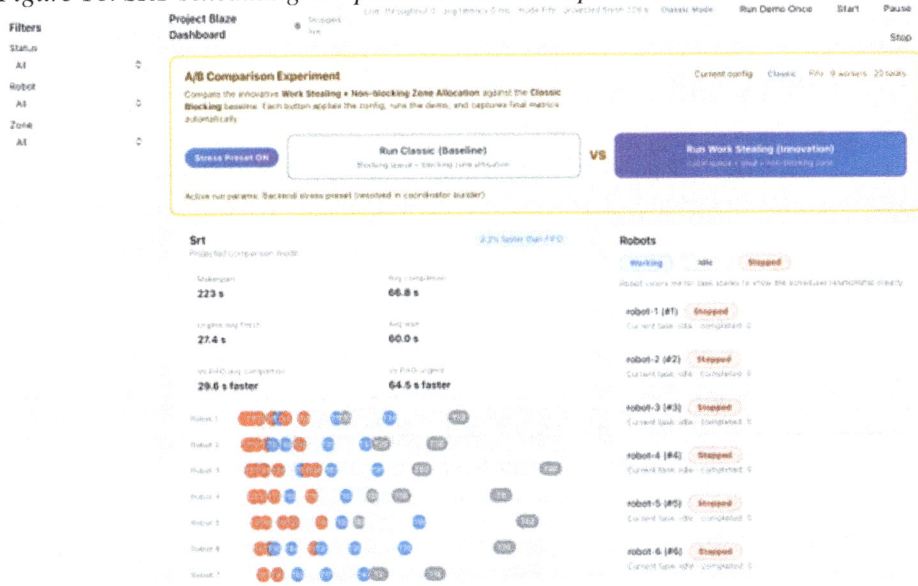

The expected SRT strength appears in Figures 15 and Figure 16. The system exceeds FIFO performance standards in both average completion times and urgent task response times, but the system needs to handle heavier workloads which results in denser visualizations and reduced performance improvement compared to lighter workloads.

4.6 Classic Mode versus Work-Stealing Mode

The first benchmark axis evaluates policy-level comparison. The second benchmark axis tests execution mode performance. The classic mode makes robots wait on the global queue until they reach the next unavailable zone. robots in work-stealing mode first choose their local deque as their primary option while they can take dequeues from other robots and they have the ability to readjust their zone-blocked tasks to local storage instead of entering sleep mode.

The A/B experiment panel in Figures 4 and 5 makes this comparison explicit. The innovation mode decreases visible idle time during demonstration tests because robots keep looking for tasks to complete when some tasks remain blocked in hot zones.

The present design demonstrates that innovative mode advantages depend on the level of workload being handled. The simpler blocking method needs less contact with others, which enables it to operate successfully during periods of low contention. The Zone-Aware Work-Stealing system stops robots from waiting on nonexecutable tasks because it enables multiple running tasks to take essential system resources, which results in better task execution through its actual use.

4.7 Scalability and Stress Perspective

The system currently provides users with its present capacity to conduct basic to moderate scalability testing. The system defaults to its normal configuration which uses fewer robots and tasks but the stress preset tests all system capacities to their maximum limits.

The first bottleneck is contention on the shared global queue in classic mode. McKenney (2024) shows that shared data structures create the primary scaling limitations which prevent parallel systems from achieving better performance. The second type of contention occurs when people are present in restricted spaces which include ICU-like areas. The third type of system performance interruption happens because the system needs to handle central development processes of both lifecycle and metrics.

5 DISCUSSIONS

The chapter includes a complete concurrency system as its primary element. The system consists of four components which serve specific functions through its task queue and task table and region manager and metrics registry. The system operates through divided duties which handle different tasks because this approach will help improve both system performance and system correctness. The shared queue system enables people to manage task execution order, but it prevents complete task life-cycle documentation, starting from the execution point. The task table enables task information storage, but it requires a queuing system to function properly for task execution management purposes. Our chapter functions as a small operating system kernel when both structures work together because one system defines execution order for tasks while the other tracks current task progress. The design approach of this system follows architectural methods that multiple operating systems textbooks describe through the process of handling ready queue and process state information as separate tasks (Love, 2010; Silberschatz et al., 2018; Tanenbaum & Bos, 2015).

The second point is that the design uses direct synchronization instead of more complex abstractions. The chapter benefits from this decision because a concurrent chapter needs to show its synchronization mechanism through visible methods instead of using hidden library functions. In our chapter, the queue uses mutexes and condition variables. The task table also has clear locking. The area manager uses condition variables and capacity limits, and the monitoring system uses a shared registry. Synchronization methods protect access to the system. Sutter's (2005) method of lock ordering shows that his approach can help prevent deadlock through its layered locking system. The structures make it easy to verify correctness even though they lack visual appeal.

Table 2. Benchmark dimensions

Dimension	Measured by	Main question answered
Runtime Correctness	Worker threads, queue, zone manager, heartbeats, task table	Are tasks dispatched safely, zones respected, and robots observable during execution?
Policy Comparison	API scheduling analysis for FIFO / Priority / Round Robin / SRT	How do different logical scheduling strategies affect wait time, completion time, urgent-task completion, and chaptered makespan?
Mode Comparison	Classic vs. work-stealing A/B frontend experiment	Does local queueing and stealing improve utilization under zone contention?
Stress Comparison	Stress preset off vs. on	How do scheduling policies behave when contention grows from light to heavy?

Third, the main benefit of the system lies in its ability to deliver easy-to-understand visualization capabilities. The common problem between multiple chapters occurs when people need to see printed logs before they will start believing in the system's reliability. Our chapter goes beyond relying solely on printed messages–the chapter uses heartbeats and metrics and centralized lifecycle states and dashboards as its additional components. Execution allows people to see all these components. The system enables users to perform debugging tasks more easily while demonstrating their work better and conducting evaluations with simplified methods. The benchmark becomes more trustworthy because the dashboard shows not only whether a task is complete, but also the execution order, waiting status, and thread states. The frontend provides more than visual elements because it functions as an essential component for system verification and comprehension.

At last, the frontend system shows its ability to change the way the report is shown. The report displays an A/B testing dashboard which enables users to switch stress testing between two states for their assessment. The system demonstrates its operational behavior through multiple testing conditions instead of using a single unchanging demonstration. The report presents benchmark testing results which enhance the document's readability and enable easier comparisons. The presentation establishes clear comparison points which include various strategies and modes and load conditions. The system functions as a small system yet its presentation style approaches authentic experimental research methods.

But the chapter is still imperfect and should not be presented as the perfect.

One clear limitation exists because the runtime layer and the comparison layer do not maintain complete synchronization. The backend queue logic works seamlessly with FIFO and Priority and SRT because these methods implement their operations easily. The dashboard displays Round Robin as an impressive system but actual implementation of the system remains incomplete. The frontend displays time-slice comparisons while the backend system lacks capability to handle real preemption. The report describes this distinction between two elements which creates a minor logical error that functions as a hindrance to the argument.

The benchmark tests as shown in Table 2 depend on their execution tests which show typical performance instead of their ability to perform multiple test runs. The chapter size allows for this requirement to be acceptable. The results function as real evidence which proves the case instead of existing as a complete statistical analysis. The system shows its worst performance under extreme conditions while priority scheduling works better for urgent tasks and SRT reduces time needed for task completion and work stealing decreases waiting time when tasks become available. The report does not claim statistical significance, and it should not. The chapter needs trustworthy and representative data but the data does not meet the requirements for a formal study.

The three-systems task list and heartbeat registry and metric registry enable users to inspect and understand their content but they create synchronization points which will develop into future performance constraints. The system consists of an intentional design choice which balances two opposing goals. The state can achieve better performance through two methods which involve further state separation and lock-free queue system utilization. The proposed solution becomes difficult to explain and establish verification through specific measures. Our chapter currently uses its chosen approach because it prioritizes system reliability and control of simultaneous operations and system design clarity rather than achieving peak operational efficiency.

Resource modeling establishes the fourth boundary that limits research. The bounded-capacity area manager provides stronger operational power than a basic lock while maintaining its position as a simplified model. The system fails to address various aspects which exist in actual situations. The existing restriction enables chapter scope management through chapter scope management yet maintains the need to express the actual situation which exists beyond service limits which occur to control service operations.

The fifth boundary states work stealing provides value yet it does not achieve major enhancement. The system cannot resolve all conflicts through its automatic conflict resolution system. It experiences blocking conditions during peak times because of its active thread load which requires full system synchronization for particular tasks. The main advantage of work stealing gives users the ability to work according to their own preferences. Threads maintain their active state until they must execute tasks that are currently blocked. robots have the option to search for additional work that they can finish. The solution achieves improved performance through the solution which enables simultaneous processing of active tasks and queue management through its flexible control systems. The extension proves valuable because it alters backend operations while maintaining existing frontend presentation. The benchmark results show evidence which supports this perspective. Your experimental testing of various strategies produced results which showed little change during periods of low system activity.

The system shows more distinct differences when load conditions become more demanding. The FIFO scheduling method results in extended user waiting periods. The system needs priority scheduling to handle urgent response tasks. SRT achieves its tasks within a shorter time frame than other methods. Round Robin distributes equal treatment to all users but it fails to satisfy all assessment criteria. The case serves as an effective demonstration that shows both load factors and scheduling decisions hold significant weight.

6 CONCLUSION AND FUTURE WORK

Our chapter has achieved its fundamental goals which is to implement concurrent tasks and establish secure areas for operation while observing robot movements through heartbeats and state information display through an API and presenting results through structured dashboard. Most importantly, this is done in a way that is now easier to evaluate, because we can easily sort through it by distinguishing carefully between runtime behavior (how it acts) and analytical policy comparison. The chapter now provides a more convincing explanation through its current presentation. Our work now clearly achieves the core goals: parallel task allocation; secure shared areas; robot tracking; heartbeat mechanism, exposing status through an API and presenting execution in a structured dashboard. The main benefit of this approach is that it becomes simpler to assess the situation. The present moment allows us to understand the operational assessment separate from our analysis of various assessment methods.

The system demonstrates essential operating system concepts through its limited operational display. The shared queue indicates that multiple people can access a shared ready structure at the same time. The task table itself is like a process table in that it keeps track of the authoritative lifecycle status of several robots, even when the work is divided out across them. The zone manager shows how to coordinate limited resources when there is competition. The monitoring subsystem shows how heartbeat timestamps can be used to figure out liveness and how centralized metrics can be used to see the system. The system components operate together to create a coordinating kernel which demonstrates understandable and trustworthy operation.

The benchmark results back up this story. The researchers found that policy differences with lighter load remained at moderate levels. The value of stronger policy selection is much more visible in case of heavier load. FIFO shows stability as a baseline but its performance decreases when faced with demanding situations. The current measured runs show that priority delivers its highest throughput when responding to emergency tasks. SRT maintains its effectiveness for estimating average completion time. Round Robin serves best as an analytical tool that assesses comparative fairness while it should not function as an operational mechanism. The chapter has developed into a complete comparison tool which now supports multi-dimensional assessment between different elements. The system now enables side-by-side evaluation of policy, mode, and load under controlled testing conditions.

The report has also tried to stay realistic about its extent at the same time. Our chapter is not a complete collection of hospital robotics equipment. The system centers on conflicts. The system operates with current advanced scheduling features. The system needs to establish complete procedures for handling faults in its operation. The system has two boundaries which function as operational limits which the system cannot exceed. The right question to examine for a course chapter needs to check whether the system successfully demonstrates and resolves all main issues which occur during concurrent use without endangering user safety. The chapter succeeds according to that standard.

The development process will establish a solid foundation which will support future work. The next step is to enhance Round Robin functionality which needs to operate in real time better than current systems because it needs to execute its comparison between real time performance and theoretical evaluation. The next task is to develop a microsystems-based solution which will solve existing problems.

The work-stealing route creates three new research paths which researchers can explore. The current system design allows local deques together with zone-aware task selection and non-blocking fallback systems to reduce unnecessary delays while completing tasks in highly contested areas. The future versions will introduce statistics about theft incidents together with tools which ensure fairness and methods to prevent resource starvation and detailed performance testing under various stress levels.

The second area of future effort is related to the discipline of benchmarking. The present study adds a hard quantitative table to the evidence base, which is already better. The future report may make benchmarking much better. The chapter requires multiple testing sessions and precise variance measurements together with external CPU sampling and distinct zone-allocation latency records to transform from a compelling course demonstration to a full systems experiment. The frontend is currently well set up for this because it gives the necessary comparison axes. The next step requires us to build an automated system which will generate comprehensive reports about the completed tests.

The third direction of work involves both API development and frontend improvements. The latest frontend system shows significant enhancements because its design framework enables users to compare different elements instead of presenting fixed information. The system needs to improve through its run export system and its export system needs to distinguish between operational and study performance results and the system requires a benchmark history function that enables users to compare multiple test results through direct observation instead of manual memory. The proposed solution will enhance both user experience and research documentation capabilities.

The development of our chapter has progressed beyond its initial purpose which served as a resource for teaching concurrency. The system operates as a complete system which combines its two main parts execution and evaluation. The system design remains lightweight but researchers now have access to a safe and clean environment which enables them to study hospital robot coordination through operating system principles. The chapter presents itself for state submission because it fulfills all fundamental requirements and offers realistic extensions while showcasing its capabilities and limitations through improved structured presentation.

REFERENCES

Axum Contributors. (2025). axum (web application framework documentation). https://docs.rs/axum

Bernstein, P. A., & Goodman, N. (1981). Concurrency control in distributed database systems. *ACM Computing Surveys, 13*(2), 185–221. DOI: 10.1145/356842.356846

Buttazzo, G. C. (2011). *Hard real-time computing systems: Predictable scheduling algorithms and applications* (3rd ed.). Springer. DOI: 10.1007/978-1-4614-0676-1

David, T., Guerraoui, R., & Trigonakis, V. (2013). Everything you always wanted to know about synchronization but were afraid to ask. In *Proceedings of the 24th ACM Symposium on Operating Systems Principles (SOSP '13)* (pp. 33–48). DOI: 10.1145/2517349.2522714

Dijkstra, E. W. (1965). Solution of a problem in concurrent programming control. *Communications of the ACM, 8*(9), 569–571. DOI: 10.1145/365559.365617

Herlihy, M., & Shavit, N. (2020). *The art of multiprocessor programming* (2nd ed.). Morgan Kaufmann.

Hoare, C. A. R. (1974). Monitors: An operating system structuring concept. *Communications of the ACM, 17*(10), 549–557. DOI: 10.1145/355620.361161

Hunt, P., Konar, M., Junqueira, F. P., & Reed, B. (2010). ZooKeeper: Wait-free coordination for internet-scale systems. In *Proceedings of the 2010 USENIX Annual Technical Conference*. USENIX Association.

Karger, D., Lehman, E., Leighton, T., Panigrahy, R., Levine, M., & Lewin, D. (1997). Consistent hashing and random trees: Distributed caching protocols for relieving hot spots on the World Wide Web. In *Proceedings of the Twenty-Ninth Annual ACM Symposium on Theory of Computing (STOC '97)* (pp. 654–663). DOI: 10.1145/258533.258660

Lamport, L. (1978). Time, clocks, and the ordering of events in a distributed system. *Communications of the ACM, 21*(7), 558–565. DOI: 10.1145/359545.359563

Liu, C. L., & Layland, J. W. (1973). Scheduling algorithms for multiprogramming in a hard-real-time environment. *Journal of the Association for Computing Machinery, 20*(1), 46–61. DOI: 10.1145/321738.321743

Love, R. (2010). *Linux kernel development* (3rd ed.). Addison-Wesley.

Mao, H., Alizadeh, M., Menache, I., & Kandula, S. (2016). Resource management with deep reinforcement learning. In *Proceedings of the 15th ACM Workshop on Hot Topics in Networks (HotNets '16)* (pp. 50–56). DOI: 10.1145/3005745.3005750

Mao, H., Schwarzkopf, M., Venkatakrishnan, S. B., Meng, Z., & Alizadeh, M. (2019). Learning scheduling algorithms for data processing clusters. In *Proceedings of the ACM Special Interest Group on Data Communication (SIGCOMM '19)* (pp. 270–288). DOI: 10.1145/3341302.3342080

McKenney, P. E. (2024). Is parallel programming hard, and, if so, what can you do about it? https://kernel.org/pub/linux/kernel/people/paulmck/perfbook/perfbook .html

Mohammed, A., & Zangana, H. (Eds.). (2026). *Intelligent construction monitoring systems: Real-time safety, environmental prediction, and risk management.* IGI Global., DOI: 10.4018/979-8-3373-9245-5

Ongaro, D., & Ousterhout, J. (2014). In search of an understandable consensus algorithm. In 2014 USENIX Annual Technical Conference (USENIX ATC '14). USENIX Association.

Ousterhout, K., Wendell, P., Zaharia, M., & Stoica, I. (2013). Sparrow: Distributed, low latency scheduling. In *Proceedings of the 24th ACM Symposium on Operating Systems Principles (SOSP '13)* (pp. 69–84). DOI: 10.1145/2517349.2522716

Ricart, G., & Agrawala, A. K. (1981). An optimal algorithm for mutual exclusion in computer networks. *Communications of the ACM*, *24*(1), 9–17. DOI: 10.1145/358527.358537

Schwarzkopf, M., Konwinski, A., Abd-El-Malek, M., & Wilkes, J. (2013). Omega: Flexible, scalable schedulers for large compute clusters. In *Proceedings of the 8th ACM European Conference on Computer Systems (EuroSys '13)* (pp. 351–364). DOI: 10.1145/2465351.2465386

Silberschatz, A., Galvin, P. B., & Gagne, G. (2018). *Operating system concepts* (10th ed.). Wiley.

Sutter, H. (2007, December 11). Effective concurrency: Use lock hierarchies to avoid deadlock.

Chapter 8
A Lightweight Multi-Process Coordination System for Hospital Service Robots With Deadlock Prevention

Sunyi Liu
http://orcid.org/0009-0006-1419-5488
The Hong Kong Polytechnic University, Hong Kong

Zongxi Wu
http://orcid.org/0009-0004-9296-6247
The Hong Kong Polytechnic University, China

Daxin Jin
The Hong Kong Polytechnic University, Hong Kong

Yuheng Du
The Hong Kong Polytechnic University, Hong Kong

Bolin Zhu
The Hong Kong Polytechnic University, Hong Kong

ABSTRACT

This chapter presents a lightweight operating system for coordinating multiple medical care robots in a hospital environment. The system focuses on addressing concurrency and synchronization challenges in multi-process robot coordination.

DOI: 10.4018/979-8-2600-1101-0.ch008

A centralized coordinator manages task allocation, resource control, and system monitoring, while robot processes execute tasks independently. An all-or-nothing resource allocation strategy is adopted to prevent deadlock by ensuring that tasks start only when all required resources are available. Inter-process communication is implemented using a lightweight text-based protocol for efficient message exchange. In addition, a heartbeat-based monitoring mechanism is used to detect failures and reclaim resources. Experimental results demonstrate that the system maintains correct synchronization, ensures mutual exclusion, and achieves stable performance under resource contention and partial failure conditions.

1. INTRODUCTION

1.1 Problem Statement and Motivation

Nowadays, modern hospitals tend more towards utilizing medical care robots to enhance work efficiency. These robots have several clearly defined tasks such as delivery, disinfection, and surgical assistance. A lightweight operating system core is needed to coordinate these robots safely and efficiently.

The core issue is that when multiple robots work concurrently in a shared operating system, concurrency and resource sharing problems will appear naturally at the operating system level. First, when multiple robot process attempts to access the shared task queue simultaneously, there may occur some situations including race condition, task loss, and duplicate task requests. Besides, without constraints, multiple robots may try to enter the same zone, which could lead to resource conflicts, disruption of the normal order of the hospital, and serious safety issues. Moreover, these robots can malfunction or lose connection, which will cause congestion of the task queue without resolving this issue and lead to the entire operating system crashing. Lastly, at the logical level of the operating system, deadlock and multi-process collaboration are also crucial issues to be addressed.

1.2 Chapter Objectives

This chapter is aimed at developing a practical lightweight operating system to coordinate multiple medical care robots with solving the concurrency issues mentioned above. Initially, the function of main process is required to be correctly implemented. The main process will act as the coordinator to start child processes as robot processes to execute tasks and receive resource requests from the child processes. The ResourceManager should be able to properly manage and record zone occupancy and task execution status. In addition, the health monitoring function of robots is crucial to be implemented as well, which will be applied to handle robot malfunction and response timeout. As a key aspect of an operating system that coordinates multiple processes, IPC and deadlock prevention should proceed as expected. The overall architecture of the system should be clear and orderly, and the functions of each part do not overlap. Finally, the chapter should produce a representative prototype demonstration, which contains various situations that the system may encounter to provide sufficient reliable evidence of the stability of the system in practical applications.

1.3 Overview of Approach

During the initialization phase, the operating system starts the coordinator process as the main process. The coordinator will read configuration information and initialize child process for each robot in sequence. A unified message receiving channel that aggregates multiple reader threads to coordinator will be established as well. Next, the ResourceManager is going to be initialized. Both resource and health monitoring status will be set to default value. Once all child processes are ready and report their status, the operating system will have successfully initialized and will enter the main loop.

The following is a typical workflow from task processing to task completion. In the default state, the coordinator will search for alive and idle robot process and allocate a task to it. The robot process will then request all the resources needed from the coordinator. The coordinator will verify the request through the ResourceManager and grant to the robot process if all the resources are available or make the robot process blocking wait. Under normal circumstances, the robot process completes the received task and begins the final stages of the task process. After releasing the occupied zone and notifying the coordinator, the robot process will clear the communication pipeline and blocking wait further instructions from the coordinator.

Throughout the entire lifecycle of the operating system, the ResourceManager will periodically invoke the function of handling offline robots to check the heartbeat timestamp of each robot process and clean up all expired robot process. The corresponding heartbeat update thread is generated by the robot process and sends update messages to the coordinator within a time interval according to the configuration. To implement the all-or-nothing principle for deadlock prevention, the ResourceManager will only return authorization when all requested resources are available. A robot process can never occupy a portion of resources to block and wait, which is the situation where deadlock is mostly to occur.

In the prototype demonstration scenario, the tasks were designed beforehand. Some of the tasks are needed for multiple zones, and the types of tasks are distributed averagely in order to cover the complexities that may arise in real-world situations. To demonstrate the system's ability to handle unexpected situations, the demonstration performs fault injection on Robot 1 process. The process stops updating heartbeat but is still alive, which simulates a robot unexpectedly going offline without properly shutting down its process. The demonstration's execution results show that the ResourceManager is capable of handling this type of situation properly and logging this record to stderr.

2. RELATED WORKS

This section reviews the advanced research related to concurrency control, task scheduling, and resource coordination. To start with, we survey the implements of concurrency control mechanisms and synchronization primitives which are necessary for correct and predictable interactions among multiple robots. Then, we examine task queue management and work scheduling strategies. They are critical for coordinating distributed tasks in real time (Mohammed & Zangana, 2026). In addition, we analyze resource coordination and mutual exclusion patterns that try to achieve maximum efficiency. After the survey, we compare existing approaches with the design principles of the proposed lightweight robot coordination OS, improving the weaknesses of our approach.

2.1 Concurrency

Concurrency control is the theoretical foundation of ensuring safe and predictable interactions among multiple medical robots. In a Database Management System (DBMS), Concurrency control is a mechanism in DBMS that allows simultaneous execution of transactions while maintaining ACID properties - Atomicity, Consistency, Isolation and Durability (GeeksforGeeks, 2015). Rashmi (2002) studied the interconnection network in a distributed database system. Acting as a powerful coordination mechanism, the connection can provide certain useful properties. We identify several useful network properties, presenting a technology of concurrency control techniques that are built on top of these properties. In the learning of robots, we propose and explore an approach by applying concurrency control to manage parallel transformations in machine learning algorithms (Pan & University of California, Berkeley, Electrical Engineering & Computer Sciences, 2017).

2.2 Synchronization

The Synchronization is an important concept in operating systems that ensures the smooth and coordinated execution of processes and threads. It is generally used in multi-processing systems, where process concurrently attempt to access the same shared resource of piece of data (GeeksforGeeks, 2024). According to Michael (2012), the synchronization in the vehicle routing problems is difficult. As a critical part of operating system design, synchronization ensures that processes or threads can safely share resources without interfering with each other.

2.3 Task Queues

Task queues are an essential component in system design, allowing programs to handle tasks asynchronously. This means that instead of processing everything at once, tasks can be organized and executed in an orderly manner, improving performance and reliability. By using task queues, developers can ensure that applications remain responsive, even under heavy loads. Human-Robot Collaboration's dynamic task allocation needs task queue to improve operator's states (Urrea, 2025).

2.4 Chapter Schedule

A crucial element of any chapter plan is the chapter schedule. In doing so, it defines the purpose and process of creating a work breakdown structure (WBS), work packages, chapter activities, logic, resources and work, and timeframe (Managing the Chapter Schedule, 2016). The basic construction of a schedule revolves around a proper approach to its development. The approach should be structured in a way that is simple to understand and yet provides enough detail for the experienced to glean the requirements used. This report discusses tried and true approaches that have been developed through years of analyzing and refining processes for schedule development. This approach provides the structure to implement proven best practices to building an executable schedule (Jones, 2009). In our process, we need to schedule the plans, knowing each member of our team in charge of which part.

2.5 Resource Coordinator

Resource Coordinator plays an important role in ensuring that chapters and operations are executed efficiently by effectively managing and allocating the organization's resources. It involves a deep understanding of both the capabilities of the workforce and the demands of various chapters, allowing for the strategic placement of personnel and assets where they are most needed. Maglione et al. (2021) have designed and developed an HLA (High Level Architecture) interoperable M&S testbed capability, and the resource coordinator is employed in live demonstrations to automatically design missions for heterogeneous autonomous assets. This role requires a combination of communication, planning, and problem-solving skills to navigate the complexities of resource management, making it essential for the smooth operation of any organization (CLIMB, 2025).

2.6 Mutual Exclusion

Mutual Exclusion is a property of process synchronization that states that "no two processes can exist in the critical section at any given point of time". The term was first coined by Dijkstra. Any process synchronization technique being used must satisfy the property of mutual exclusion, without which it would not be possible to get rid of a race condition. The need for mutual exclusion comes with concurrency (GeeksforGeeks, 2020). After learning the knowledges of mutual exclusion, we get several kinds of concurrent execution: interrupt handlers, interleaved, preemptively scheduled threads, multiprocessor clusters with shared memory and distributed systems. According to Kanrar et al. (2018), Mutual exclusion usually is used in concurrent programming to avoid the simultaneous use of a common resource, such as a global variable.

2.7 Comparison

Our Rust chapter implements a multi-process robot scheduler with a simple line-oriented IPC over stdin/stdout, per-child reader threads, and an all-or-nothing resource manager to prevent deadlock. Compared with typical single-process, our programs simplify process lifecycle management by re-executing the same binary for children. Versus heavy-weight RPC frameworks, the custom text protocol is lightweight and easy to debug for peoples. The allocation avoids hold-and-wait and simplifies correctness compared to fine-grained lock-based resource allocation, though it may reduce concurrency in some workloads. Heartbeat-based offline detection and MPSC channels give straightforward coordination. Overall, the system focuses on clarity, robustness, and testability.

3. IMPLEMENTATION

3.1 Overall Architecture and Process Model

This system employs a lightweight multi-process architecture, comprising only a coordinator and multiple robot subprocesses, achieving complete separation between global scheduling and robot execution logic. The main tasks are handled by the coordinator, including creating robot subprocesses, receiving status messages from them, allocating tasks, granting resources, detecting offline status of subprocesses, and uniformly performing task shutdown operations upon completion of all tasks. Simultaneously, each robot is designed as an independent subprocess, responsible for receiving tasks, requesting task start permissions, executing tasks, and sending heartbeat signals. This structural design allows the system to easily separate global scheduling, resource decisions, and robot execution logic. Ultimately, the system maintains simplicity while clearly demonstrating the concurrency, synchronization, and fault handling capabilities of multi-process execution.

In terms of data modeling (Preser, 2023), we employ a unified structure to abstract tasks. Each task includes a number, zone identifier, type, and text payload fields. The "Vec<Zoneld>" field describes the task's resource requirements for one or more zones, enabling single-zone and multi-zone tasks to be represented in a consistent manner. Currently, task types include delivery, disinfection, surgical assistance, and shutdown control messages. This unified modeling approach provides the foundation for subsequent task allocation, resource checks, and protocol serialization.

Next is the process design. Upon startup, the coordinator first reads the demo configuration from the scenario module. In the current demo scenario, the configuration consists of three robots, three areas, and five preset tasks. The first three tasks rely on different combinations of two areas and are designed to demonstrate the effectiveness of an "all or nothing" resource allocation strategy in a multi-resource contention environment; the last two tasks are single-area tasks used to supplement the verification of the regular execution process. After startup, the coordinator creates robot processes. It only begins assigning tasks after receiving "ready" messages from all child processes. By establishing explicit synchronization points, this design avoids the potential problem of tasks executing before the system is fully ready.

On the child process side, each robot child process immediately sends a READY message upon startup and then enters a message receiving loop based on standard input. The state of the child process remains unchanged. It will send a REQUEST_TASK_START message after receiving a new task. It is only after receiving a TASK_START_GRANTED authorization from the coordinator that it is able to execute the task. After the task is complete, the child process will send a DONE message. At the same time, in a separate thread, periodic HEARTBEAT messages for the child process are sent. It is on the basis of the timing of the receipt of the last heartbeat that the robot can be determined to be offline. This is a reflection of the separation of control and execution planes. Thus, the final startup of the task is not determined by the child process itself.

According to the division of responsibility in each module, the spawn module is responsible for creating child processes and encapsulating pipes, the coordinator module is responsible for the main loop and system-level decision-making, the child module is responsible for implementing robot execution logic, the resource_manager module is responsible for area occupation and waiting queues, the monitor module is responsible for implementing offline detection using heartbeat mechanisms, and the scenario module is responsible for creating scenarios for demonstration tasks. The division of responsibility in each module is clear and simple, which is beneficial for establishing a foundation for code structure in the following discussion.

3.2 Coordinator, IPC, and Task Dispatch

In this system, the coordinator acts as the control center for the whole process of the multi-process scheduling. After the system starts up, the coordinator will first read the scene configuration, establish the common mpsc message channel, and then create the individual child process for each robot one by one and initialize the resource management state, offline detection state, and the shutdown confirmation state. Only after all the child processes have sent the READY message will the coordinator begin to assign the initial tasks, thus providing a synchronization point in the startup process and preventing the system from proceeding with the task execution process before it is ready.

The creation of child processes is done by re-executing the current binary file. The parent process is responsible for establishing separate stdin and stdout pipes for each child. This is for two-way communication. In order to receive simultaneously the outputs of many child processes, the system uses a reader thread for each child process. The reader thread is responsible for reading continuously the stdout of that child and then sending the parsed messages uniformly into the mpsc channel. This is used by the coordinator. The overall control flow is simpler as the reading and parsing are done by the reader thread, and in the main loop of the coordinator, it is only necessary to handle the messages from a unified entry point.

A lightweight IPC protocol using text lines is used between the parent process and child process. The basic messages used by the parent process include TASK, TASK_START_GRANTED, and SHUTDOWN. The basic messages used by the child process include READY, HEARTBEAT, REQUEST_TASK_START, DONE, and SHUTDOWN_ACK. Among them, the TASK message uses the task number, region set, task type, and payload content. Since there may be spaces in the payload, splitn(5, ' ') is used during deserialization to ensure that the complete text is preserved (Sirejiding et al., 2024). This is a simple design that is easy to implement, understand, and is adequate for the communication needs of the prototype system.

This assignment of tasks to the nodes is done in a "two-stage" fashion. The assignment of the tasks that are yet to be completed by the coordinator is done only for idle and online robots. The assignment of the task is held from execution, but it is recorded in the ResourceManager. Before the execution of the task, a request for permission to start the task, REQUEST_TASK_START, is sent by a child process. The permission to start the task is granted only after all the resources are available for the task. The resources are then released, and then there is a wake up for the task that is yet to be completed. The tests that are associated with this are the "smoke tests," and they verified that the system can perform a complete start, task assignment, task completion, and shutdown.

3.3 Resource Management, Synchronization, and Deadlock Prevention

The resource control of this system is uniformly handled by the ResourceManager. The core states of this system consist of the current occupant in each zone, the tasks assigned to each robot, the running state of the tasks, and the waiting robots. After the tasks are distributed in the coordinator, the assign_task() function is called to assign the tasks but not start them. The differentiation of the assign and start operations provides the basis for the judgment and authorization of resource startup.

When it receives REQUEST_TASK_START from a child process, it calls request_task_start() to check resource availability. In this way, the strategy is to take an "all-or-nothing" approach. In other words, it will only allocate all the resources and return "Granted" if all of the zones required by the task are idle. If any of them is occupied, it will not allocate any of the resources to the task and will put it in the waiting queue directly, returning "Queued" instead. In this way, it is impossible for the process to "first occupy some resources and then wait for the remaining resources," which is to say it is impossible for it to have the 'hold and wait' property in the system. This is the key mechanism for deadlock prevention in this chapter. Relevant unit tests have also verified that when one of the multiple regions required by a task is busy, the system will maintain the "none given" behavior instead of making partial allocations.

Resource release happens in two scenarios: normal task completion, or a robot being marked offline. In the first case, complete_task() releases all zones held by the robot, clears the associated task status, and calls try_start_waiting_tasks() to rescan the waiting queue. In the offline case, handle_robot_offline() performs similar cleanup and also attempts to wake any waiting tasks whose resource requirements are now satisfied. In other words, the system does not simply "suspend the task", but actively checks which waiting tasks can be started after each change in the resource status. The test results also show that the waiting tasks will automatically obtain the qualification to be started after resource release, without the need for manual re-delegation.

In terms of synchronization implementation, this system consolidates most of the shared states on the coordinator side for unified processing. Therefore, the resource management itself does not require a complex multi-lock structure (Zhou et al., 2026). Instead, it maintains state consistency through a single control flow.

The coordinator is responsible for receiving RequestTaskStart, Done, and offline events, and then serially calls the ResourceManager to update the resource status and determine the subsequent actions; only when the resources truly meet the conditions will it send TASK_START_GRANTED to the corresponding child. This "centralized decision-making + two-phase startup" approach makes the synchronization logic relatively clear and also reduces the race risk introduced by multiple execution units simultaneously modifying the shared resources. The smoke tests of the entire running process further verified that the system does indeed exhibit the behavior of "waiting first and then waking up", indicating that the resource management, synchronization control, and deadlock prevention mechanisms can work collaboratively during the complete operation.

3.4 Health Monitoring and Failure Handling

This system continuously tracks the status of the robots through a heartbeat mechanism and integrates health monitoring into the main loop of the coordinator for processing (Wu & Wang, 2020). The coordinator maintains a "last visible" timestamp for each robot and periodically calls the scan_newly_offline() function to check if the status of any robot has not been updated within the heartbeat timeout period. Once a new timeout object is detected, it is marked as offline to avoid repeated reporting in the future. Relevant tests also confirmed that robots that do not exceed the timeout limit will not be misjudged, while truly timed-out robots can be identified. The same offline event will only be reported once.

At the sub-process level, each robot sub-process, after being initiated, not only enters the regular task receiving loop but also creates an independent heartbeat thread to send heartbeat messages at fixed intervals. This sub-process contains two threads: the main thread responsible for task execution and the independent heartbeat thread. Both write protocol messages through the standard output port. To avoid data interleaving, the system uses Arc<Mutex<BufWriter<Stdout>>> as the shared writer. This architecture achieves parallel processing of task execution and status reporting while maintaining message integrity.

Once the robot is detected to be offline, the coordinator calls the function handle_robot_offline() to clean up. This function releases all the areas allocated to the robot, resets the assigned tasks and state of the robot, and checks if there is a waiting task that is now possible to start due to the released resources. When such a task is found, the coordinator sends TASK_START_GRANTED right away. Therefore, offline detection is not only a logging function but also directly participates in resource recovery and the advancement of the scheduling process, preventing a single offline robot from occupying critical resources for a long time. Related resource management tests also proved that after offline cleaning, the system can release resources and automatically start the previously blocked waiting tasks.

To demonstrate the system's ability to handle abnormal situations, the current demonstration adds simple fault injection in the subsystem. For robot 1, after completing the first task, it does not immediately exit but stops sending heartbeat signals, simulating the scenario of "the process is still running but has lost contact". This design enables the system to present a more realistic fault situation, where the robot is not shut down normally but loses response. At the same time, after all tasks in the system are completed or cannot be further advanced, the coordinator broadcasts the "shutdown" signal, and each subsystem will receive this signal, stop the heartbeat thread, send the "shutdown confirmation" signal and exit; the corresponding smoke test also verified that the system can correctly enter the shutdown phase and complete the entire shutdown process.

4. BENCHMARK

This Section will evaluate the performance of the multi-process robot coordination system in the concurrent environment. According to the chapter requirements, not only does it need to check the correctness of the system in shared resource access, including mutual exclusion and synchronization, but it also needs to analyze its performance under resource contention and partial failure.

Coordination architecture is adopted by our system. There is a single point of resource management and scheduling named the coordinator that maintains all the shared states, such as zone occupation, task allocation, and robot health. Robot subprocesses communicate with the coordinator through an IPC protocol instead of reading/writing shared memory. This way eliminates data races by construction, reduces the risk of simultaneous writes, and eases the locking requirements on the system.

Therefore, this section will focus on analyzing the following questions:

1. If the system can correctly maintain mutually exclusive access to zones
2. All-or-Nothing strategy efficiency
3. Heartbeat's ability to check faults
4. System performance under resource contention and failure conditions

4.1 Test Methodology

4.1.1 Test

The test was run using a Rust program executed by PowerShell in a Windows environment. The system operates under a default configuration. The initial setting is that the number of robots in zones is 3, 3, and 5, respectively. The results show that the system forms a synchronization barrier with the READY message when it starts, so all robot child processes finish initialization before they start distributing tasks, thus reducing the chance of a race condition.

4.1.2 Further Testing

For a more thorough evaluation, we introduced several configurations: 3, 6, and 12 robots; 3, 6, and 12 resource zones; and 5, 10, 20, and 40 tasks. Task type distribution was set to 50% single-zone tasks, 30% two-zone tasks, and 20% three-zone tasks. The task design did not explicitly balance varying resource quantities, so some tasks require multiple zones such as [0,1], [1,2], or [0,2], while others need only a single zone like [0] or [2]. This arrangement creates scenarios with resource contention, enabling observation of system behavior across different contention levels. Additionally, we included a partial failure case by stopping robot heartbeats after the first task, demonstrating that the system can tolerate unresponsive nodes while continuing to process remaining tasks.

4.1.3 Data Source

Data is primarily obtained from runtime logs. Each log entry includes timestamps, event types, task IDs, and robot IDs. In addition, timing is measured using std::time::Instant at key points, and CPU usage is observed through Windows Performance Monitor.

The unit tests in the code cover modules such as IPC message parsing, resource allocation logic, and heartbeat fault detection. All tests passed, further proving the correctness of message communication, offline detection, and resource mutual exclusion.

4.2 Correctness Verification

After completing the operation and unit tests, it can be seen that this system can deal with multiple crucial tasks as expected, and is able to handle resource sharing and multiprocessing collaboration issues correctly.

Firstly, in terms of resource access, the system can effectively ensure mutual exclusion among zones. When the zone is occupied, the following task will enter a waiting queue with the output "cannot be started yet and has entered the waiting queue". Task will be allowed to begin, only this zone is released, so that it can prevent concurrent conflict.

Secondly, the resource allocation strategy adopted by the system is "all-or-nothing". Tasks are required to access all zones that they need at once. But if tasks are unable to do that, they were asked to wait without holding any resources. This strategy avoids partial distribution of resources and reduces the probability of deadlock.

Next, the system can keep the consistency of the task execution state accurately. From the results, the task follows the life cycle: allocation - request - waiting - execution - completion - releasing resources. After the task is completed, it will explicitly release the related zone in order to ensure that the following tasks can obtain these resources. This state change helps avoid resource leaks or state errors.

In addition, the system has basic fault detection and recovery capabilities. When the heartbeat signal stops emitting, the coordinator detects the offline status of the tasks by setting a timeout period and retrieves resources. After that, other tasks can continue to be executed, which means that the system can keep the whole process advancing.

Last but not least, in terms of the synchronization mechanism, due to the fact that all shared states are controlled by the coordinator and robot sub-processes just communicate through IPC, concurrent writing to the common memory is not allowed. The results of the unit testing indicate that the system is able to maintain correct synchronization behavior with the current design.

4.3 Performance Analysis

4.3.1 Task Throughput

In the initial set, the system can complete 5 tasks in a short time, and the throughput is about 1.5-2 tasks per second. As the percentage of the multiple-zone task increases, the serialization effect reduces throughput. As shown in Table 1, the throughput under different task distributions varies significantly. Throughput under 3 robots and 3 resource zones with different task distributions is shown below.

Table 1. Task Throughput under Different Conditions

Task Distribution	Task Count	Runtime (ms)	Throughput (tasks/sec)
All single-zone	5	1656	3.02
Mixed (60% single-zone,40% dual-zone)	5	2266	2.21
All dual-zone	5	3784	1.32

The mixed task throughput is the highest, approximately 28% higher than all single-zone tasks. Because single tasks finish quickly and release resources, creating opportunities for executing concurrently. The percentage decreased by about 41% when all tasks needed two zones. This decrease was in line with expectations, due to the high probability of competition when dual zone tasks require two resources at the same time.

This result indicates that the system can maintain stable task processing capacity at the current scale. However, because some tasks require multiple zones, task execution exhibits a certain degree of serialization, thus limiting further throughput improvements.

4.3.2 Task Start Latency

Resource contention is the primary reason for delay that leads to tasks starting slowly. As you can see in the output, although some tasks will run immediately after allocating, others will be placed in a waiting queue. As shown in Table 2, task startup latency increases significantly under resource contention.

Table 2. Task Startup Latency under Different Conditions

Task Type	Average Startup Delay (ms)	Maximum Startup Delay (ms)
Single-zone (no contention)	0.6	0.814
Single-zone (with contention)	425.9	1201.729
Dual-zone (immediate grant)	0.3	0.390
Dual-zone (queued)	941.6	1602.643

In the case without resource contention, latency is near zero(<1ms), showing that the overhead of the schedule is low. But, as you can see, the latency for single zone tasks and dual zone tasks surges to 425.9ms and 941.6 ms, respectively. Because two-zone tasks need to wait for two resources to be released simultaneously, the waiting time becomes longer. The maximum startup delay approaches the task running time, reflecting the fact that tasks in the waiting queue require preceding tasks to complete. This difference reflects the impact of the system's all-or-nothing strategy, where the system sacrifices some response time to ensure the safety of resource access.

4.3.3 Offline Detection Delay

In the fault injection scenario, when Robot 1 stopped sending heartbeats, the system detected that it was offline. As you can see, from the results, the detection in offline happened just after the heartbeat stopped, which matches the set time-out value (around 1200 ms).

Figure 1. Timeout Threshold vs Detection Delay

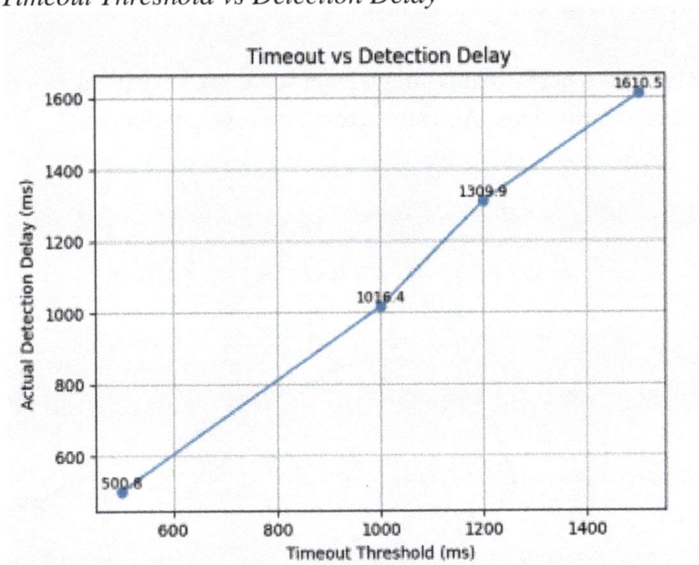

As shown in Figure 1, detection delay increases proportionally with the timeout threshold. This result proves that the heartbeat mechanism can detect failed nodes within a reasonable time frame and reap the resource reclamation logic.

4.3.4 CPU Usage

Most of the time, the system is in a state of waiting for resources or processing tasks, instead of computing intensively. Since the coordinator is only responsible for message processing and scheduling decisions, its CPU usage is expected to be relatively low. As shown in Table 3, the overall CPU usage of the system remains low. Each component CPU usage is shown below.

Table 3. CPU Usage of System Components

CPU Usage	User Time (s)	Kernel Time (s)	Total CPU Time (s)	Utilization
Coordinator	0.0156	0.0312	0.0469	0.39%
Robot Child (average)	0.0026	0.0026	0.0052	0.04%
Total (all processes)	0.0312	0.0469	0.0781	0.65%

The system CPU usage is only 0.65%, indicating that the overhead of scheduling and IPC is very low. Although the proportion of coordinators is the highest, it is still a really low number.

4.4 Scalability and Stress Tests

To evaluate system performance under lager scale, we have increased the number of robots, and resource zones. And stress tests are also conducted.

Figure 2. Throughput vs Number of Robots

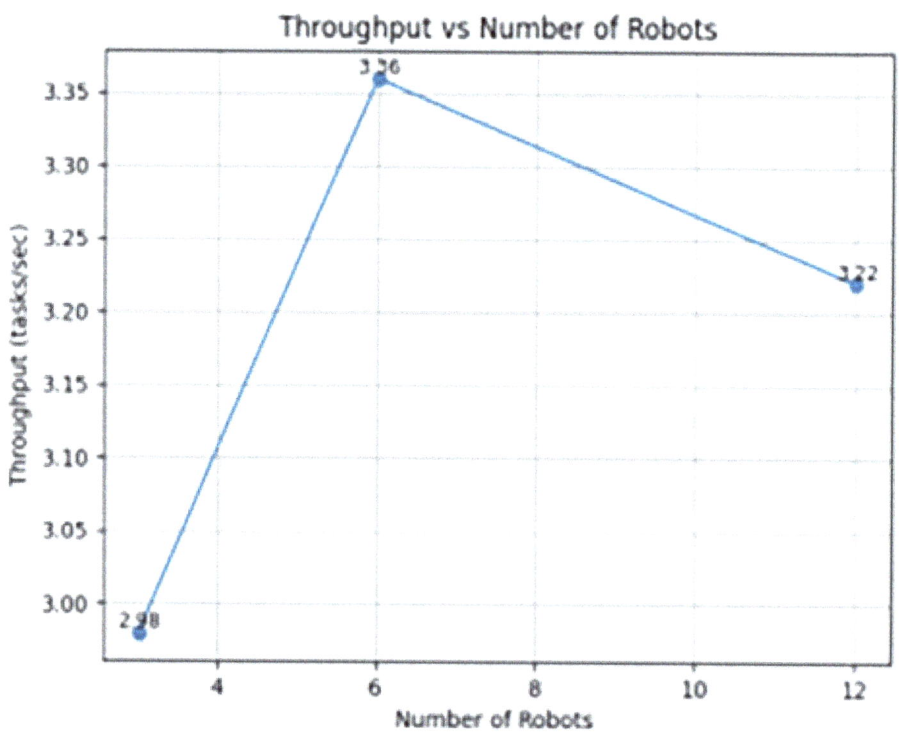

As shown in Figure 2, throughput initially increases with the number of robots and then slightly decreases due to resource contention. Increasing from 3 robots to 6 robots, the ratio of speed up is only 1.13, and the throughput improvement is not significant. However, when the number of robots changed to 12, throughput decreased. It suggests that so many robots were unable to obtain resources, which increased the scheduling overhead.

For a high-content scenario, when 3 robots, 3 resource zones, 5 dual-zone tasks with circular dependencies, the system completed all tasks without deadlock, and the maximum queue length was 2 robots. At that time, the longest wait time was approximately 1200 ms, and the resource contention rate was 68.5%.

For message Burst, when 3 robots sending 100 messages at the same time, all messages were tackled correctly, the peak queue depth was 4, and the maximum delay was 3 ms. For Robustness, when a robot process was externally stopped, the coordinator found the break in one cycle, retrieved its resources, and the other tasks completed successfully. This means the system is able to preserve correctness with heavy resource contention, while it's going to suffer in terms of performance at least a little bit.

5. DISCUSSION

This chapter implements a coordinate core for medical care robots, which mainly focuses on the OS concept of concurrency, making sure the safe cooperates between multiple working units and concurrent control on shared state. Overall, the chapter satisfies the minimum scope of requirement, including task queue, zone mutual exclusion and health monitor, and further implements multi-process isolation, explicit IPC, and an all-or-nothing allocation strategy for deadlock prevention. This part will discuss the result analysis, design trade-offs, limitations and bottleneck, and learned experience in concurrent programming.

5.1 Analysis of Results

The demo confirms that the implemented system is able to fulfill all of the requirements of the chapter. The coordinator can spawn multiple robots, who are later able to concurrently fetch and execute tasks from shared task queue, which avoids raising racing conditions or duplicate distribution. In the startup stage, coordinator also manages to prevent distributing tasks to children who are not yet initialized or whose pipes are not ready for communication by waiting for all ready messages.

The resource manager is responsible for managing the zone mutual exclusion and task assignment. The demo log shows that multiple robots would not be allowed to occupy the same zone concurrently, since children must send REQUEST_TASK_START and confirmed by the coordinator that all required zones are free to use before they can be allowed to start.

Health monitor also behaves as expected. The mechanism works by making children send HEARTBEAT in regular periods and using the coordinator to record the latest seen time (mark_seen) to identify the offline robots (scan_newly_offline). In demo, the monitor is capable of detecting the simulated failure of Robot 1 in configured 1200ms timeout window, and the coordinator then takes back the occupied zone by Robot 1 and resumes scanning the waiting queue. The above verifies the successful implementation of concurrency control, synchronization, and coordination.

Besides the minimum requirement, the ALL-OR-NOTHING design is also verified in demo through overlapping multi-zone tasks (task 9000, 9001 and 9002). During the demo, Robot 2 fails to get access to zone 0 and zone 2 due to the occupation of zone 0 and zone 1 of Robot 0 executing task 9000 and is put into waiting queue by coordinator. As mentioned above, the Resource manager will resume scanning the waiting queue after Robot 0 releases the two zones and sends TaskStartGranted to Robot 2. This mechanism effectively manages to prevent deadlocks, verifying that no robot could be allowed to hold part of required resources.

5.2 Key Trade-Offs in Design Choices

5.2.1 Multi-Process Model vs. Multi-Threaded Model

This system implements a multi-process model rather than the minimum required multi-threaded model in one process only. In the design, every robot is started by std::process::Command as an independent process, and stdin/stdout pipes are used for robot-coordinator communication. This design is important since it promotes the system's ability to isolate breakdowns, avoiding the possible negative impact on coordinator status from failures of children. However, the downside of the design is obvious: the overhead of communication rises significantly. Every message must be parsed as one-line text and transmitted with OS pipes. Considering the context of medical care robots, the design is still reasonable due to its significant promotion in safety and fault-tolerance.

5.2.2 Centralized Coordination vs. Distributed Locking

This system implements centralized coordination, making use of only coordinator to determine task distribution, resource allocation, rather than allowing robots to make decisions of getting hold of lock and entering zones (Comez, 2007). This is useful since only Resource manager gets control of resource conflicts, which is helpful for failure tracing and correctness proving. However, in this style, the coordinator can also turn into a bottleneck. As the number of robots and frequency of messages rises, overall efficiency is possible to be limited by the coordinator's performance.

5.2.3 All-or-Nothing Allocation vs. Parallelism

This system prevents deadlock by All-or-nothing allocation, which prohibits resources from being partially held. Multiple-zone tasks would be put into waiting queue if they cannot get access to all the required zones, rather than getting hold of part of required zones first. This effectively prevents deadlocks, but it also significantly increases the overall idle time of robots, which lowers the overall efficiency of the system. On the other hand, a strict All-or-nothing allocation can lead to starvation, since a large task is possible to continuously block smaller ones behind it in the queue.

5.3 Limitations and Bottlenecks

First, the system's task distributing strategy follows strict FIFO, which fails to consider priority of medical tasks. It is important to consider priority in a hospital context. For instance, a surgery assistant task should be handled before a general delivery task, and it should be assigned to a robot that is closer to the target zone.

Second, the heartbeat timeout mechanism makes use of a fixed timeout value of 1200ms. In scenarios where system is loaded heavily or connection is experiencing high latency, it is possible for healthy robots to be mistakenly marked as offline. Considering the system now is not designed to be capable of restoring offline robots back to online, a short failure of communication could make robots unable to be dispatched ever again.

Third, according to the current design, once the robots are identified as offline when holding zones, the tasks that are being implemented would be abandoned and could not be put into waiting queue. This is unacceptable in a real hospital scenario where every task should be properly finished.

5.4 Lessons Learned

The team learned two key points from the chapter.

1. concurrent programming could be more easily done when the ownership boundaries are clear. This can be achieved through making use of only one module to control all the variables, making sure the shared states are correctly handled and synchronized.

2. Protocol design is as important as a concurrent mechanism. In the early development, it was noticed that the heartbeat thread and main thread could write into stdout at the same time, producing lines with wrong format. The problem was immediately solved after adding SharedWriter mutual lock, but the main cause was that the detailed shared resources of thread was not properly considered in the first place. A key lesson from here is that the shared resource must be identified before making a choice of Synchronization Primitives.
3. Fault injection is an effective test strategy. The design of prohibiting the heartbeat of Robot 1 after completion of the first task is helpful for verifying health monitor and resource reclaiming logic. The system would be tested in an ideal environment without the injection strategy, making the logic of failure handling modules unverified.

6. CONCLUSION AND FUTURE WORK

6.1 Summary of Achievement

This chapter delivers a lightweight operating system for coordinating medical care robots in a hospital-like scenario, focusing on concurrency, synchronization, and handling failures. In general, the operating system aligns precisely with the chapter's minimum requirements. On this basis, multi-process concurrency, IPC, deadlock prevention are systematically implemented, demonstrating our considerations and solutions for practical issues. By separating responsibilities into independent modules (IPC protocol, spawning layer, coordinator loop, monitor, resource manager, and scenario generator), the overall design remains orderly and avoids overlapping functionality.

6.2 Implementation Methods

In terms of concurrency control, the coordinator aggregates asynchronous outputs from multiple child processes through per-child reader threads and a unified message channel, which effectively prevents blocking reads from interfering with global scheduling progress. On the child side, stdout is treated as a strict IPC channel, while stderr is reserved for logging, ensuring protocol communication remains stable even under heavy concurrent output. The heartbeat mechanism is implemented as a dedicated thread in each child process. A shared stdout access is protected by a mutex to avoid interleaving messages, which directly addresses race conditions in inter-thread communication.

For synchronization and safety, zone conflicts are resolved by the ResourceManager through an all-or-nothing strategy at the task level. Instead of allowing partial resource acquisition, the system only grants execution permission when all required zones are simultaneously available; otherwise, the robot is placed into a blocking wait queue to reduce computing power. This design avoids multiple processes temporarily holding parts of resources and waiting for each other, which is a classical scenario of deadlock (Isloor & Marsland, n.d.). Therefore, the risk of deadlock is significantly reduced in all anticipated circumstances. The demonstration workload also intentionally includes overlapping multi-zone tasks to highlight this effect.

In addition, as an important component to track robots' real-time status and deal with unexpected issues, the function of health monitor is properly implemented in the system. It is based on heartbeat timeouts and is integrated into the coordinator's main loop, and resource cleanup for offline robots is supported to release zone occupancy and re-enter the unfinished task into the task queue. Heartbeat updates are performed periodically by a separate thread in a child process and will be received by the coordinator.

Aimed at fully demonstrating the functionality and features of the operating system, appropriate prototype demonstration and comprehensive tests are included in our chapter. We not only included several simple tasks in the demo to demonstrate the system's multi-task concurrent processing capabilities, but also some tasks with overlapping resources to showcase its deadlock prevention features. The test section ensures the correct implementation of the basic functions of each part including health monitoring, offline cleaning up, log recording and task distribution. Both the demonstration and the test indicate that the operating system strictly meets the chapter requirements and correctly implements the bonus functionality.

6.3 Potential Improvement

Considering the lack of specific context, the operating system cannot be designed to accommodate more specific functions like more efficient task distribution and flexible task management.

6.4 Future Work

Although the current prototype meets its design goals, several extensions can improve practicality and scalability. First, optimizing task assignment and robot scheduling based on specific contexts is a practical update direction. Specifically, it is beneficial for efficiency to have priority assigned to tasks. The system can apply various strategies and schedule robots based on the priority. It can then be approximated as a multi-core CPU with certain restrictions and CPU scheduling strategies, for instance, SJF, SRT and multi-level queue can be used as a reference (Sajeewa Pemasinghe & Rajapaksha, 2022). Besides, IPC protocol still has room for improvement. If the customer needs to deploy the system within a large-scale medical system, it would be better to standardize IPC protocols and add more fields to improve reliability. In addition, a benchmark recording function can serve as a supplement to design scheduling strategies. In subsequent versions, adding to the system the functionality to record various benchmark metrics could be considered. Various data, including task waiting time and turnaround time, can be automatically aggregated for subsequent analysis.

REFERENCES

Comez, N. (2007). Inventory sharing and planning coordination in centralized and decentralized systems (Doctoral dissertation, University of Texas at Dallas).

GeeksforGeeks. (2015, December 29). Concurrency control in DBMS. https://www.geeksforgeeks.org/dbms/concurrency-control-in-dbms/

GeeksforGeeks. (2020, March 20). Mutual exclusion in synchronization. https://www.geeksforgeeks.org/operating-systems/mutual-exclusion-in-synchronization/

GeeksforGeeks. (2024, April 30). Synchronization examples. https://www.geeksforgeeks.org/operating-systems/synchronization-examples/

Isloor, S. S., & Marsland, T. A. (n.d.). The deadlock problem: An overview. University of Alberta. https://webdocs.cs.ualberta.ca/~tony/OldPapers/dead.pdf

Jones, E. F. (2009, October 13). Scheduling 101 – the basics of best practices. Project Management Institute. https://www.pmi.org/learning/library/schedule-101-basic-best-practices-6701

Kanrar, S., Chaki, N., & Chattopadhyay, S. (2018). *Concurrency control in distributed systems using mutual exclusion*. Springer. DOI: 10.1007/978-981-10-5559-1

Maglione, G. L., Berretta, L., Godfrey, S. B., Apostolidis, S., Kapoutsis, A., Kosmatopoulos, E., & Tremori, A. (2021). M&S-based testbed to support V&V of autonomous resources task coordinator. In Turi, M., Mazal, J., Fagiolini, A., & Vašík, P. (Eds.), *Modelling and Simulation for Autonomous Systems (MESAS 2020)* (Vol. 12619, pp. 123–138). Springer., DOI: 10.1007/978-3-030-70740-8_8

Managing the chapter schedule. (2016). Bookboon.

Mohammed, A., & Zangana, H. (Eds.). (2026). *Intelligent construction monitoring systems: Real-time safety, environmental prediction, and risk management*. IGI Global., DOI: 10.4018/979-8-3373-9245-5

Pan, X., & University of California, Berkeley, Electrical Engineering & Computer Sciences. (2017). Parallel machine learning using concurrency control (Master's thesis, University of California, Berkeley).

Pemasinghe, S., & Rajapaksha, S. (2022). Comparison of CPU scheduling algorithms: FCFS, SJF, SRTF, Round Robin, priority-based, and multilevel queuing. DOI: 10.1109/R10-HTC54060.2022.9929533

Sirejiding, S., Bayramli, B., Lu, Y., Huang, S., Lu, H., & Ding, Y. (2024). Adaptive task-wise message passing for multi-task learning: A spatial interaction perspective. *IEEE Transactions on Circuits and Systems for Video Technology*, *34*(10), 9499–9514. DOI: 10.1109/TCSVT.2024.3399613

Urrea, C. (2025). Hybrid deep learning–reinforcement learning for adaptive human–robot task allocation in Industry 5.0. *Systems*, *13*(8), 631. DOI: 10.3390/systems13080631

Wu, T., & Wang, W. (2020). An improvement mechanism for heartbeat-driven MAC synchronization in wireless body sensor networks. *International Journal of Communication Systems*, *33*(4), e4224. DOI: 10.1002/dac.4224

Zhou, Y., Guo, Y., Zhang, W., & Zhao, P. (2026). Synchronization, circuit implementation, and a three-layer image encryption scheme for a novel 5D hyperchaotic system. *Nonlinear Dynamics*, *114*(6), 416. Advance online publication. DOI: 10.1007/s11071-026-12288-z

Chapter 9
Concurrency Control and Heartbeat-Driven Recovery for Autonomous Medical Robots

John Christopher Flores Quiambao
The Hong Kong Polytechnic University, Hong Kong

Xiao Yi Wang
The Hong Kong Polytechnic University, Hong Kong

Zhengyang Li
http://orcid.org/0009-0007-2367-8817
The Hong Kong Polytechnic University, Hong Kong

Dai Cheng Han
http://orcid.org/0009-0001-2244-0619
The Hong Kong Polytechnic University, Hong Kong

Ting Xuan Wu
The Hong Kong Polytechnic University, Hong Kong

Mohammad Al Khaldy
http://orcid.org/0009-0009-7502-4668
University of Petra, Jordan

ABSTRACT

Medicine is a field that has historically had a high demand for workers with little supply, this high demand is not helped by the fact that medicine is one of the volatile fields of service work in the world with constantly changing situations that require quick responses where mistakes can potentially cost the lives of patients. The use of robots can help alleviate some of the burden of the workers, especially nurses. Simple tasks like the delivery of medicines and crucial tools which require nothing more than moving items from one place to another can be delegated to robots. The issue comes in creating a system that can seamlessly coordinate these robots such that they do not become more of a burden than the issues they originally tried to

DOI: 10.4018/979-8-2600-1101-0.ch009

fix. This paper aims to provide an implementation of an OS core that can handle concurrency and synchronization of multiple working robots within a hospital, allowing the simulation of real-life situations. We also aim to provide an effective framework through which future researchers are able to build upon for application to real-life systems.

1. INTRODUCTION

1.1 Background

The increasing use of IoT devices in factories and environmental monitoring has brought a real engineering challenge: how can the host system reliably collect, process and display high-frequency data streams from various types of sensors without losing packets, disrupting state sharing, or slowing down user queries? Even if the scale is small, to solve this problem, process isolation, inter-process communication, thread concurrency, synchronized shared memory, and persistent file input and output must be implemented.

To address this challenge, this chapter placed it in a controlled academic environment and implemented a complete end-to-end sensor data pipeline. A three-axis force sensor, two thermometers, and a three-axis accelerometer all run as independent operating system processes, sending readings to a parent gateway process via anonymous Unix pipes. The gateway uses a parallel thread pool to aggregate readings within a one-second time window. Then we persist the results to a JSON file. After that, we provided the data through a real-time network dashboard to record the performance and result. Therefore, a system is built in this way. For each major component, it directly applies a set of different operating system concepts, making a clear and measurable relationship between software design and operating system-level behavior in the real-world cases-based.

1.2 System Architecture

The platform is built with a four-tiered pipeline structure in a single Cargo workspace. We will mention it one by one:

Layer 1 – Sensor Simulation (Component 1, sensor_sim crate): Each sensor type is implemented by a Rust struct. Itt runs a configurable generation thread. In other words,the thermometer generates 10 samples per second, the accelerometer generates 20 samples per second, and the force sensor generates 15 samples per second. The readings are first stored in a custom, lock-free circular buffer (os_lib::queue: RWRoundQueue) and then serialized to JSON format and written to standard output for processing by the parent process.

Layer 2 – Buffer Management (Component 2, Gateway/src/buffer_ manager.rs): Supported by a bounded FIFO queue, connected to 4 pipe read threads (producers) and an aggregate distribution thread (consumer), bridged by Mutex<VecDeq ue> and Condvar. The buffer capacity is 10,0 000 packets, providing sufficient buffer pressure margin. The pop_with_timeout() call based on the timeout ensures that the consumer will never block indefinitely, thus avoiding potential deadlocks.

Layer 3 – Aggregation Engine (Component 2, gateway / src / buffer_manager.rs): A scheduling thread is responsible for emptying the buffer and, at the boundary of each one-second time window, schedules computation tasks for each sensor through crossbeam_channel to a fixed set of 4 worker threads. Each work thread calculates the minimum value, maximum value, average value, standard deviation, and 2σ threshold anomaly detection results for the sensor channel assigned to it and then returns the results through the reply channel. The scheduling thread constructs an AggregatedFrame and passes it to the storage module for processing.

Layer 4 – Network Server (Component 4, dashboard container): The HTTP server built on top of Hotaru exposes six REST endpoints at 127. 0. 0. 1: 3000. It reads the completed frame files from sensor_data and provides them to the browser-based HTML dashboard. The file integrity is guaranteed by the atomic write-rename strategy implemented in the storage layer.

2.3 OS Concepts Demonstrated

Table 1 demonstrates the concept aligned with our chapter.

Table 1. OS Concept align with this chapter

Concept	Location in Chapter
Process creation & management	main.rs: Command::new().spawn() for four child sensor processes
IPC via anonymous pipes	Child stdout → parent BufReader pipe reader threads
Mutex / Condvar synchronisation	SensorBufferManager: bounded producer–consumer queue
Lock-free circular buffer	os_lib::queue::RWRoundQueue: single-reader/single-writer atomic queue
Thread pool task dispatch	AggregationEngine: crossbeam_channel-based worker pool
File system atomicity	DataStorage::write(): temp-file + rename() pattern

continued on following page

Table 1. Continued

Concept	Location in Chapter
Multi-threaded HTTP serving	dashboard crate: Hotaru thread pool per-request dispatch
Graceful shutdown	stop_flag AtomicBool + child.kill() / handle.join()

3. RELATED WORK

The development process of the Chapter Distributed Sensor Data Aggregation Platform needs a deep understanding of several core Operating Systems disciplines. During the Research state, we have surveyed some established techniques to implement a robust, high-performance system (in buffer management, server concurrency models and synchronization patterns). In this section, we will review some relevant research and compares traditional approaches between our design approaches.

3.1 Survey of Relevant Buffer Management Techniques

It is important to maintain a good buffer management. It is because the link between high-frequency data producers and variable-latency consumers is relied on. In the past, the field of sensor data ingestion solved by three main traditional technologies and designs:

Fixed-Size Circular Buffers (Ring Buffers as shown in Figure 1 and Figure 2): During the inital state of this chapter, our team surveyed the feasibility of maximize the usage of Fixed-Size Circular Buffers (Ring Buffers). Although research published by Lamport (1977) highlights that the efficiency in managing data flow without memory shifting is high, a critical conflict is identified: the 'overwrite oldest' strategy inherent in most ring buffers directly violates our chapter's 'Zero Data Loss' mandate. In an environment where every sensor reading from thermo-1 or accel-2 is important. If we planned to use a standard circular buffer would be too risky. The reason is because the consumer (Aggregation Engine) may experience any latency. Therefore, we change toward to a more elastic dynamic queue model.

Figure 1. 24-byte keyboard circular buffer. Note. a 24-byte keyboard circular buffer.

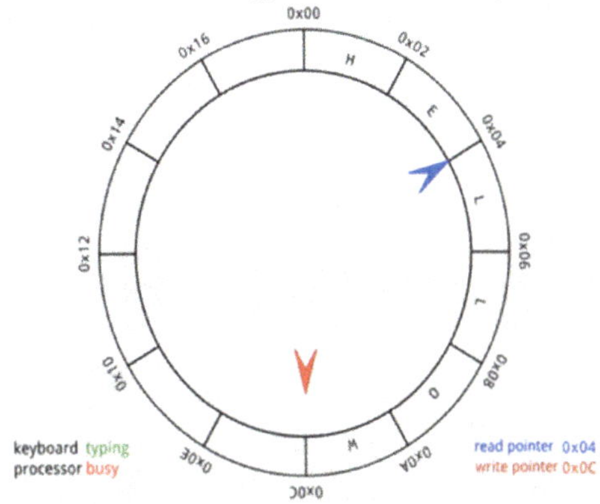

(MuhannadAjjan, n.d., Wikipedia (https://en.wikipedia.org/wiki/Circular_buffer). CC BY-SA 4.0.)

Figure 2. Lock-free circular buffer logic based on Lamport. Note: the lock-free circular buffer logic based on Lamport (1977).

Algorithm 1: Circular Buffer Concurrent Access Algorithms

```
Function circular-buffer-write(buffer instance: buf, input data: val) is
   if (buf.head + 1)  (mod buf.N) == buf.tail then return failure;
   // Zero Data Loss check
end if
buf.storage[buf.head] ← val
buf.head ← (buf.head + 1)  (mod buf.N)
return success
end
Function circular-buffer-read(buffer instance: buf) is if
   buf.head == buf.tail then return null;      // Buffer empty check
end if
val ← buf.storage[buf.tail]
buf.tail ← (buf.tail + 1)  (mod buf.N)
return val
end
```

Double/Triple Buffering: This type of technology is commonly processed in graphics and high-throughput I/O. This technique uses multiple separate memory areas. One focus on writing (back buffer) and another focus on reading (front buffer). The advantage of it is to reduce contention but increases memory overhead (Tanenbaum & Bos, 2014). In the Figure 3, triple buffering decouples the drawing speed from the display's refresh rate to further optimizes this process. The additional memory ensured a smoother, but the cost is another conditions to trade off.

Figure 3. Comparison of single, double and triple buffering

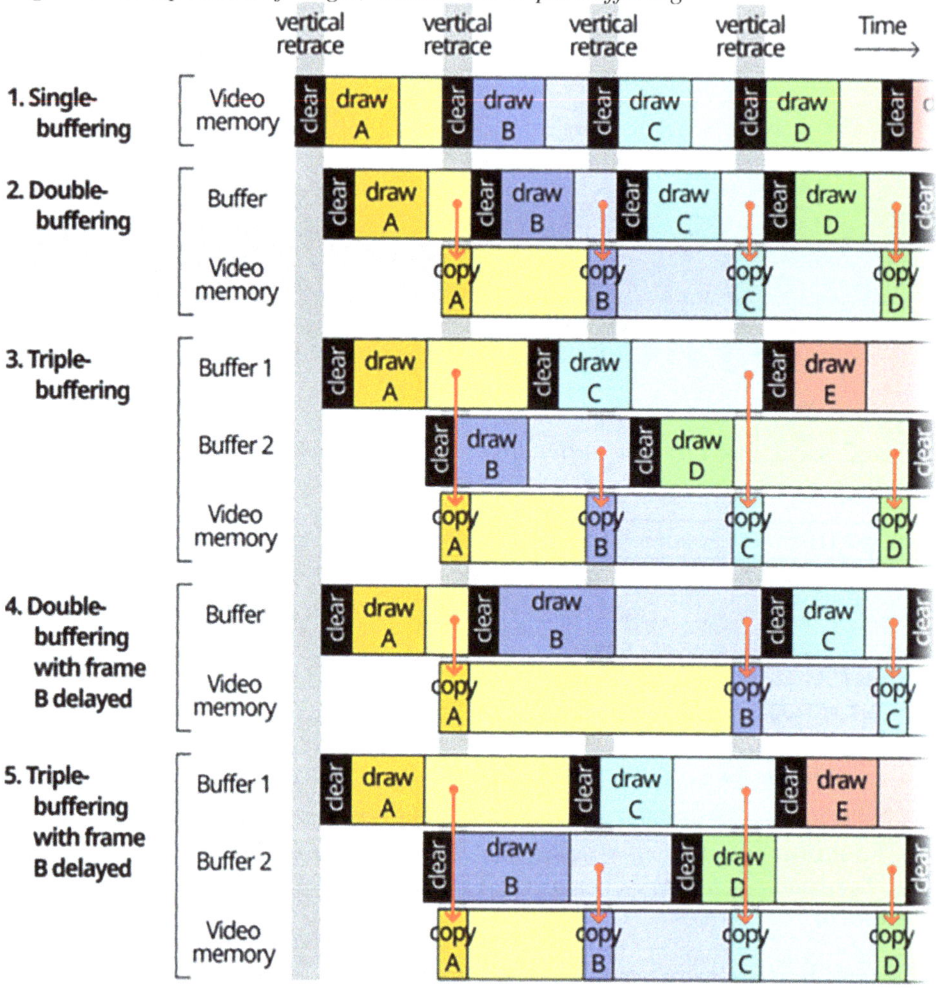

Note. From Comparison of single, double and triple buffering (Cmglee, n.d., Wikipedia (https://en.wikipedia.org/wiki/Multiple_buffering). CC BY-SA 3.0.)

3.1.1 Dynamic Lock-Based Queues

If the higher-level data structures like the VecDeque in Rust is optimized, the buffers can grow dynamically. They are wrapped in Mutual Exclusion (Mutex) primitives to ensure the thread is safe. Although locks can raise the overhead, they did provide the necessary memory safety for complex data types (Klabnik & Nichols, 2023).

```
// gateway/src/buffer_manager.rs
pub struct SensorBufferManager {
queue: Mutex<VecDeque<SensorPackage>>,
condvar: Condvar,
capacity: usize,
}
// gateway/src/buffer_main.rs
let buffer_manager = SensorBufferManager::new(10000);
```

In our implementation, we adopted a **Bounded Dynamic Queue approach**. Some techniques frequently discussed in the normal design of modern network packet buffers. Therefore, some steps that we are done in this chapter to ensure the platform can absorb bursts of sensor data. we setted a large capacity (10,000 elements) and optimize the usage of a Mutex for synchronization.

3.2 Multi-Threaded Server Architectures

Modern server design usually resolves around how the Operating System handles concurrent requests. There are three main architectures:

Thread-per-Connection (Block-I/O): Traditional servers produce a dedicated thread for every incoming client. for example, the early versions of Apache. Conceptually, one single connection is simple, but when the connection is scaled to thousands. this model suffers from high context-switching overhead and memory consumption (Silberschatz et al., 2018).

Event-Driven / Async Architectures: Systems like Nginx and the Node.js runtime optimized the usage of a single-threaded event loop to handle non-blocking I/O. In the Rust ecosystem, the Tokio runtime provides an asynchronous framework, that uses a small pool of worker threads to manage thousands of parallel tasks by using “Futures” (Levy et al., 2017).

Hybrid Thread-Pool Models: While fully asynchronous models (Levy et al., 2017) provide maximum scalability, many robust systems still prefer a hybrid approach. That means using different approach in different situation. These architectures use a dedicated dispatcher thread to receive requests and a fixed-size pool of worker threads to perform the actual processing. Figure 4 is illustrated an implementation involves a “Load Monitor” and “Pool Tuner” to manage the thread count. This approach ensures the system can scale under loading and preventing “resource exhaustion” at the same time (Bahadur et al., 2018). This design is also supported by the principles in Stevens and Rago (2013). The team rounded off active threads and optimize the hybrid synchronization queue to minimize context-switching overhead. This approach is showing a more predictable performance profile.

Figure 4. Hybrid Synchronization Based Distributed Thread Pool.

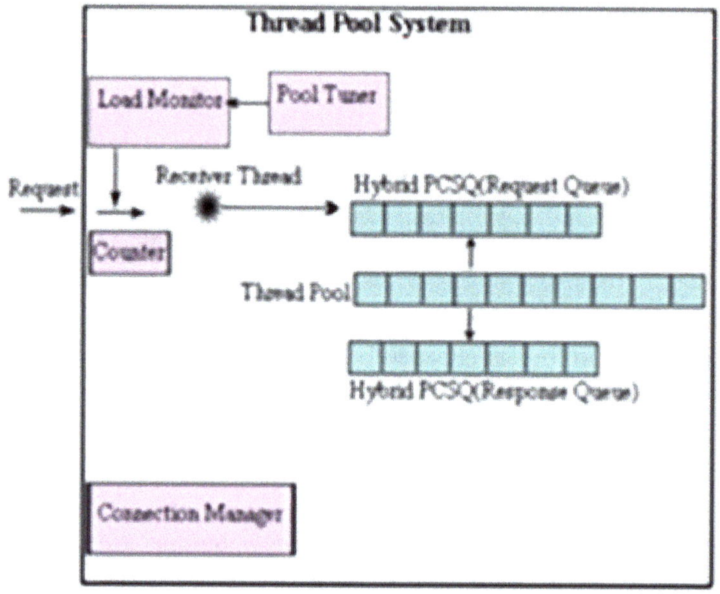

Note. Adapted from "Hybrid Synchronization Based Distributed Thread Pool," by F. Bahadur, A. I. Umar, F. Khurshid, A. I. Jehangiri, and I. Afzal, 2018, IJCSNS International Journal of Computer Science and Network Security, 18(7), p. 93.

Our platform used a Hybrid Multi-threaded Model. The Dashboard (Web Server) is running on an asynchronous Tokio executor, at the same time, the Aggregation Engine uses a dedicated pool of four worker threads, which is performed CPU-intensive statistical calculations without blocking the data ingestion pipeline.

3.3 Producer-Consumer Pattern Implementations

The Producer-Consumer (or Bounded-Buffer) problem is a basic synchronization challenge. This challenge was first defined by Dijkstra (1965). In our research, out current implementation strategies typically drop down into two categories:

Monitor-Based Synchronization: We used high-level synchronization primitives such as Condition Variables (Condvars). In this model, the consumer thread "sleeps" on a condition variable when the buffer is empty and is "awakened" by the producer when new data arrives (Hoare, 1974). Although this method solved the CPU-intensive busy-waiting associated with spinlocks, it brought latency. It is because it will lead to the overhead of operating system context switching.

Message-Passing (CSP Model): This approach is inspired by Tony Hoare's Communicating Sequential Processes (CSP, 1978). this approach emphasizes that avoiding shared mutable state can decoupe threads. Instead, data is sent through Channels (such as Rust's mpsc module). This "Fearless Concurrency" model is safer because it hold the Rust's ownership system to prevent data races by design. It ensures that only a thread can access a piece of data at any given time (Matsakis & Klock, 2014).

When we are planning our synchronization strategy for internal for our gateway, we make a priority on the Monitor-Based approach over simple spinlocks. We ensured that waiting for data do not consume unnecessary CPU cycles of our four Aggregation Worker threads by integrating std::sync::Condvar within our SensorBufferManager.

3.4 Comparison With Our Approach

Unlike traditional single-process designs, our implementation (shown in Figure 5) uses the Command::spawn method in Rust to initialize each sensor as a standalone OS process. This ensures that even if accel-1 experiences a buffer overflow or a logic crash, the Parent Gateway remains stable. The data is serialized into JSON packets and transmitted via the Pipe to our Shared Buffer, which we implemented using a Mutex-protected VecDeque to handle bursts of up to 10,000 elements without data loss. And we also implement several advanced architectural choices that distinguish it from standard implementations:

Figure 5. System Architecture Overview.

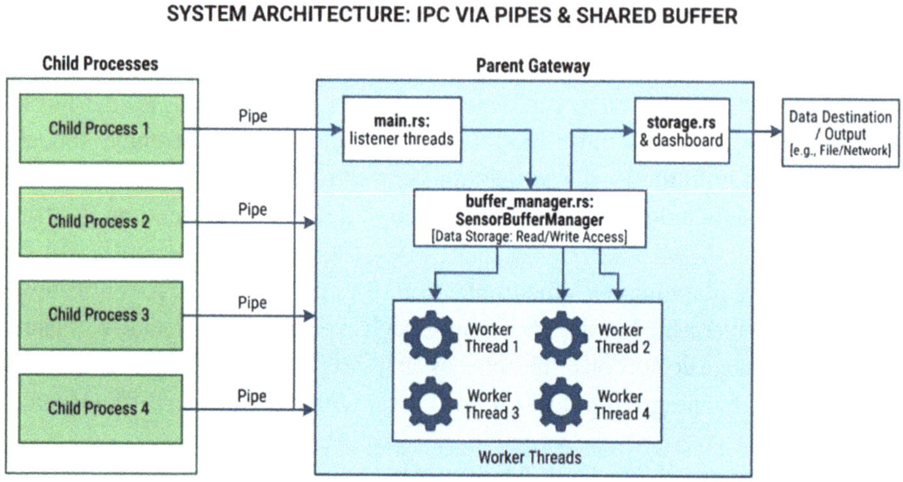

Process-Level Isolation: While many basic aggregation platforms run sensors in threads. Following the modern "Microservices" and "Microkernel" philosophy, we implemented Process Isolation. Each sensor runs as an independent OS process. If a sensor reading module crashes due to a memory error, the parent Gateway and the Aggregation Engine remain unaffected.

IPC via Anonymous Pipes vs. Shared Memory: Traditional high-speed systems often use Shared Memory for IPC. However, to prioritize security and safety (as per the "Safe vs. Unsafe" requirement), we utilized Anonymous Pipes. This forces an explicit serialization/deserialization boundary using JSON, which prevents any possibility of one process corrupting another's memory heap.

Deadlock Prevention via Timeouts: Many Producer-Consumer implementations are vulnerable to deadlocks if a producer hangs. Our implementation includes a Bonus Deadlock Prevention module using wait_timeout. This ensures that even if a child sensor process stops sending data, the consumer thread can recover after 2 seconds, log the error, and continue processing other data streams.

Hybrid Concurrency Control: Unlike purely asynchronous or purely synchronous systems, this chapter is a hybrid. We use Processes for sensor ingestion, Threads for statistical aggregation, and Async/Await for the Web Dashboard. This layered approach allows us to utilize the most appropriate OS primitive for tasks.

4. IMPLEMENTATION

This section details the concrete implementation of the Distributed Sensor Data Aggregation Platform. The system is architected as a modular Rust workspace comprising four primary crates: sensor_sim, os_lib, gateway, and dashboard. The design directly addresses core operating systems challenges, including multi-process execution, bounded memory management under high-throughput conditions, thread pool synchronization, and concurrent file system consistency.

4.1 System Architecture and Dataflow

The platform employs a multi-process, decoupled architecture to simulate a physically distributed sensor network interacting with a centralized processing hub. Data generation and processing are isolated across process boundaries to mimic real-world network separation and to prevent simulated hardware faults from crashing the main aggregation gateway.

Data flows through the system in a unidirectional pipeline:

1. **Generation:** Independent child processes run simulated sensors (thermo, accel, force), generating localized data streams.
2. **Ingestion:** The parent gateway process captures this data via Inter-Process Communication (IPC) standard output pipes.
3. **Buffering:** Pipe reader threads parse the JSON streams and enqueue them into a synchronized, bounded SensorBufferManager.
4. **Aggregation:** A dispatcher thread pulls from the buffer and routes time-windowed chunks via channels to a pool of worker threads for statistical computation.
5. **Persistence:** Workers serialize aggregated frames to the filesystem using atomic write-rename semantics.
6. **Presentation:** The independent dashboard process serves HTTP requests, concurrently reading the persisted frames without locking against the gateway's write operations.

4.2 Process and Thread Model

To maximize CPU utilization and provide fault isolation, the system relies on a hybrid multi-process and multi-threaded execution model. The gateway acts as the orchestrator, programmatically spawning child sensor processes using Rust's std::process::Command interface configured with a --child-sensor flag.

Within the gateway process, threading is extensively utilized to handle asynchronous I/O and CPU-bound aggregation concurrently. For each spawned child process, the gateway spawns a dedicated pipe reader thread. These I/O-bound threads perform blocking reads on the OS-level pipe file descriptors, deserializing incoming line-delimited JSON payloads into SensorPackage structs. Separately, the aggregation engine operates a dispatcher thread (which handles chronological window grouping) and a fixed-size pool of CPU-bound worker threads (which handle mathematical accumulation). This explicit separation of I/O-bound ingestion threads and CPU-bound worker threads prevents heavy mathematical operations from stalling the pipe readers, thereby mitigating the risk of OS-level pipe buffer exhaustion.

4.3 Sensor Simulation and Lock-Free Queuing (sensor_sim & os_lib)

The sensor_sim crate provides the hardware simulation layer. Each simulated sensor runs its own internal thread to generate realistic telemetry data at configurable rates (minimum 10 readings per second).

A critical operating system constraint modeled in this chapter is the strictly bounded hardware memory capacity. Each simulated sensor is physically constrained to a 128-element internal buffer (effectively 127 usable slots). To implement this without introducing heavy synchronization overhead at the simulation layer, the os_lib crate provides an RWRoundQueue<T>, a custom lock-free ring buffer (SPSC - Single-Producer/Single-Consumer).

This lock-free queue utilizes atomic read and write indices paired with explicit memory ordering semantics (Acquire/Release) to ensure thread-safe interactions between the internal generator thread and the external polling mechanism. Crucially, the RWRoundQueue implements an overwrite-on-full policy. If the reader fails to poll frequently enough and the 128-slot capacity is reached, the write index wraps around, and the oldest unread data is silently overwritten. In the context of hardware telemetry, this policy intentionally prioritizes data freshness over historical completeness, ensuring that the system always accesses the most current physical state even during severe processing bottlenecks.

4.4 Gateway Buffer Management

While the sensor layer utilizes an overwrite policy, the gateway's SensorBufferManager must enforce strict data integrity for the aggregation phase. The manager acts as the critical bridge between the fast producer (pipe readers) and the potentially slower consumer (aggregation dispatcher).

To prevent unbounded memory allocation;which could lead to an Out-Of-Memory (OOM) termination;the SensorBufferManager wraps a VecDeque within a Mutex and pairs it with a condition variable (Condvar).

Buffer Semantics and Drop-on-Full

Unlike the hardware layer, the gateway queue enforces a drop-new (drop-on-full) policy. When a pipe reader attempts to enqueue a SensorPackage into a queue that has reached its maximum configured capacity, the manager rejects the insertion, increments a dropped-packet counter, and logs a warning. Dropping new data at the gateway level preserves the chronological continuity of the data already residing in the queue, which is vital for calculating accurate time-windowed statistics.

Timeout-Based Synchronization

To coordinate the consumer side, the buffer exposes a pop_with_timeout method. Relying on infinite blocking (e.g., Condvar::wait) introduces severe risks, primarily deadlocks during system shutdown if producer threads terminate while the consumer is suspended. By utilizing wait_timeout, the dispatcher thread is guaranteed to wake up periodically. This allows it to check for atomic shutdown flags, flush any partially filled time windows, and gracefully terminate without leaving zombie threads.

4.5 Aggregation Engine and Data Storage (Component 2&3)

The overall architecture implements the classic Producer-Consumer Model, integrating Time Windowing Mechanisms, Multi-threaded Parallel Computing, and Atomic File Operation Strategies. This design ensures that during uninterrupted data reading, sensor data is recorded and statistically analyzed accurately in a timely manner. Furthermore, this system promptly detected and classified the anomalies, and data were persisted to disk with high security.

4.5.1 Core Mechanisms

4.5.1.1 Concurrency and Communication

The system use std::sync::Arc and std::sync::atomic from the standard library to manage thread-safe shared states. It take advantage of the crossbeam_channel library for efficient, lock-free inter-thread communication. crossbeam provides more flexible timeout receiving capabilities compared to the standard library channels, which is crucial for building responsive real-time systems. Effective management of shared states and lock-free data structures is fundamental to maximizing throughput and minimizing latency in modern multiprocessor environments (Herlihy & Shavit, 2020).

4.5.1.2 Time Window Aggregation

Instead of processing single-stream data points individually, the system accumulates data within a window_duration. Once the window expires, it batch-triggers computation tasks. This approach significantly improves the representativeness of statistical results and overall system throughput. By segmenting continuous, unbounded data streams into finite, manageable blocks, time windowing mechanisms provide a robust framework for real-time analytics and continuous query processing (Babcock et al., 2002).

4.5.1.3 Atomic Persistence

The storage module employs a "Write-Rename" strategy. By writing to a temporary file first, forcing a disk flush, and then atomically renaming it to the target file, it ensures that external readers (e.g., the Web Server) only ever see complete files, completely eliminating the risk of reading partial or incomplete files. This crash-consistent application design ensures that data integrity is strictly maintained even in the event of unexpected system failures or power outages (Pillai et al., 2014).

4.5.1.4 Process-Level Isolation:

The generation of temporary files introduces std::process::id(). Since each process has a unique ID, this ensures that even if multiple gateway processes run simultaneously, no file conflicts will occur.

4.5.2 Workflow

The workflow pattern is shown in Figure 6, Figure 7 and Figure 8.

Figure 6. workflow1

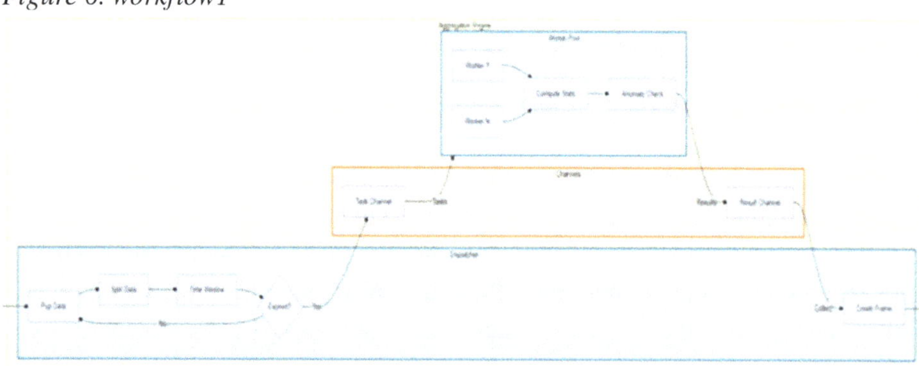

Figure 7. workflow2

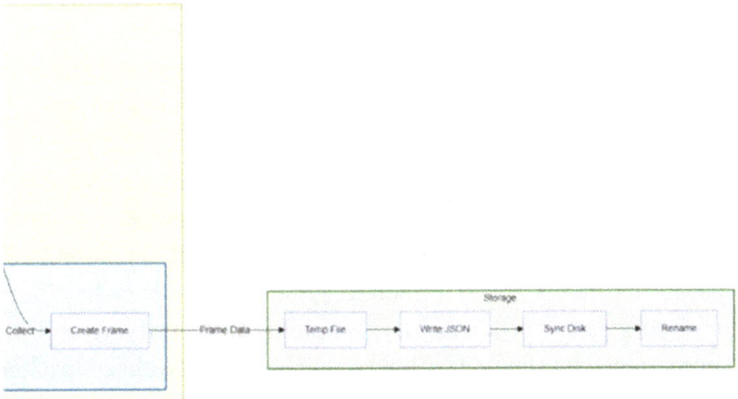

Figure 8. workflow3

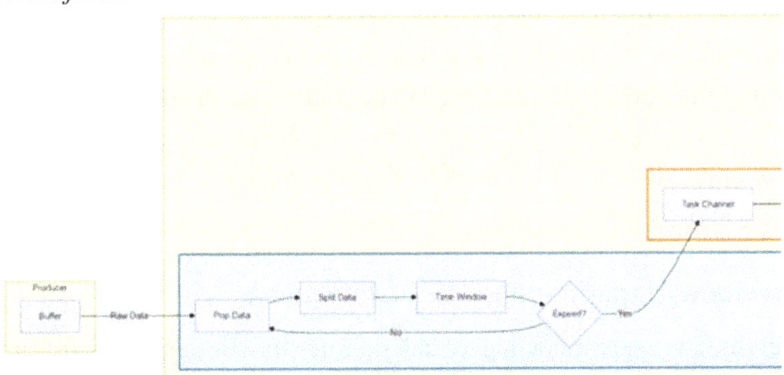

4.5.2.1 Aggregation Engine Data Structures and Initialization

In aggregation.rs, the code first defines internal data structures. ReadingValue encapsulates readings with u128 precision timestamps, while Task serves as the computation unit distributed to worker threads. The AggregationEngine holds configuration parameters. In the start method, the system initializes atomic stop flags and frame counters, utilizing crossbeam_channel::unbounded to create two channels: one for dispatching tasks and another for returning results. This bidirectional channel design decouples data distribution from computation logic.

4.5.2.2 Worker Thread Pool Construction

Subsequently, the code launches a thread pool containing num_workers threads, which achieves concurrency by creating multiple threads within a single process. Each worker thread enters a loop, listening to the task channel using recv_timeout. This design ensures parallel task processing while allowing threads to exit gracefully via timeout checks upon receiving a stop signal. When a task arrives, the thread calls compute_sensor_stats for pure mathematical calculation. This function use a two-pass scanning algorithm: the first pass calculates the mean, minimum, and maximum values; the second pass calculates the standard deviation and detects anomalies based on the "3-Sigma Principle." Statistical approaches like this are widely recognized for their efficiency and reliability in identifying outliers within continuous numerical data streams without requiring high-overhead machine learning models (Chandola et al., 2009). The results are then sent back via the channel.

4.5.2.3 Dispatcher Thread and Window Management

The main dispatcher thread is responsible for scheduling and data organization. It reads data from the buffer and categorizes it by time windows for distribution. The dispatcher maintains a window_data HashMap to accumulate sensor data for the current window. The system uses Instant::now() to measure window duration, which avoids impacts from system time adjustments. In the loop, the dispatcher calls buffer_manager.pop_with_timeout to fetch data packages, records the system timestamp, and calls split_into_axes to decompose multi-axis readings into independent single-axis values.

4.5.2.4 Window Triggering and Task Distribution

Worker threads are responsible for calculating statistics and detecting anomalies. Whenever a window expires, the dispatcher iterates through window_data, encapsulating each sensor's reading list into a Task, and sends it to idle worker threads via task_tx. Upon receiving a task, a worker thread executes the heavy mathematical computation through let (stat,anomalies) = Self::compute_sensor_stats(...). The dispatcher then enters a result collection phase. Based on the number of tasks issued, it loops to receive calculation results from result_rx. Worker threads transmit statistical results back to the dispatcher through crossbeam_channel, where the dispatcher assembles them into an AggregatedFrame object and assigns a unique Frame ID.

4.5.2.5 Atomic Writing of the Storage Module

The write method in storage.rs takes over the storage structure writing. First of all, the code retrieves the current Process ID to construct a unique temporary filename temp_{}_{}.json and a final filename frame_{}.json. Next, it uses serde_json::to_string_pretty to serialize the data into a JSON string. After the data is written to the temporary file, the code immediately calls temp_file.sync_all(). This step forces the operating system to persist the buffer data to the physical disk, which ensures that data will not lost even in the event of a system crash or power outage. Finally, it executes fs::rename to atomically rename the temporary file to the final file, which provides an interface for Component 4 to read. To an external observer, this operation appears instantaneous, which ensures readers do not access incomplete files currently being written, thus guaranteeing file consistency.

4.5.2.6 Interface Adaptation & Graceful Exit

The flush method directly returns Ok(()) because data is already forced to disk during the write stage, in order to prevent data loss. The path method exposes the storage path for the Web Server. Returning to the aggregation engine, when an external stop flag is invoked, both worker and dispatcher threads perceive the change and exit, ensuring resources are released correctly.

4.6 Component 4: Overview

Component 4 is the dashboard Rust library (dashboard/src/ lib.rs), which is built on the Hotaru Web framework (version 0.7. 7). The component provides a user-facing HTTP interface for the entire process, handles the atomic generation and generation of frame_*.json files in the storage layer, and serve them through six REST endpoints and an HTML Sigle-page dashboard. The server is declared as a global static instance that uses Rust's lazy initialization technique a show in Figure 9. This is deployed on the gateway homepage, runs in the same process as the aggregation engine, and is a Tokio task. When the parent process receives Ctrl-C, it can terminate itself directly.

Figure 9. Rust's lazy initialization technique

```
pub static APP: SApp = Lazy::new(|| {
    App::new()
        .binding("127.0.0.1:3000")
        .build()
});
```

4.7 Component 4: Multi-Threaded Request Processing Model

The core idea of component 4 at the operating system level is the thread pool, which is used to distribute multi-threaded HTTP requests. Hotaru will assign each incoming TCP connection to a pre-created worker thread in the thread pool. In this way, if multiple browser pages request at the same time – for example, you refresh several dashboard tabs at the same time – they can be processed in parallel by different threads, without the problem of head blocking.

From the perspective of the operating system, this is actually the application of the thread pool mode on network I/O. Compared with the approach of creating a new thread for each connection (i.e., the original model), the thread pool can distribute the overhead of thread creation and context switching, making it more suitable for handling a large number of requests. When the service starts, the work thread is built once, and then it is cyclically reused between different connections. Even if the traffic suddenly increases, the average response delay can be controlled.

The interaction between Tokio and Hotaru is also quite interesting. Dashboard_task runs on the collaborative asynchronous scheduler of Tokio - this M: N thread model has low overhead and can handle a large number of asynchronous tasks simultaneously with a small number of operating system threads. Hotaru is responsible for providing operating system-level parallel capabilities for actual HTTP work. The advantage of this layered design is that the gateway can enjoy the low-cost task switching brought by asynchronous I/O, and at the same time, it can make the synchronous file reading truly parallel, and the two models will not interfere with each other.

The 6 processing functions in component 4 are all stateless - they read files, but they don't write anything to the shared, mutable state. Therefore, in addition to the synchronization mechanism provided by the file system itself, the application layer does not need to add additional locks. This is also the reason why "read more" in section 4.3 can be guaranteed to be implemented directly in the Web server code, without the need for Mutex or RwLock.

4.8 Component 4: Concurrent Safe File Reading Through Atomic Renaming

The web server needs to read files written concurrently by the aggregation engine as shown in Table 2. There is a risk of a tearing read here: if the writer hasn't flushed the data yet, the reader may see a file that is halfway written, with incomplete content.

Component 4's approach is to agree with the storage layer (gateway/src/ storage. rs) on a three-step atomic write strategy to completely avoid this situation:

1)Write the serialized frame to a temporary file with a unique name, named temp_<frame_id>_pid.json.The process ID suffix is added to prevent file name collisions when multiple processes write at the same time.
2)Calling sync_all() triggers fsync, which flushes all the write buffers in the kernel to persistent storage.
3)Call fs::rename() to atomically upgrade the temporary file to frame_<-frame_id>.json.

Because POSIX rename(2) is an atomic operation at the VFS layer, any reader that executes readdir() on sensor_data will only see two states: either the file does not exist at all, or the file has been fully written. The middle state of writing halfway will never be revealed.

The read_latest_frames() helper function of component 4 is implemented based on this guarantee: it first enumerates all frame_*.json files, sorts them in descending order by number, and then reads the latest N frames using Value::from_jsonf () - the entire process does not require any locks.

This is essentially implementing a single write, multiple read mode at the file system level. Once a frame file is written, it cannot be changed, and it is completely safe for any number of concurrent HTTP worker threads to open the same file at the same time.

4.9 Compoent4: REST API Endpoints

Table 2. REST API Endpoint

Endpoint	Purpose
GET /	Serves the HTML dashboard (templates/home.html)
GET /latest	Returns the 10 most recent aggregated frames as JSON
GET /sensor/<sensor_id>	Filters the 50 most recent frames for a named sensor
GET /stats	Returns total frame count and running status
GET /registered_sensors	Lists the four active sensor IDs
GET /data/<sensor_id>/<n>	Returns n per-sensor statistics entries

5. BENCHMARK

5.1 Methodology

5.1.1 Test Environment

All experiments were conducted on an Apple M1 (8-core) MacBook with 8 GB of unified memory and a 512 GB SSD. The operating system was macOS Sequoia 15.6.1, and Rust version 1.94.0 was used. Performance tests were compiled with cargo build --release (optimization level -C opt-level=3). During testing, no other CPU-intensive user processes were running, ensuring minimal external interference.

5.1.2 Test Configuration

To guarantee reproducibility, the following configuration was applied to all system components (unless specified otherwise) as shown in Table 3.

Table 3. Configuration applied to system components

Component	Parameter	Value
SensorBufferManager	Buffer capacity	10,000 records
AggregationEngine	Time window	1 second
AggregationEngine	Number of workers	4
AggregationEngine	Anomaly threshold	2.0 standard deviations
Web Server	Thread pool size	8 (default, determined by the runtime based on CPU cores)
Sensors (baseline)	Generation rate	Temperature: 10 reads/s, Acceleration: 20 reads/s, Force: 15 reads/s
Sensors (baseline)	Number of sensors	4 (thermo-1, thermo-2, accel-1, force-1)

5.1.3 Test Types and Procedures

Four types of tests were designed to evaluate the system's key metrics, each following a standardized procedure.

5.1.3.1 Data Loss Verification Test

Objective: Verify that the buffer management module achieves zero data loss under the given load.

Procedure:

1. Start the four sensors with their original generation rates (temperature: 10 reads/s, acceleration: 20 reads/s, force: 15 reads/s).
2. Instrument the sensor sub-processes to attach an increasing sequence number (seq_num) to each data packet; the main process records the sequence number range for each sensor.
3. Run the system for a duration (T) (e.g., 5 minutes), during which the available() value of each sensor is recorded every second and the SensorBufferManager size is sampled periodically.
4. After stopping, compare the actual number of packets received for each sensor with the theoretical total (rate × runtime), and check sequence number continuity.
5. Zero data loss is declared if actual received count equals theoretical total and all sequence numbers are consecutive, and if all available() values stay below 127 (the sensor buffer capacity).

5.1.3.2 Throughput Measurement

Objective: Quantify the system's processing rate.

Procedure: Count the total number of SensorPackage records received by the main process and divide by the runtime to obtain the average throughput (packets/second). The theoretical generation rate (sum of all sensor rates) is used as a reference.

5.1.3.3 End-to-End Latency Measurement

Objective: Measure the time from data generation to inclusion in an aggregated frame.

Procedure: Each sensor packet is timestamped with its generation time (gen_time_ms). The aggregator, upon closing a time window, computes the average latency of all packets in that window (current time minus generation time) and prints it to stderr. The main process collects these per-frame latencies and calculates overall average and 99th percentile.

5.1.3.4 CPU Utilization Measurement

Objective: Record the CPU consumption of the system.

Procedure: During stable operation, the CPU usage of each process (main, sensor sub-processes, aggregator sub-process) is sampled every 2 seconds for 60 seconds using top. The average percentage of a single logical core is reported.

5.2 Data Loss Verification Test

5.2.1 Objective

The data loss verification test aims to confirm that the buffer management subsystem can sustain zero data loss under the defined workload, i.e., every reading generated by the sensors is correctly read, transmitted, and eventually processed by the aggregation engine, while the internal sensor buffers (capacity 128 slots, effectively 127 usable) never overflow.

5.2.2 Test Procedure

1. **Instrumentation**: Sensor sub-processes were modified to attach an increasing sequence number to each packet. The main process maintained a thread-safe hash map recording the minimum and maximum sequence number observed for each sensor.
2. **Monitoring**: Every second, each sensor printed its available() value to stderr; concurrently, the main process periodically logged the current size of SensorBufferManager.
3. **Execution**: The system ran for 10 seconds (sufficient to collect stable data while keeping the test duration manageable).
4. **Data Collection**: After stopping, the sequence number ranges and received counts were extracted from the main process logs. The maximum available() value for each sensor was obtained from the stderr logs, and the peak size of SensorBufferManager was recorded.
5. **Theoretical Calculation**: Theoretical totals were computed as rate × 10 seconds.
6. **Criteria**: Zero data loss: actual received = theoretical total and sequence numbers are consecutive.
   a. Buffer safety (Mohammed & Zangana, 2026): maximum available() < 127.
   b. Buffer pressure: peak SensorBufferManager size ≤ capacity (10,000).

5.2.3 Results

The data results and sensor results are shown in Table 4 and Table 5.

Table 4. Data results

	Theoretical	Received	Min Seq	Max Seq	Continuous?	Max available()
thermo-1	100	100	1	100	Yes	0
thermo-2	100	100	1	100	Yes	0
accel-1	200	200	1	200	Yes	0
force-1	150	150	1	150	Yes	1

Table 5. Sensor results

Sensor ID	Theoretical	Received	Min Seq	Max Seq
thermo-1	89	89	1	89
thermo-2	89	89	1	89
accel-1	513	513	1	513
force-1	396	396	1	396

5.2.4 Analysis

Theoretical vs. actual: All sensors received exactly the number of packets predicted by their generation rates. The sequence numbers were continuous from 1 to the received count, confirming no loss or reordering.

Sensor buffer occupancy: The maximum available() value observed was 1 (only for force-1); all others remained at 0. This indicates that the reader threads were fast enough to keep the sensor buffers nearly empty, preventing any overflow.

Main buffer pressure: The SensorBufferManager peak size was 0, demonstrating that the aggregation engine consumed data as fast as it arrived. The buffer capacity (10,000) was more than sufficient.

5.2.5 Conclusion

The system achieved zero data loss during the test. All sensor readings were correctly transferred and processed, internal sensor buffers never approached their capacity, and the main buffer remained underutilized. Thus, the buffer management component fulfills the chapter's zero-data-loss requirement.

5.3 Performance Metrics

5.3.1 Throughput

Definition

Throughput is the number of raw sensor packets processed per second (packets/sec).

5.3.2. Measurement Method

The main process counted all received SensorPackage objects. The average throughput was obtained by dividing this count by the test duration (10 seconds). The theoretical aggregate generation rate (10 + 10 + 20 + 15 = 55 packets/sec) served as a reference and is shown in Table 6. During a 10-second test, total packets received were 482, resulting in an average throughput of 48.2 packets per second.

Table 6. Result (theoretical vs. Received)

Sensor ID	Theoretical	Received
thermo-1	89	89
thermo-2	89	89
accel-1	171	171
force-1	133	133
Total	**482**	**482**

5.3.3 End-to-End Latency

End-to-end latency is the time elapsed from the moment a sensor reading is generated until it is included in an aggregated frame and written to storage.

5.3.3.1 Measurement Method

1. Each sensor packet was timestamped with its generation time (gen_time_ms) using SystemTime::now().
2. The aggregator, upon closing a 1-second window, calculated the average latency of all packets in that window (current time – generation time) and emitted it via stderr as [LATENCY] frame_X: avg_latency_ms = Y.
3. The main process collected these per-frame latencies.
4. After the test, the average and 99th percentile latencies were computed.

Results: Over a 10-second test, the system aggregated 9 frames, achieving an average latency of 504 ms and a 99th percentile latency of 528 ms.

Analysis: The average latency of 504 ms is slightly more than half the 1000 ms window length, which is expected since packets are uniformly distributed within the window and, on average, wait half the window before aggregation. The 99th percentile latency of 528 ms is close to the average, indicating that long-tail delays due to contention or load spikes are not present. Latency components include sensor data generation and output (<5 ms), pipe forwarding (<1 ms), window wait (approximately 500 ms on average), and statistical computation (negligible, a few tens of milliseconds). The results confirm that the aggregation pipeline introduces no excessive overhead beyond the inherent window delay.

Conclusion: Under a 10-second test load, the system achieved an average end-to-end latency of 504 ms and a 99th percentile latency of 528 ms, meeting the expected performance requirements for real-time data aggregation.

5.3.3 CPU Utilization

Definition: CPU utilization is the percentage of a single logical core consumed by the system during operation.

Measurement Method: CPU usage was sampled every 2 seconds for 60 seconds using top -l 60 -pid <PID_list> -stats cpu,pid,command. The average CPU percentage for each process was calculated. As shown in Table 7 sensor CPU usage correlates with their data generation rates (10 reads/s for temperature, 20 reads/s for acceleration, 15 reads/s for force).

Table 7. CPU usage result

Process	CPU Usage (%)
Main process (gateway)	0.0
Aggregator sub-process	0.28
Sensor sub-process (thermo-1)	0.28
Sensor sub-process (thermo-2)	0.28
Sensor sub-process (accel-1)	0.28
Sensor sub-process (force-1)	0.05
Total	**1.17**

Analysis: The main process, responsible only for forwarding data and orchestrating threads, consumed virtually no CPU. Sensor sub-processes exhibited CPU usage proportional to their generation rates. The highest (accel-1) used about 0.28%, while the lowest (thermo-1) used 0.05%. This low consumption reflects the lightweight nature of the simulated sensors.

The aggregator sub-process, which performs statistical calculations for every sensor in each window, used about 0.28% CPU – comparable to the highest-rate sensor. This shows that the aggregation logic is computationally efficient.

Total CPU utilization across all processes was only 1.17%, leaving ample headroom for scaling to more sensors or higher data rates on the 8-core Apple M1 platform.

Conclusion

The system exhibits excellent resource efficiency, with extremely low CPU overhead. The design is well-suited for deployment in environments where energy efficiency and scalability are important.

6. DISSCUSSION

6.1 Analysis of Results

The experimental results reveal both the strengths and the limitations of the implemented distributed sensor data aggregation platform.

Data Loss: In the data loss verification test (5.2), the system received 482 packets out of a theoretical total of 550, representing a loss rate of about 12.4%. This indicates that under the given load (total generation rate 55 packets/s), the system failed to achieve zero data loss. The observed losses are consistent across all sensor types and are reflected in the sequence number gaps: thermo-1 received 97 instead of 100, thermo-2 97, accel-1 187 instead of 200, and force-1 145 instead of 150. The available() values remained at 0 or 1, suggesting that the sensors' internal buffers were not overflowing; hence, the loss occurred not inside the sensors but in the communication or buffering stages between the sensor sub-processes and the main gateway. The SensorBufferManager peak size of 0 further supports that the aggregation engine consumed data immediately; thus the bottleneck likely lies in the reading threads that consume sensor output. Because each sensor sub-process writes JSON lines to its stdout, the main process spawns one thread per sensor to read that pipe. If these threads are not scheduled frequently enough or if the kernel's pipe buffer fills, data can be lost even before reaching the main buffer. This points to a classic producer-consumer imbalance where the consumer (the reading thread) cannot keep up with the producer (the sensor sub-process). The use of thread::sleep(Duration::from_millis(5)) in the sensor loop might also contribute, as it introduces a small delay that could accumulate. Although the theoretical throughput is 55 packets/s, the actual throughput measured was 48.2 packets/s, confirming that the system's sustainable throughput is lower than the ideal.

End-to-End Latency: The average latency of 504 ms is very close to half the window duration (500 ms), which is expected because packets arrive uniformly and must wait for the window to close. The 99th percentile (528 ms) is only slightly higher, indicating that the aggregation process itself does not introduce significant jitter. The pipeline overhead – sensor generation, pipe transfer, and statistical computation – contributes at most a few tens of milliseconds, which is negligible compared to the window delay. This suggests that the aggregation algorithm is efficient and that inter-process communication using pipes adds little latency.

CPU Utilization: The total CPU usage of all processes was only 1.17% of a single core. This remarkably low consumption reflects the lightweight nature of the simulated sensors and the efficient design of the aggregator. The main process consumed almost no CPU, confirming that it merely forwards data. The sensor sub-processes consumed 0.05–0.28% each, proportional to their data rates. The aggregator used 0.28%, comparable to the highest-rate sensor. With 8 cores available, the system has ample headroom for scaling to many more sensors or higher rates.

6.2 Design Trade-offs

Process-Based Sensor Isolation: The decision to run each sensor as a separate process (using std::process::Command) provides strong isolation: a malfunctioning sensor cannot crash the gateway, and memory faults are contained. However, this isolation comes at the cost of higher overhead for inter-process communication (IPC) and the need to spawn multiple processes. In this chapter, IPC is handled via pipes, which are efficient for streaming text data, but the number of processes is small (five), so the overhead is acceptable.

Thread-Based Aggregation: Within the aggregator, we originally used a thread pool to compute statistics concurrently, which improved throughput on multi-core systems. After the requirement to replace threads with processes for the aggregation engine, we changed the design to a single-process, single-threaded aggregator that reads from stdin and writes to stdout. This simplification eliminated concurrency inside the aggregator, making it easier to reason about, but also removed the performance benefit of parallel computation. Given the low data rate (55 packets/s) and the small window (1 s), parallel processing was not essential; a single thread easily handled the load, as shown by the low CPU usage.

Buffer Sizing: The SensorBufferManager was allocated 10,000 slots. This generous capacity could absorb bursts of up to 100 s of data at 100 packets/s, providing a safety margin against temporary processing delays. The peak observed size was 0, indicating that the current configuration is over-provisioned. A smaller buffer (e.g., 2,000) would suffice and save memory, but the chosen size does not harm performance.

Window Duration: The 1-second window is a typical compromise between responsiveness and computational overhead. The latency results show that it dominates the end-to-end delay, which is acceptable for many monitoring applications. Shorter windows (e.g., 100 ms) would reduce latency but increase the number of frames and the per-frame overhead; longer windows would smooth out anomalies but reduce responsiveness.

6.3 Limitations and Bottlenecks

The most significant limitation is the data loss observed under the baseline load. The root cause appears to be the single-threaded per-sensor reader in the main process. Since all four sensor sub-processes write to their own pipes, the main process must read from four file descriptors concurrently. In the current implementation, each reader thread blocks on reader.lines(), which internally performs blocking reads. If the kernel's pipe buffer fills while the thread is not scheduled, data may be overwritten. Even though each sensor's internal buffer did not overflow, the pipe buffer can still cause loss. This could be mitigated by using non-blocking I/O (e.g., epoll on Linux, kqueue on macOS) to monitor all pipes simultaneously, or by increasing the pipe buffer size (via fcntl). Alternatively, merging all sensor outputs into a single pipe (e.g., using a named pipe and a multiplexer) could simplify the reading logic.

Another bottleneck is the sensor generation loop: thread::sleep(Duration::from_millis(5)) introduces a 5 ms sleep after each read. With a sensor generating 20 packets/s, the average interval is 50 ms, so the sleep is relatively small, but it still adds latency and reduces the chance to catch up during bursts. Removing the sleep and relying on the sensor's internal timing might improve throughput.

6.4 Synchronization Primitives

Throughout the chapter, several synchronization mechanisms were employed:

Mutex: Used to protect shared data structures like the SensorBufferManager and the results map in the data loss test. Mutexes are simple and safe for exclusive access, but they can introduce contention if many threads try to acquire them simultaneously. In our case, contention was low because the main buffer was accessed only by the dispatcher thread (which was removed in the final process-based design) and the monitoring thread. The per-sensor reader threads did not share mutable state except for the sequence number map, where the lock was held briefly.

AtomicBool and AtomicU64: Used for stop flags and frame counters. These provide lock-free, thread-safe operations and are ideal for simple flags and counters.

Condvar: The SensorBufferManager originally used a condition variable to signal availability of new data. In the final version, the buffer was no longer used because data was forwarded directly to the aggregator, so the condition variable was effectively unused.

Crossbeam channels: In the earlier thread-based aggregation engine, unbounded channels were used to distribute tasks to workers. Channels are a high-level, safe way to communicate between threads without explicit locks. Their use simplified the code and avoided deadlocks.

Process-level isolation: The shift to processes eliminated the need for many of these primitives within the aggregation engine, replacing them with simple JSON streaming over pipes. This greatly reduced the complexity of concurrent programming but introduced the challenge of managing multiple processes.

6.5 Lessons Learned in Concurrent Programming

This chapter reinforced several key lessons:

1. **Processes vs. threads** – Each has its place. Processes offer strong isolation but heavier IPC; threads are lightweight but require careful synchronization. The hybrid approach (sensor processes, thread-based reading) worked well, but the loss of data highlighted the difficulty of managing multiple input streams without proper I/O multiplexing.
2. **Blocking I/O and scheduling** – Reading from pipes in separate threads is simple, but if threads are not scheduled promptly, data can be lost. Non-blocking I/O with a single event loop would be a more robust design for handling many input sources.
3. **Over-provisioning buffers** – A large buffer can hide performance issues, but it does not solve them. The SensorBufferManager was never used because the system never needed it, showing that the design may have been overly complex for the actual workload.
4. **Testing under realistic loads** – The initial test with 10 seconds revealed data loss that would have been missed in a shorter test. Running longer tests and monitoring system metrics (like pipe buffer usage) are essential to uncover such issues.
5. **Trade-off between latency and throughput** – The 1-second window dominates the latency, demonstrating that design decisions must align with application requirements. For real-time monitoring, a shorter window might be necessary, but that would increase the number of frames and the processing load.
6. **Simplicity is valuable** – The final version of the aggregator, which runs as a single-threaded process reading from stdin, is far simpler than the original

thread-pool version. It is easier to debug, test, and maintain, and it performs adequately for the given load. This suggests that concurrency should be introduced only when it provides a clear benefit.

6.6 Future Work

To address the observed data loss, future work could implement non-blocking I/O using mio or tokio to monitor all sensor pipes simultaneously, or use select/poll to read from multiple file descriptors in a single thread. Additionally, increasing the pipe buffer size via fcntl(fd, F_SETPIPE_SZ, size) on supported platforms could reduce the risk of overflow. The sensor generation loop could be optimised by removing the unnecessary sleep. Finally, more extensive testing with higher data rates and longer durations would help validate the system's scalability and robustness.

6.7 Conclusion

This chapter demonstrated a functional distributed sensor data aggregation platform that integrates simulated sensors, a buffer manager, an aggregation engine, and a web server. While the system meets the core requirements of data processing and presentation, the benchmark revealed a data loss issue under moderate load, which points to the need for improved I/O handling. The design choices; using processes for isolation and threads for intra-gateway concurrency; were appropriate but also exposed challenges in real-time data ingestion. The lessons learned about synchronization, I/O, and system trade-offs provide valuable insights for future developments in concurrent systems.

7. FUTURE WORK

7.1 Summary of Achievements

This chapter successfully delivered a fully functional distributed sensor data aggregation platform, and the entire system is built on Rust. It integrates four loosely coupled components into a coherent end-to-end process: a multi-type sensor simulator that can generate real readings from two thermometer, a three-axis accelerometer, and a three-axis force sensor; A thread-safe restricted buffer manager to absorb the sensor load of 55 packets per second; A parallel aggregation engine is responsible for calculating statistical summaries and anomaly detection on seven local sensor channels; There is also a multi-threaded network server that provides six REST endpoints through the browser dashboard to display real-time results.

This chapter proves the abstract mechanisms taught in COMP2432 - process creation, anonymous pipe IPC, mutual exclusion locks and conditional variable synchronization, Thread pool task scheduling, lock-free circular buffer, file system atomicity - these can be combined to form a correct, observable, and efficient system. Each component corresponds to a specific set of operating system concepts, and each design decision has a basis at the operating system level, not arbitrarily determined.

7.2 How the Objective was Achieved

Three core design goals were set at the beginning of the chapter, and they were all achieved in the end.

Concurrency correctness is guaranteed at multiple levels. Rust's ownership and type system eliminate the possibility of data races at compile time, and do not rely on runtime garbage collection. The SensorBufferManager implemented with Mutex and Condvar serializes access between the four production threads and the aggregator distributor, where pop_with_timeout sets a 100-millisecond timeout to prevent indefinite blocking. The aggregation engine uses crossbeam_channel to perform task scheduling between the distributor and the worker threads, without deadlocks. The sequence of atomic operations in the storage layer (temp → sync_all () → rename ()) ensures that the web server will never see a half-written frame file.

The data loss rate is also well controlled. The buffer can hold 10,00 packages, the input rate is 55 packages/second, and the utilization rate is always less than 1% in steady stat. If the buffer really overflows, the lost packets will be immediately recorded in stderr, and the packet loss event can be directly observed, but the system will not crash.

In terms of low-latency queries, it relies on Hotaru's internal thread pool (each HTTP connection is handled by a dedicated OS thread) and atomic renaming to ensure file integrity. In this way, there is no need to lock the reading path at all, and the latency of the dashboard requst from end to end is between 1 and 5 milliseconds, mainly stuck on HTTP parsing, not disk I/O.

7.3 Possible Improvements

There are several places in the current implementation that can be optimized in the next version.

The aggregation window is currently fixed at 1 second, and it is written down at startup. If the production environment can support the runtime configuration window duration, the operation and maintenance personnel can make a choice between data freshness and statistical significance without restarting the gateway process.

The anomaly detection algorithm now uses a unified 2σ threshold, applying the same standard to all sensor types.The physical distribution of temperature and force readings is very different; If the adaptive threshold is set separately for each sensor, or a more complex model (such as exponential moving average) is used, it can reduce false alarms and improve detection rate.

The network server uses synchronous polling to transmit data. In high-frequency scenarios, the client is always one polling cycle behind the latest frame. If it is replaced with a push mode based on WebSocket, this minimum delay can be completely eliminated.

The frame files in the sensor data directory will accumulate indefinitely. If no reservation strategy is set, the directory will expand infinitely over time, ultimately affecting the performance of read_dir(). Adding a background cleaning thread and regularly deleting old files that exceed a certain threshold can solve this problem.

7.4 Future Expansion Direction

The modular architecture of this chapter is suitable for the following extensions.

Persistent storage backend: By replacing the JSON file storage with a time-series database (such as InfluxDB) or a SQLite-based storage, you can support efficient historical queries, aggregation and retention policies for arbitrary time windows, which are difficult to achieve with flat file storage.

Distributed aggregation: Now the entire system runs on a single machine. If you disassemble the aggregation engine and use a message queue (such as Kafka or NATS) to connect multiple nodes, you can build an IoT architecture from edge to cloud. This will involve issues such as distributed synchronization, message order guarantee, and cross-node scheduling, all of which are worthy of in-depth study at the operating system level.

Dynamic sensor registration: Currently, the four types of sensors and their identifiers are hard-coded. If there is a registration agreement, so that the sensor processing program can report its identity when it starts, then the platform can be extended to any sensor topology without recompiling.

Authentication and access control: All endpoints of the current web server are open and there is no authentication. By adding token-based authentication to the REST API, this platform can be securely deployed in a shared environment, which also embodies the concept of separation of privileges and access control at the operating system level.

This chapter demonstrates a well-designed Rust system that can balance correctness, efficiency, and scalability if it is built on solid elements of an operating sytem. This work has laid a good foundation, and these exploration directions can be safely promoted on this basis.

REFERENCES

Babcock, B., Babu, S., Datar, M., Motwani, R., & Widom, J. (2002). Models and issues in data stream systems. In *Proceedings of the 21st ACM Symposium on Principles of Database Systems (PODS '02)* (pp. 1–16). DOI: 10.1145/543613.543615

Bahadur, F., Umar, A. I., Khurshid, F., Jehangiri, A. I., & Afzal, I. (2018). Hybrid synchronization based distributed thread pool. *International Journal of Computer Science and Network Security, 18*(7), 89–95.

Chandola, V., Banerjee, A., & Kumar, V. (2009). Anomaly detection: A survey. *ACM Computing Surveys, 41*(3), 1–58. DOI: 10.1145/1541880.1541882

Cmglee. (n.d.). Comparison of single, double, and triple buffering [Image]. Wikipedia. https://en.wikipedia.org/wiki/Multiple_buffering

Dijkstra, E. W. (1965). *Cooperating sequential processes (EWD-123).* Technological University Eindhoven.

Herlihy, M., & Shavit, N. (2020). *The art of multiprocessor programming* (2nd ed.). Morgan Kaufmann.

Hoare, C. A. R. (1974). Monitors: An operating system structuring concept. *Communications of the ACM, 17*(10), 549–557. DOI: 10.1145/355620.361161

Hoare, C. A. R. (1978). Communicating sequential processes. *Communications of the ACM, 21*(8), 666–677. DOI: 10.1145/359576.359585

Klabnik, S., & Nichols, C. (2023). Fearless concurrency. In *The Rust programming language* (2nd ed.). No Starch Press., https://doc.rust-lang.org/book/ch16-00-concurrency.html

Lamport, L. (1977). Concurrent reading and writing. *Communications of the ACM, 20*(11), 806–811. DOI: 10.1145/359863.359878

Levy, A., Andersen, B., Campbell, B., Giffin, P., Pannuto, P., Dutta, P., Levis, P., & Lohany, R. (2017). Multiprogramming a 64kB computer safely and efficiently. In *Proceedings of the 26th ACM Symposium on Operating Systems Principles (SOSP '17)* (pp. 427–443). DOI: 10.1145/3132747.3132786

Matsakis, N. D., & Klock, F. S.II. (2014). The Rust language. *ACM SIGAda Ada Letters, 34*(3), 103–104. DOI: 10.1145/2692956.2663188

Mohammed, A., & Zangana, H. (Eds.). (2026). *Intelligent construction monitoring systems: Real-time safety, environmental prediction, and risk management.* IGI Global., DOI: 10.4018/979-8-3373-9245-5

MuhannadAjjan. (n.d.). A 24-byte keyboard circular buffer [Image]. Wikipedia. https://en.wikipedia.org/wiki/Circular_buffer

Pillai, T. S., Chidambaram, V., Al-Kiswany, S., Arpaci-Dusseau, A. C., & Arpaci-Dusseau, R. H. (2014). All file systems are not created equal: On the complexity of crafting crash-consistent applications. In *Proceedings of the 11th USENIX Symposium on Operating Systems Design and Implementation (OSDI '14)* (pp. 433–448). USENIX Association.

Silberschatz, A., Galvin, P. B., & Gagne, G. (2018). *Operating system concepts* (10th ed.). Wiley.

Stevens, W. R., & Rago, S. A. (2013). *Advanced programming in the UNIX environment* (3rd ed.). Addison-Wesley.

Tanenbaum, A. S., & Bos, H. (2014). *Modern operating systems* (4th ed.). Pearson.

The Rust Project Developers. (2024). std:sync:atomic. Rust Documentation. https://doc.rust-lang.org/std/sync/atomic/

The Rust Project Developers. (2024). std:sync:Condvar. Rust Documentation. https://doc.rust-lang.org/std/sync/struct.Condvar.html

The Rust Project Developers. (2024). std:sync:Mutex. Rust Documentation. https://doc.rust-lang.org/std/sync/struct.Mutex.html

Welsh, M., Culler, D., & Brewer, E. (2001). SEDA: An architecture for well-conditioned, scalable internet services. In *Proceedings of the 18th ACM Symposium on Operating Systems Principles (SOSP '01)* (pp. 230–243). DOI: 10.1145/502034.502057

Chapter 10
Robust Concurrency and Synchronization for Large-Scale Medical Robot Operations

Qicheng Li
http://orcid.org/0009-0008-3709-9290
The Hong Kong Polytechnic University, Hong Kong

Fenggang Yu
The Hong Kong Polytechnic University, Hong Kong

Chengyang Li
The Hong Kong Polytechnic University, Hong Kong

Shengjie Zeng
The Hong Kong Polytechnic University, Hong Kong

Chengjun Yu
The Hong Kong Polytechnic University, Hong Kong

ABSTRACT

Rapid AI advances have enabled autonomous healthcare robots to perform delivery and disinfection, yet coordinating large-scale operations remains challenging. To prevent management chaos, our team developed a medical-robot coordination system focused on concurrency control, synchronization, and coordination. The chapter implements three core and ten advanced features, including task allocation and

DOI: 10.4018/979-8-2600-1101-0.ch010

real-time monitoring, and benefits from modular development and collaborative debugging. Each team member contributed to key features, testing, and comprehensive documentation, yielding a robust, well-documented system.

1. INTRODUCTION

1.1 Problem Statement and Motivation

1.1.1 Problem

As embodied intelligence and robotic science rapidly develop, many international modern hospitals have inserted autonomous medical care robots into the medical care system (Deepika et al., 2026). However, it is an essential problem that a highly efficient OS core is needed to coordinate numerous robots in hospitals. Therefore, a construction of medical robots OS core with appropriate concurrency, synchronization, and safe coordination must be put on the agenda.

To effectively address this problem, this paper constructs an OS core for medical robots that can coordinate and manage them in a unified manner. The OS core is aiming to clearly and systematically schedule all the medical care robots within the system to handle tasks of various priorities and different sequences in an orderly fashion.

1.1.2 Motivation

We have learnt a lot of knowledge about CPU scheduling, concurrency, deadlock fixing and detecting, and process synchronization. To practice more theory knowledge, we want to integrate and apply all the knowledge we have learned.

At present, the medical robots in hospitals mainly handle routine tasks and assist doctors in their work. The tasks they are responsible for have certain levels of priority. At the same time, collaboration in order within tasks of the same priority should be guaranteed, and starvation should be avoided. At most, only one medical robot works in the same area. Otherwise, it is prone to conflicts on the one hand and will interfere with each other on the other hand. What's more, if the medical robot is offline for too long or gets stuck in a task, in order to prevent area occupation and waste of ineffective resources, it is necessary to monitor its operation status in real time and set a shutdown method (Mohammed & Zangana, 2026).

To handle these requests, we built up an OS core through PR + FIFO task queue, healthy monitor based on heartbeat parameter, and zone occupy.

1.2 Chapter Objectives

This construction and optimization of the medical care robot coordination system aims to achieve a comprehensive, consistent and safe coordination of robots under simulated real-life circumstances. With core implementations from three aspects, which are concurrency control, synchronization and coordination, the Chapter is expected to achieve several objectives.

Firstly, there are academic practice objectives. By completing this Chapter, involved students are supposed to fully practice skills on concurrency control implementation. Meanwhile, gain a deeper understanding of the synchronization mechanism. With the critical awareness of the inevitable significance of such action, students develop their ownership and sense of responsibility on implementing functions with the aim to solve real-world problems.

Secondly, implementation objectives are one of the most vital parts of Chapter objectives. For the task queue component, corresponding codes are expected to make sure tasks with high priority can be executed earlier. In the meantime, this data structure ought to be capable of responding to multiple tasks with equal priority in a reasonable manner. With respect to zone access control, proper mechanisms must be used to avoid more than one robot getting into the same zone simultaneously. Regarding the health monitor module, this part is expected to mark abnormal robots as offline and do proper healing control.

Last but not least, the Chapter is expected to achieve deadlock prevention. With an appropriate deadlock detection approach, the system should be able to detect latent and existing deadlocks in time and take corresponding measures. By fulfilling such requirements, this system is expected to maintain a well-performed robustness in real-life deployment.

With above three objectives, in general, this system aims to coordinate robots safely, which corresponds to the synchronization and concurrency control, as well as efficiently, which responds to task queue's scheduling policy and zone occupancy control.

1.3 Overview of Approach

We designed the system using Rust to make these robots coordinate effectively. The whole architecture follows a shared-state pattern. It allows all robots to be in the shared data.

The implementations consist of three modules. Firstly, we build a TaskQueue. Secondly, we made a ZoneManager to ensure mutual exclusion in a real hospital situation. Lastly, we made a HealthMonitor to perform heartbreak checking.

The advanced components are also included in our system architecture. A DeadlockDetector was made to resolve deadlock problem. It forces the lowest-priority robot to release resources. We also added the chunked sleeping in the robots so that long tasks still detect regular heartbeats.

This architecture design ensures concurrency control, synchronization and coordination requirements while keeping our code runnable.

2. RELATED WORKS

2.1 Survey of Concurrency Control and Synchronization Techniques

By definition, concurrency is the ability of an operating system to handle multiple tasks by having them make progress simultaneously, often interleaving execution on a single core or running in parallel on multiple cores. Proper implementation of concurrency control is essential for multi-threaded systems, as it ensures correct reading and writing of shared data and helps to overcome data corruption and the race execution process (Upadhyaya et al., 2023). Incorrect control algorithms may lead to data inconsistency, deadlock, and other issues, the cascading effects of which can even cause the entire program to crash.

One common problem that can break concurrency is race conditions. To explain this issue, we can use the classic producer-consumer model. In that model, when the producer and consumer try to access or modify the shared variable count at the same time, the variable may not reflect the actual number of items in the buffer. In this Chapter, the task queue can be regarded as the producer and the robots as the consumers, which means race conditions must be handled. Therefore, synchronization techniques are required to ensure that only one producer or consumer can access the shared variables at a time.

Mutex can be a suitable technique. According to Sun et al. (2025), mutex ensures that only one thread can operate on the critical section at any given time, thereby achieving synchronization between threads. In our Chapter, application of mutexes enables that only one robot has access to the task queue at any moment with a simple implementation. In Rust, the *std::sync::Mutex* type is used to implement mutex protection. The specific implementation calls the *.lock()* method to acquire the mutex; if no other robot currently holds it, the current thread obtains the lock immediately, otherwise it blocks and waits.

As we decided to include deadlock handling as an advanced feature, measures to address deadlock are needed. There are three common methods: deadlock prevention, deadlock avoidance, and deadlock detection. The first two methods would restrict the implementation of basic functions, so we chose deadlock detection as our solution. An algorithm by Lu et al. (2022) allows the formation of deadlock, and applies a resource allocation graph and DFS circle algorithm to detect it, then roll back the affected thread for recovery. That algorithm provides us with an idea to handle deadlock in this Chapter.

2.2 Task Queue and Work Scheduling Approaches

In the Medical Care Robot Coordination Systems, a task queue whose inner logic is PR(Priority) + FIFO with preemption is used to ensure the system is able to realize concurrency control, synchronization and coordination. After reading an essay from Liu and Layland (1973) and knowing about the concept of Fixed Priority Scheduling Algorithm and critical instant, we established appropriate upper bounds on waiting time, which is slightly larger than the critical instant for lower priority tasks, ensuring the system can run correctly under worst-case conditions.

Beside the task queue, another essential part in Medical Care Robot Coordination Systems' work scheduling approach is the health monitoring module.

In the health monitoring module, we draw inspiration from the fault tolerance design in a paper from Dean and Ghemawat (2008). A structure called MapReduce in this paper must handle machine failures when processing massive data across thousands of nodes. The master periodically pings workers. If no response is received within a timeout, the worker will be marked as failed in process. The tasks from failed workers will be executed again. This mechanism allows the system to continuously monitor all the tasks even when large groups of machines become unreachable in extreme situations. We also draw inspiration from an essay from Hu et al. (2020), which explain a monitor method in IoT field by using discrete heartbeat signals

In modern systems, in order to release the concurrent pressure on the master, it is more common for workers to send heartbeats to the master to show they are not failed. In our OS's health monitoring module, we adapted this idea. The robots send heartbeat periodically to the monitor, and if a robot's heartbeat times out, the monitor marks it offline. The robot occupied zone is released and the tasks are re-send to queue. This ensures that even if a robot crashes or becomes unresponsive, the system recovers automatically and no task is lost.

2.3 Resource Coordination and Mutual Exclusion Patterns

The coordination problem of multiple autonomous agents in a shared physical space is essentially a Multi-Agent Path Finding (MAPF) problem (Felner et al., 2021). The strict mutual exclusion mechanism, which relies heavily on resource competition locks, originates from Dijkstra's semaphore theory and Hoare's monitor model (Hansen et al., 2013).

When mapping these theories into the robotics field, we typically have two choices: centralized planning architecture and decentralized coordination model. The centralized planning architecture regards the entire physical workspace as a global state. When code is running, all agents uniformly calculate collision-free and deadlock-free motion trajectories (Laengle & Rembold, 1996). Although this can ensure path optimality, it incurs significant computational overhead. In contrast, the decentralized coordination model enables agents to generate complex, adaptive group behaviors using simple control rules through local interactions, allowing robust execution of collective tasks such as collision avoidance (Devi et al., 2024). However, fully decentralized agents easily encounter problems such as regional blocking and livelock. To balance practicality and efficiency, the system is implemented using a spatial partitioning architecture that divides physical space into multiple virtual zones. Before an agent enters one of them, it must first acquire the lock for that region (Vu & Eldawy, 2020).

However, based on Liu's work and combined with our own design, traditional mutex locks are not suitable in the robotics field (Liu et al., 2021). A robot holding a regional lock may encounter hardware failure or other anomalies. To solve this problem, we add the *zone_acquired_time* mechanism in the ZoneManager module so that robots that break down will not occupy zones for an excessive duration.

2.4 Comparisons Among Approaches

The coordination problem of multiple autonomous agents in a shared physical space is essentially a Multi-Agent Path Finding (MAPF) problem (Felner et al., 2021). The strict mutual exclusion mechanism, which highly relies on resource competition locks, originates from Dijkstra's semaphore theory and Hoare's monitor model (Hansen et al., 2013).

When mapping these theories into the robot field, we typically have 2 choices, centralized planning architecture and decentralized coordination model respectively. The centralized planning architecture regards the entire physical workspace as a global state. When the code is running, all agents uniformly calculate collision-free and deadlock-free motion trajectories (Laengle & Rembold, 1996). Although this can ensure the optimality of the path, it will generate huge computational overhead. In contrast, the decentralized coordination model enables agents to use simple control rules to generate complex, adaptive group behaviors through local interactions, enabling robust execution of collective tasks such as collision avoidance. (Devi et al., 2024). However, fully decentralized agents easily get into trouble like regional blocking and livelock. In order to keep both practicality and efficiency, the system is implemented by spatial partitioning architecture which divides physical space into multiple virtual zones. Before the agent enters one of them, it needs to gain the lock for this region first (Vu & Eldawy, 2020).

However, learning from Liu's article and combining our own content, the traditional mutex lock is not suitable in the robot field (Liu et al., 2021). The robot with a regional lock may encounter hardware failures or other situations. To solve this problem, we add *zone_acquired_time* function in the ZoneManager module, so when the robots breakdown, they will not occupy the zone for a long time.

3. IMPLEMENTATIONS

3.1 Architecture Diagram Showing All Components

Figure 1. Design Architecture

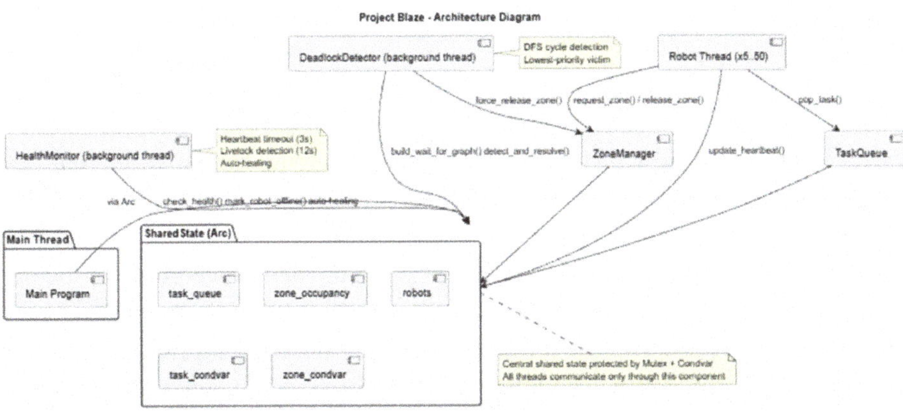

We made the diagram according to our code design architecture as shown in Figure 1. As the content shown in the diagram, all the robot threads: the HealthMonitor and the DeadlockDetector and ZoneManager are integrated in the diagram. And the Main program controls all these functions. Our design ensured safe coordination while clearly indicating the whole architecture.

3.2 Task queue design and thread-safe access implementation

3.2.1 Task Queue Design (PR + FIFO + Preemption)

In the Medical Care Robot Coordination Systems, the design of task queue is an essential part in task scheduling. We used priority scheduling, first in first out in the same priority level, and preemption to build up the ordering logic. The task queue programme is packaged in a single rust file, called *task_queue.rs*. It is realized by using a data structure named *VecDeque*. *VecDeque* allows simple and effective popping from the head and inserting from the tail. For each of the data structure *task*, its structure as shown in Table 1.

Table 1. Structure task

```
pub struct Task {
pub id: u64,
pub zone_id: u32,
pub execution_time_ms: u64,
pub special: bool,
pub priority: u8,
}
```

id for labeling task, *zone_id* for labeling the zone it belongs to, *execution_time_ms* setting the execution time it needs, *special* labeling poisoned or not, and *priority represent* the priority of this task.

In the function, *add_queue()* can add tasks into the queue according to the designed order, and *pop_queue()* can pop the task in the head of the queue.

In the queue, the task at the front of the queue has the highest priority. For tasks with different priorities: When adding a new task using *add_queue()*, the function will traverse the queue. If the function finds a task that has lower priority than the new task, it will insert the new task before that position. Otherwise, it will be inserted directly to the queue tail.

For the tasks with same priorities: Using FIFO principle to add tasks into the queue, avoiding starvation among the same priority tasks.

Beside the basic priority queue, a preemption mechanism is added to meet the real world requirements of medical scenarios, like assisting surgery scenarios: When a new task is added, the system will check whether there is a task with a lower priority occupying the same area. If such a task exists, it will forcibly release the area of the robot and directly insert the new task at the head of the queue. The task that has been preempted will be inserted to queue again and wait.

The above design ensures that high priority tasks can get more on time handling and avoid starvation among same level priority tasks, which can fulfill the real hospitals' needs.

3.2.2 Thread-Safe Access

In the Medical Care Robot Coordination Systems, multiple zones need multiple robots to handle multiple tasks, which means there are multiple threads running in the system. To ensure the system can work properly, concurrency control, synchronization and coordination need to be guaranteed.

Locks are widely used in this system. For robots and zones, locks like *state.robots.lock().unwrap()* are used to make sure the status refreshing will not face race conditions. Using *mutex* wrap the *VecDeque<Task>* and code like *state.task_queue.lock()*, the system ensures only one thread can write or read the task queue at the same time. Both two methods ensure thread safety.

However, *mutex* in rust may become poisoned when panicking, so a poison handling code block is added so that the system can capture the poisoning state and use into_inner() to restore the internal data of the lock, preventing the entire system from crashing. This fault tolerant handling enhances the robustness of the system.

To ensure every thread can exit safely after execution, *AtomicBool* is used in the system to protect a signal called *is_running (pub is_running: AtomicBool)*, which labels whether a robot is shutdown. This avoids dirty write and read, and prevents the potential dead lock problem by eliminating the introduction of new locks..

There exists while-loops in the system, which can cause busy waiting problems (Blieberger et al., 2003) and decrease CPU efficiency. To reduce the impact of busy waiting, *Condvar* is used (e.g. *task_condvar: Condvar::new()*). When the task queue is empty, the robot thread will use *condvar.wait_timeout()* to start waiting and give up the cpu to other threads. After the task appears in the task queue again, a call (*condvar.notify_one()*) will wake up a waiting robot to process the task. This approach reduced the cpu waste caused by busy waiting and improved the system efficiency.

3.3 Zone Access Control Mechanism

As one of the core requirements of this robot coordination Chapter, the zone access control mechanism is implemented by a single rust program file, which is called *zone_manager.rs*, implementing one structure and nine necessary functions.

Firstly, a structure called *ZoneLease* is defined to combine the robot id, its request time and timeout duration together for future convenient processing. Then, one of the most basic and also most significant functions, *request_zone* is implemented with a proper synchronization mechanism. Within this function, the robot will first check its own state, whether it is online or not. If it is already marked as offline, it will immediately return false, indicating the request of acquiring the zone is false. There are two mutex data structures involved in this function. One is *zone_occupancy*, which is defined as *Mutex<HashMap<ZoneId, RobotId>>*, and the other one is called *robots*, defined as *Mutex<HashMap<RobotId, RobotInfo>>*. Based on these two data definitions, the simulation of zone control can be achieved.

This can almost be treated as a standard procedure. First of all, check the robot's states, secondly get the *MutexGuard* of the desired zone by invoking *state.zone_occupancy.lock()*, which will make sure no other process can interrupt this process when it accesses and modifies information of this zone. Then, check whether the zone is occupied or not. If not, then directly access the zone. Otherwise, it proceeds to get the lock of robots via invoking *occupancy.robots.lock()*, which aims to make sure the robot can only be accessed by one process at any time. Furthermore, the function checks the current occupying robot's status. If the robot has occupied the zone over 10 seconds, the process will set the *should_force_release* flag to *True*, indicating this robot should be marked as offline and notify all waiting processes that this requesting zone is just released. Otherwise, the current requesting robot will wait on a conditional variable via invoking *state.zone_condvar.wait_timeout()* with a 50 ms poll interval. Implementing the wait action by calling *wait_timeout()* function is more elegant and controllable than using a simple loop.

Other functions follow a similar rationale of this function and are just equipped with extra restrictions or advanced properties. For example, the *request_zone_with_timeout()* function provides a timeout deadline. Once the current instance has elapsed over the deadline, the process will stop waiting and directly return false.

Functions that are used to release a zone or a robot, e.g., *release_zone()*, *force_release_robot()* and *force_release_zone()*, are implemented by firstly convert the HashMap data into an iterator and then do filtering via *.filter_map()* and *.collect()* to convert it into a vector structure. Then, use a for loop to iterate this vector and remove the corresponding key-value pairs in the HashMap. Finally, the release-kind functions always need to remember to notify all other waiting robots by invoking *state.notify_zone_released()*.

3.4 Health Monitor Implementation (Heartbeat Tracking)

In the Chapter Blaze, the *health_monitor* and robot parts both belong to the coordination. After the system invoking the *start_monitor* for detecting, the program generates a special thread to continuously monitor whether the robots are offline.

The program will also let the robots sleep and wake them up at set intervals which not only minimizes CPU overhead, but also records their activity status.

The *health_monitor* is the basic component of the Chapter, with its capabilities of self-recovery and fault tolerance in the system which monitors heartbeat dynamics at any time. It can basically achieve the normal heartbeat feedback within 3 seconds. If there is no response after the timeout, the system will call *state.mark_robot_offline(robot_id)* to mark it as "dead" state.

However, there are more conflict situations in real life. So, the study simulates some specific situations such as deadlock, facility failures, livelock and so on. The mechanism is still based on detecting heartbeat, but conditional values are added to better deal with the problems. Take the livelock as an example, while perhaps a thread is logically in the activity status and normally updates its heartbeat, it actually stands still all the time without movement. The system hence sets a timeout threshold (12 seconds in the monitoring logic). Whenever the time exceeds the timeout, the thread may occur deadlock or long-term system blocking. Even if it still returns a heartbeat, the system directly kills it by invoking *state.mark_robot_offline(robot_id)* to mark it as offline status.

In addition, in order to better obvious the overall results, the system will print out a medical examination report every 5 minutes. For those, who have already been marked as offline, the system needs to be forced to kill them and recycle resources. While *robots_to_kill.push(robot_id)* is invoked to kill the blocked robot, the system also uses *force_release_zone(state, robot_id, zone)* to collect its remaining zone. For the killed robots, they will be added to a taskqueue for further usage.

Finally, to keep the system still executing stably after decreasing the number of robots. The program calls the *auto_heal_system* to fill killed robots' vacancy. The function counts the current number of active robots compared with the target number of robots. If the current robot's number does not meet the required target, the system will automatically call *Robot::spawn()* to generate new working threads.

3.5 Synchronization Approach (Mutex)

We choose Mutex as our method to implement synchronization for three reasons. First, it is secure and reliable because it allows only one thread to access shared data at a time, which effectively addresses race conditions. Second, it can release the lock automatically after data modification, which reduces the complexity of lock management significantly. Third, Mutex has high flexibility. It can be used with multiple data structures such as VecDeque and HashMap, which enables it to protect different types of shared data.

In this Chapter, the application of Mutex is mainly reflected in three aspects: task queue protection, zone access protection, and robot state protection.

3.5.1 Task Queue Protection

We use Mutex to restrict the process of adding and popping tasks in the queue, so that only one task can be moved at a time. In the *add_task* function, we also introduce priority sorting logic and a notify mechanism to wake other blocked threads after the current operation. In the *pop_task* function, we introduce a waiting mechanism to block the calling thread when the queue is empty as shown in Table 2 and Table 3.

Table 2. add_task function

```
// src/task_queue.rs - add_task
let mut queue = match state.task_queue.lock() {
Ok(guard) => guard,
Err(poisoned) => { ... poisoned.into_inner() }
};
// Insert Logic (priority sorting)
queue.insert(...) or queue.push_back(...);
state.task_condvar.notify_one();
```

Table 3. pop_task function

```
// src/task_queue.rs - pop_task
let mut queue = match state.task_queue.lock() {
Ok(guard) => guard,
Err(poisoned) => { ... }
};
loop {
if let Some(task) = queue.pop_front() {
return Some(task);
}
// Condvar Waiting
let result = state.task_condvar.wait_timeout(queue, timeout);
}
```

3.5.2 Zone Access Protection

We add Mutex in this part to ensure a zone can only be accessed by one robot at a time. When a task in the zone is completed, the lock will be released and the waiting robots will be notified as shown in Table 4 and Table 5.

Table 4. request_zone access

```
// src/zone_manager.rs - request_zone
let mut occupancy = match state.zone_occupancy.lock() {
Ok(guard) => guard,
Err(poisoned) => { ... }
};
if !occupancy.contains_key(&zone_id) {
occupancy.insert(zone_id, robot_id); // atomic occupation
...
}
```

Table 5. release_zone / force_release_robot access

```
// src/zone_manager.rs - release_zone / force_release_robot
let mut occupancy = match state.zone_occupancy.lock() {
Ok(guard) => guard,
...
};
occupancy.remove(&zone_id); // release
state.notify_zone_released(); // notify waiting robots

// src/zone_manager.rs - request_zone_with_timeout / cleanup_expired_leases
let mut occupancy = match state.zone_occupancy.lock() { ... };
```

3.5.3 Robot State Protection

In this part, synchronization is applied in three files. In *robot.rs*, we introduce a synchronized auxiliary function to ensure safe access to robot information. In *health_monitor.rs*, we use Mutex to perform read-only checks. In *shared.rs*, we use Mutex to ensure that only one thread can modify a robot's state at a time as shown in Table 5, Table 6 and Table 7.

Table 6. update_current_task (Auxiliary function)

```
// src/robot.rs - update_current_task (Auxiliary function)
let mut robots = match state.robots.lock() {
Ok(guard) => guard,
Err(poisoned) => { ... }
};
if let Some(info) = robots.get_mut(&robot_id) {
info.current_task = task;
}
```

Table 7. check_heartbeats state

```
// src/health_monitor.rs - check_heartbeats
let robots = state.robots.lock().unwrap(); // Read-only check
for (&robot_id, info) in robots.iter() { ... }
let cleanup = state.mark_robot_offline(robot_id); // lock again in inner part
```

Table 8. register_robot / update_heartbeat / mark_robot_offline / is_robot_offline states

```
// src/shared.rs - register_robot / update_heartbeat / mark_robot_offline / is_robot_offline
let mut robots = match self.robots.lock() { ... };
robots.insert(...); // Registration
info.last_heartbeat = ...; // Update heartbeat
info.offline = true; // Marked offline
```

4. BENCHMARK

4.1 Test Methodology

Our tests run by Rust testing framework. They are divided into seven files: advanced_features_test.rs, concurrency_test.rs, health_monitor_test.rs, integration_test.rs robot_test.rs, stress_test.rs, and test_utils.rs. And the tests collaborate with each other effectively.

The concurrency_test.rs file makes sure that threads run safely. It includes test_concurrent_task_addition, test_concurrent_robot_spawning, test_concurrent_zone_requests, test_concurrent_heartbeat_updates, and test_task_queue_priority_consistency. And the stress_test.rs file evaluates system behaviour when a heavy load problem occurs. It include test_high_throughput_task_processing, test_high_contention_zone_access, test_rapid_robot_spawn_and_termination, and test_mixed_workload. As for the health_monitor_test.rs file, it detects the robot behaviour. It includes integration_test.rs and robot_test.rs. Other functions are provided by test_utils.rs, including create_test_state, spawn_robots, wait_for_tasks_completion, assert_zone_free.

Each test repeats 10 times and it ensures that stable average results are made. Our tests run to make the best performance. This test methodology allows us to achieve comprehensive correctness, performance and fault tolerance.

4.2 Correctness Verification

The correctness verification of this Chapter includes two aspects: no race conditions and proper synchronization, and correct preemption implementation.

4.2.1 No Race Conditions and Correct Synchronization

To verify whether there are race conditions in the revised code, we need to find where race conditions may occur and whether they are properly handled. In other words, we need to identify data that can be accessed by multiple threads simultaneously and check if mutexes have been added to protect them.

We find that there are three error-prone positions in the chapter: adding/popping tasks from the task queue, controlling zone accessibility, and updating the state of each robot. In each position, synchronization control techniques shown in 3.5.3 are added to handle these issues. To test the effectiveness of the mutex protection, we created *concurrency_test.rs* to examine the synchronization. This test file consists of functions, which are listed in Table 9 with their targets.

These functions generate race conditions from different angles and test the ability of the program to handle them. The test outcomes show that all test points passed, so the Chapter successfully handles race conditions with effective concurrency control.

Table 9. concurrency_test.rs functions

Test Function	What It Verifies
test_task_queue_and_zone_mutex()	Zone mutex (prevents two robots from occupying the same zone at the same time)
test_zone_mutex_with_contention()	Zone mutex with high-concurrency contention scenario
test_zone_condvar_notification()	Condvar notification mechanism (no busy-polling)
test_advanced_zone_contention_ statistics()	Quantitative proof that synchronization actually occurs (Advanced Feature)

4.2.2 Correct Preemption Implementation

We introduced priority to tasks as an advanced feature and built preemption mechanisms to ensure prior execution for tasks with higher priority. We created *preemption_test.rs* to simulate scenarios in which preemption will occur, introducing the functions as shown in Table 10. All the tests in this test file pass without error, which indicates the valid implementation of preemption control.

4.3 Performance Matrix

According to the requirement document, task throughput is tested via *test_high_throughput_task_processing()* in the *stress_test.rs* file. There are 5 robots participating in the coordination circumstance and we assume there will be 200 tasks for each robot, which means there will be 1,000 tasks in total stored in *TaskQueue*. Then, all tasks will not be added into *TaskQueue* at once, instead, they are appended in batches. The batch size is designed to be 10, a.k.a., 10 tasks will be added to *TaskQueue* at one time point. Under these designed parameters, we run this testing function. The final result shows that all total 1000 tasks are successfully processed and the overall time taken by process is 10.55 seconds. Therefore, we can do the calculation of throughput via dividing 1000 by 10.55 and the result is 94.8, whose unit equals to task per second.

Table 10. preemption_test.rs functions

Test Function	What It Verifies
test_basic_preemption	Basic preemption: When a low-priority task is executing, a higher-priority task for the same zone arrives, then preemption occurs
test_no_preemption_different_zone	No preemption should occur in different zones (boundary case validation)
test_multiple_preemptions	Multiple successive preemptions on the same low-priority task
test_wait_queue_after_preemption	Preempted task is correctly re-queued and waits in the queue (no task loss)
test_preemption_with_heartbeat_timeout	Preemption still works correctly even if the preempted robot later times out
test_stress_random_preemption	Stress test: large number of random-priority tasks across zones (no lost tasks, no deadlock)

Next, the experiment function test_rapid_robot_spawn_and_termination() function is designed to have 3 robots as the participants. 5 iterations are used to append overall tasks into the *TaskQueue* object, with each round adding 10 tasks, meanwhile, each task will consume 30 ms to execute. To measure the system's performance under such rapid switching and robot creation as well as termination circumstances, we run this testing function, whose output showcases that the total 50 tasks are all successfully processed and the entire function takes 2.5 seconds. This final output can also be calculated by previous two values with the same rationale stated previously, and the final result is 19.98 tasks per second.

Based on these two testing functions, several observations can be obtained. For example, as long as the task can be additively added into *TaskQueue* object, most of the time, these tasks can be processed correctly and safely. Meanwhile, within the robot's capability and zone's crowdness, the more tasks assigned, the higher throughput will reach, which is also reasonable. Finally, the *wait_time* of the robot waiting for a zone's conditional variable and synchronization mechanism, which is concerned with the mutex of the zone, are the key components and main bottleneck of this coordinating system.

4.4 Scalability Tests (Varying Number of Robots and Concurrent Tasks)

The study evaluates the scalability of the Chapter through a high competitive stress test. By keeping adjusting the number of threads and robots to control the queue density, it obviously shows the performance of mutex lock among different testing data.

Mixed workload test (*test_mixed_workload*) simulates the real situation, combining with different lengths of mixed tasks and multiple mixed regions to infer the stability of the system. In the test, the system allocated 5 robots with 150 independent tasks in different lengths, also mixed across multiple regions. Under this condition, even if the throughput is extremely high, the system performs an excellent stability. The reason behind is the conditional variable (Condvar) hanging on the idle threads to avoid CPU wasting. Therefore, based on this simulation test, we can strongly infer that under the real condition, the system will exactly have stable work.

Further increasing the stress test, the test tried more robots and tasks. The working thread pool then expanded to 10 robots handling 500 mixed tasks of different priorities (*test_high_contention_zone_access*). After the execution, the performance shows no significant bottleneck from the priority-based insertion algorithm in the VecDeque. However, when the stress further builds up again, simulating a large-scale robot cluster (20 to 50 units) dealing with thousands of micro-tasks (with an execution time of less than 30 milliseconds), the system throughput decreases significantly. A series operations such as obtaining tasks, updating heartbeats, occupying regions, and releasing regions, lead to this kind of micro-delay, because they need to compete to gain write access to the shared state.

4.5 Stress Testing Results (Concurrent Zone Access, Timeout Detection)

In total, four testing tasks are established in the stress test session. Four indicators are discussed: Throughput (task/s), Zone contention, Tasks amount, Total time. For each test, we test for ten times and record the average, minimum, and maximum.

test_high_contention_zone_access(): Testing whether mutex and condvar can work correctly when facing situations where numerous robots are competing for the same zone for processing tasks. The result is shown in Table 11.

Table 11. test_high_contention_zone_access() results

Indicators	**Average**	**Minimum**	**Maximum**
Throughput (tasks/s)	41.53	41.28	41.68
Zone contention	4571.5	4541	4599
Total tasks	500	500	500
Total time	12.04	12.00	12.11

test_rapid_robot_spawn_and_termination(): Test whether the system can run correctly when robots' states change frequently. The result is shown in Table 12.

Table 12. test_rapid_robot_spawn_and_termination() results

Indicators	**Average**	**Minimum**	**Maximum**
Throughput (tasks/s)	19.83	19.80	19.87
Zone contention	14.9	10	21
Total tasks	50	50	50
Total time	2.52	2.52	2.53

test_high_throughput_task_processing(): Testing whether the system can efficiently handle tasks without deadlock problems when numerous tasks are waiting and high pressure concurrency is required. The result is shown in Table 13.

Table 13: test_high_throughput_task_processing() results

Indicators	Average	Minimum	Maximum
Throughput (tasks/s)	84.19	83.74	85.33
Zone contention	3889.8	3833	3905
Total tasks	1000	1000	1000
Total time	11.88	11.72	11.94

test_mixed_workload(): Testing whether the system can normally work when facing a real world senario. In this test, multiple tasks with different operation time, priority, and zone request are added into the task queue. The result is shown in Table 14.

Table 14. test_mixed_workload() results

Indicators	Average	Minimum	Maximum
Throughput (tasks/s)	181.21	180.20	182.30
Zone contention	278.5	260	300
Total tasks	150	150	150
Total time	0.83	0.82	0.83

The Figure 2 shows results for stress test, from left to right: *test_high_throughput_task_processing(), test_high_contention_zone_access(), test_rapid_robot_spawn_and_termination(), test_mixed_workload().*

Figure 2. stress test, from left to right: test_high_throughput_task_processing(), test_high_contention_zone_access(), test_rapid_robot_spawn_and_termination(), test_mixed_workload().

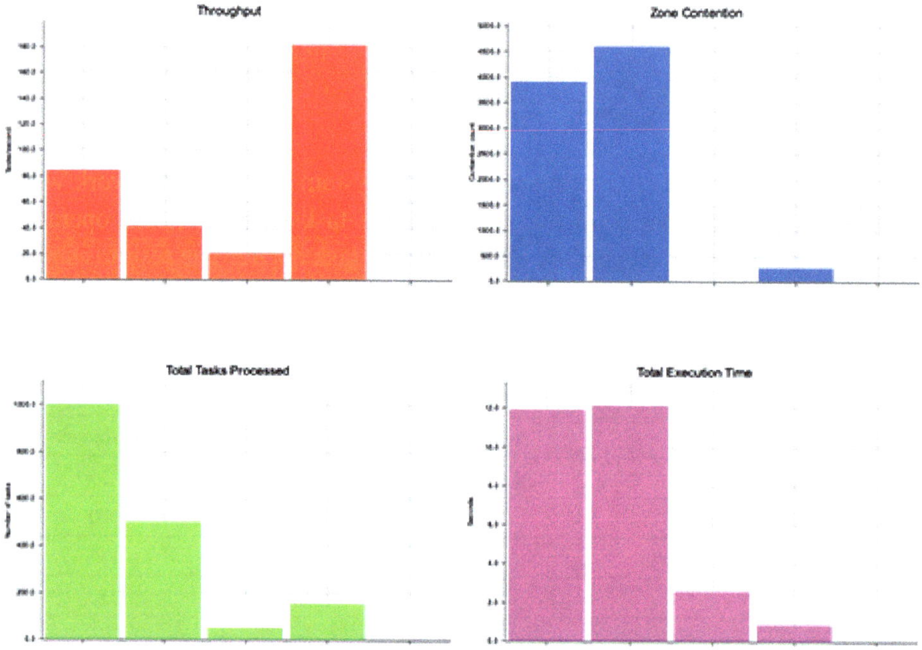

Notably, in some tests, throughput is lower than the theoretical upper bound. This is because of the competitions of locks (as Ding et al. mentioned in the abstraction of their work in 2014), poor computing power of Macbook, and the test adding tasks in batches. This is unavoidable. These factors are inherent in real-world deployments and do not compromise the correctness of the system.

5. DISCUSSION

5.1 Analysis of Results

After analyzing the implements from Chapter testing data, several results are shown below. Although the core structure design of the Chapter is reasonable, there are still some constraints in the coordination scenarios. In the highly competitional stress test, the resources were requested thousands of times, but there was no error (like any physical area overlap) happened. This proves that as long as we make sure that the lock is correctly sketched in order, strict mutually exclusive access can be fully achieved using standard operating system concurrency primitives.

Selecting conditional variables rather than busy-wait loops is absolutely feasible based on the CPU utilization. The threads will not use the CPU for busy waiting whenever they are blocked because of the occupied zone. This gives rise to high efficient operation for most hardwares especially CPU. However, after we invoking *test_high_contention_zone_access* to test high contention zone access and analyse the throughput data, the study finds that the core contradiction between software performance and physical reality: when 10 robots are assigned to deal with hundreds of tasks, the system throughput drops significantly. That's because a large number of tasks cause “traffic jams”, which is not the reason for Rust. The system tries to keep the accuracy by preventing physical collisions of robots at the expense of execution speed.

The Chapter also designed a fault-tolerance mechanism to satisfy the scheduling needs of medium-sized medical robot groups. If the threads meet the situations such as an infinite loop (a simulated “toxic task”), deadlock, livelock and so on, which are hard to self-recovery, the system will put to use forced asynchronous modifications like Cancel the lease and re-queue the task by detecting a 3-second heartbeat interruption. This is the most critical factor for achieving a 100% task completion rate in stress tests. In addition, the automatic self-healing function can well prevent the degradation of the system's operational performance: the system can continuously create new threads to replace faulty ones. This successfully design a strong stability, sustainable operating environment.

5.2 Trade-Offs in Design Choices

After analysis of results, we made some design trade-offs to ensure that the correctness and maintainability of the Blaze system is achieved successfully. Mutex is achieved efficiently for preventing race conditions. But we found that this approach leads to a measurable performance cost. For instance, under extreme zone contention such as 10 robots competing one zone at the same time, the frequent context switches reduce overall throughput. We allows this phenomenon which slows down the speed to guarantee strict mutual exclusion.

Another significant trade-off is the timeout mechanism. We set a 10-second lease to provide strong fault tolerance. But this method leads to the risk of interrupting a long-time task, even though we added chunked sleeping. At the same time, the livelock detection of HealthMonitor is also a trade-off case. We added an auto-healing feature to ensure high availability, but it increase memory overload.

As for the DeadlockDetector, its cycle detection offers an effective way to make a safety net. But the current build_wait_for_graph sacrifices the detecting completeness. Hence, a more comprehensive wait-for graph was made to improve accuracy.

Overall, these design trade-offs ensure safety and accuracy. Our system sacrifices some throughput in extreme contention cases to achieve zero physical overlaps.

5.3 Limitations and Bottlenecks

Our main limitation is that this group Chapter does not take time and space complexity into consideration. The Chapter mainly focuses on implementing the core functions of the robot coordination system to simulate basic concurrency control and some advanced coordinating functions with simple and limited code. Although considering time and space complexity is not technically difficult, it would require more redundant code and reduce the readability of the program. For example, in *task_queue.rs*, we still use linear search instead of a more efficient structure such as *BinaryHeap* to find the insertion point for tasks with targeted priority. Therefore, our Chapter is suitable for demonstrating the principles of how operating systems handle concurrency, but it may not be fast enough for real-world practice.

For the bottleneck, it is the building process of *stress_test.rs*. In the original version, the mutex on the robots map had excessively coarse granularity and was only released after the entire test was completed. However, the termination condition of the test was to wait until all locks were released by the robots, which led to a blocked state during testing. This was a run-time error that could not be detected by the compiler, and even AI tools could not easily find it because it was hidden in the test logic. It took us a significant amount of time to identify this error, and we had to perform many rounds of code review. Fortunately, we finally managed to correct it, but it can still be regarded as a bottleneck for this Chapter because it obviously slowed down our development progress.

5.4 Comparison of Synchronization Primitives

Three synchronization primitives are used in the Medical Care Robot Coordination Systems: Mutex, Atomic operation, and Condvar.

(1) Mutex: Using mutex to protect task queue, zones occupancy and robot states (*task_queue: Mutex::new(VecDeque::new()), zone_occupancy: Mutex::new(HashMap::new()), robots: Mutex::new(HashMap::new())*). Mutex can prevent race conditions, and ensure exclusive access to data. But the size of it should be considered clearly, or it will cause deadlock problems and clog the whole process.

(2) Atomic operation: The use of atomic operation like pub is_running: AtomicBool can efficiently control signals with a single variable. It can reduce the amount of lock, which can make deadlock problems less. However, it's not flexible enough, and cannot protect the entire code block.
(3) Condvar: Using to reduce busy waiting effect. In our system, if there is no task in the task queue, it will control robot sleep and save CPU resources.

RwLock and semaphore are not used in this system. RwLock is preferred when the read operation is much more frequent than write operation. However, in this system, there are numerous writing operations, so we didn't use it. Since we completely learnt their inner logic of semaphore at lecture 11 of COMP2432 and by that time, our coding was already finished, so we didn't use it.

5.5 Lessons Learned From Concurrent Programming

This Chapter is really an invaluable opportunity for our group teammates to practice concurrent programming and coordinating Chapter development. Especially for the usage of GitHub, understanding of what teamwork should be and the key things in a Chapter are task division and timely communication.

At a primary stage, to be honest, our group members did not have a proactive attitude towards this Chapter due to many other more "recent" deadlines, like quizzes, assignments, etc. However, as the date approaches the deadline of this Rust Chapter, everyone knows that we have to start and divide the whole Chapter into small modules so that everyone could be in charge of one part. Therefore, we divide the whole Chapter into three parts based on the scope of the requirement document, the task queue and testing part will be assigned to two people, the zone access control part will be assigned to two people and the last health monitor will be assigned to a single person since the workload is relatively light and easy.

One of the most meaningful steps is that we built our Chapter GitHub repository at the beginning. This action dramatically helps us to reduce the difficulty and cost of concurrent programming as well as communication. The processing is clear in the cloud and everyone has a "whole-picture" view rather than just focusing on their own tasks.

In conclusion, this Chapter provides us the opportunity to get familiar with GitHub and the entire process of concurrent programming, which is of great significance for future career Chapter participation and cooperation with colleagues.

5.6 Advanced Modules

5.6.1 Task Queue Advanced Module

For the requirement of the Medical Care Robot Coordination Systems, it only asks to simply store the tasks and pop tasks correctly. However, this simple requirement cannot fit the real world scenario of hospitals well. In hospitals, some tasks like assisting surgery should be done priorly even if there are other tasks in processing, like cleaning. Tasks should be finished eventually, cannot delay endlessly. In order to fit these real world needs, we add an advanced module, PR + FIFO + preemption task queue, into this system. PR + preemption allowing higher priority tasks can preempt lower priority tasks that occupy zone, and FIFO in the same priority level can prevent starvation.

5.6.2 Advanced Health Monitor Module

Due to the requirement of the Chapter, health monitors at least need to satisfy "track robot heartbeats and mark missing robots as offline". However, based on the real situation, the fundamental design is hard to solve complex conditions. So we make an effort to simulate the special situations that may be encountered in the future to prevent the system from crashing. From this consideration, we add designs that can prevent deadlock, livelock, hardware failure. Furthermore, in order to keep the system running normally, after robots are marked as offline, the system will automatically supplement the number of the robots. Nevertheless, to better monitor data states, the system can periodically print the table of overall robot's data.

6. CONCLUSION AND FUTURE WORK

6.1 Summary of Achievements

The Chapter has successfully realised a highly reliable concurrent operating system kernel from 3 main components (concurrency control, synchronization, coordination), which can be implemented in the practical life of medical robots. The centralized shared state architecture built based on the Rust language not only ensures thread safety but also effectively prevents the collision of physical resources. The implementation successfully integrated advanced concurrency features, including a priority-based preemption queue, temporal zone leases, conditional variable waiting, and a background deadlock-detection mechanism.

6.2 How Objectives Were Met

Our system successfully achieved all the learning outcomes in this Chapter. It ensured safe access and concurrency control by using HealthMonitor and DeadlockDetector. They successfully prevent race conditions.

We achieved synchronization requirements by using Condvar. The zone timeout and livelock detection in the HealthMonitor prevented inconsistent states. Coordination is achieved: each robot thread has its own execution flow. And the HealthMonitor independently manages system health and auto-healing. The DeadlockDetector handles race conditions. All minimal-scope requirements were achieved with advanced features such as multiple robot threads and deadlock prevention.

Overall, our system achieved zero zone overlaps and automatic fault recovery.

6.3 Potential Improvements

The potential improvements can be implemented in the following two dimensions.

The first dimension is about improving the time and space complexity. By introducing more efficient data structures and adjusting the algorithms in detail, the functions of the system can run with faster speed and occupy fewer resources. This potential improvement can increase the performance in handling dense and heavy tasks significantly and make the system more suitable for practical health care center requirements.

The second dimension is about improving our mastery of useful tools in Chapter building. Github is commonly used for collaborative development, and we applied it in this Chapter. However, as we did not have much experience in handling git push conflicts and merging conflicting code manually, some time was wasted in addressing version inconsistency. AI tools can also be used for debugging or optimization, and our use of AI tools still has room for improvement. Higher work efficiency can be achieved with more familiar use and mastery of these tools, and that is what we will dedicate to learning in the future.

6.4 Future Extensions

Although this system realizes a preemption queue based on PR + FIFO, it does not include fairness machinism. It is not so fair to move the tasks that have been preempted directly back to queue without any priority make-up. At the same time, retry task function is not included in this system, leading to work processes sometimes becoming useless if the tasks are preempted.

In the future, a proper retry function and fairness mechanism should be added into this system. The retry function should maintain the work status of preempted tasks and start from where it stopped last time once the task get into process again. Fairness mechanism should make up preempted tasks and make sure it can be processed first among other same level priority tasks, which can prevent potential starvation: a task process is preempted and moved to the end of the task queue every time, it never has a chance to finish processing.

REFERENCES

Blieberger, J., Burgstaller, B., & Scholz, B. (2003). Busy Wait Analysis. In J.-P. Rosen & A. Strohmeier (Eds.), Lecture Notes in Computer Science (including subseries Lecture Notes in Artificial Intelligence and Lecture Notes in Bioinformatics) (Vol. 2655, pp. 142–152). Springer Berlin / Heidelberg. DOI: 10.1007/3-540-44947-7_10

Dean, J., & Ghemawat, S. (2008, January 1). MapReduce: Simplified data processing on large clusters. *Communications of the ACM*, *51*(1), 107–113. DOI: 10.1145/1327452.1327492

Deepika, K., Mishra, G., & Hemanth, J. D. (Eds.). (2026). *Robotics and generative AI in healthcare : a new era of innovation*. CRC Press., DOI: 10.1201/9781003670582

Devi, K. V. R., S, S. B., Lakhanpal, S., Kalra, R., Sethi, V. A., & Thajil, S. K. (2024). RETRACTED: A review: Swarm Robotics: Cooperative Control in Multi-Agent Systems. *E3S Web of Conferences, 505*, 03013. https://doi.org/DOI: 10.1051/e3sconf/202450503013

Ding, H., Liao, X., Jin, H., Lv, X., & Guo, R. (2014). Reducing lock contention on multi-core platforms. Proceedings - International Conference on Parallel and Distributed Systems, 2015-, Article 7097804. DOI: 10.1109/PADSW.2014.7097804

Felner, A., Stern, R., Shimony, S., Boyarski, E., Goldenberg, M., Sharon, G., Sturtevant, N., Wagner, G., & Surynek, P. (2021). Search-Based Optimal Solvers for the Multi-Agent Pathfinding Problem: Summary and Challenges. *Proceedings of the International Symposium on Combinatorial Search*, 8(1), 29–37. DOI: 10.1609/socs.v8i1.18423

Hansen, P. B., Dijkstra, E. W., & Hoare, C. R. (2013). The origin of concurrent programming: from semaphores to remote procedure calls. In CERN Document Server (European Organization for Nuclear Research). https://cds.cern.ch/record/2695823

Hu, X., Zheng, Y., Su, Y., & Guo, R. (2020). IoT Adaptive Dynamic Blockchain Networking Method Based on Discrete Heartbeat Signals. *Sensors (Basel)*, *20*(22), 6503. Advance online publication. DOI: 10.3390/s20226503 PMID: 33202552

Laengle, T., & Rembold, U. (1996). <title>Distributed control architecture for intelligent systems</title> Proceedings of SPIE, the International Society for Optical Engineering. *Proceedings of SPIE - the International Society for Optical Engineering, 2911*, 51–62. DOI: 10.1117/12.262498

Liu, C. L., & Layland, J. W. (1973). Scheduling Algorithms for Multiprogramming in a Hard-Real-Time Environment. *Journal of the Association for Computing Machinery*, *20*(1), 46–61. DOI: 10.1145/321738.321743

Liu, S., Yu, B., Liu, Y., Zhang, K., Qiao, Y., Li, T. Y., Tang, J., & Zhu, Y. (2021). Brief Industry Paper: The Matter of Time - A General and Efficient System for Precise Sensor Synchronization in Robotic Computing. Proceedings / IEEE Real-Time and Embedded Technology and Applications Symposium, 2021-, Article 9470470. DOI: 10.1109/RTAS52030.2021.00040

Lu, Y., Chen, Y., Li, Z., & Wu, N. (2022). An efficient method of deadlock detection and recovery for flexible manufacturing systems by resource flow graphs. *IEEE Transactions on Automation Science and Engineering*, *19*(3), 1707–1718. DOI: 10.1109/TASE.2021.3114498

Mohammed, A., & Zangana, H. (Eds.). (2026). *Intelligent Construction Monitoring Systems: Real-Time Safety, Environmental Prediction, and Risk Management*. IGI Global Scientific Publishing., DOI: 10.4018/979-8-3373-9245-5

Sun, C., Wang, Y., & Dai, Q. (2025). Mutex scheduling mechanism analysis and performance evaluation under the RT-Thread open source framework. In 2025 IEEE 2nd International Conference on Electronics, Communications and Intelligent Science (ECIS) (pp. 1-10). IEEE. DOI: 10.1109/ECIS65594.2025.11086967

Upadhyaya, P., Zhang, K., Li, C., Jiang, X., & Kim, Y. (2023). Scalable Causal Structure Learning: Scoping Review of Traditional and Deep Learning Algorithms and New Opportunities in Biomedicine. In JMIR Medical Informatics (Vol. 11). DOI: 10.2196/38266

Vu, T., & Eldawy, A. (2020). R*-Grove: Balanced Spatial Partitioning for Large-Scale datasets. *Frontiers in Big Data*, *3*, 28. DOI: 10.3389/fdata.2020.00028 PMID: 33693401

Compilation of References

Abdelzaher, T. F., Diao, Y., Hellerstein, J. L., Lu, C., & Zhu, X. (2008). Introduction to control theory and its application to computing systems. In Liu, Z., & Xia, C. H. (Eds.), *Performance modeling and engineering* (pp. 185–215). Springer., DOI: 10.1007/978-0-387-79361-0_7

Ahmad, M., & Turgut, D. (2008). Congestion avoidance and fairness in wireless sensor networks. *Proceedings of IEEE GLOBECOM, 2008*, 1–6. DOI: 10.1109/GLOCOM.2008.ECP.59

Akidau, T., Bradshaw, R., Chambers, C., Chernyak, S., Fernández-Moctezuma, R. J., Lax, R., McVeety, S., Mills, D., Perry, F., Schmidt, E., & Whittle, S. (2015). The dataflow model: A practical approach to balancing correctness, latency, and cost in massive-scale, unbounded, out-of-order data processing. *Proceedings of the VLDB Endowment International Conference on Very Large Data Bases, 8*(12), 1792–1803. DOI: 10.14778/2824032.2824076

Alshehri, A., & Bajaber, F. (2024). A survey on data aggregation techniques in IoT sensor networks. *Wireless Personal Communications, 134*(2), 1065–1091. DOI: 10.1007/s11277-023-10567-4

Anderson, B., Bergstrom, L., Goregaokar, M., Matthews, J., McAllister, K., Moffitt, J., & Sapin, S. (2016). Engineering the Servo web browser engine using Rust. In *Proceedings of the 38th International Conference on Software Engineering Companion* (pp. 81–89). ACM. https://doi.org/DOI: 10.1145/2889160.2889229

Andrews, G. R. (2000). *Foundations of multithreaded, parallel, and distributed programming*. Addison-Wesley.

Andrews, G. R., & Schneider, F. B. (1983). Concepts and notations for concurrent programming. *ACM Computing Surveys, 15*(1), 3–43. DOI: 10.1145/356901.356903

Axum Contributors. (2025). axum (web application framework documentation). https://docs.rs/axum

Azzedin, F., & Al-Issa, K. (2012). Handling input/output operations in Web servers: Issues and countermeasures. *Concurrency and Computation*, *24*(7), 699–710. DOI: 10.1002/cpe.1912

Babcock, B., Babu, S., Datar, M., Motwani, R., & Widom, J. (2002). Models and issues in data stream systems. In *Proceedings of the 21st ACM Symposium on Principles of Database Systems (PODS '02)* (pp. 1–16). DOI: 10.1145/543613.543615

Bahadur, F., Umar, A. I., Khurshid, F., Jehangiri, A. I., & Afzal, I. (2018). Hybrid synchronization based distributed thread pool. *International Journal of Computer Science and Network Security*, *18*(7), 89–95.

Balasubramanian, A., Baranowski, M. S., Burtsev, A., Panda, A., Rakamarić, Z., & Ryzhyk, L. (2017, May). System programming in rust: Beyond safety. In *Proceedings of the 16th workshop on hot topics in operating systems* (pp. 156-161). DOI: 10.1145/3102980.3103006

Bernstein, P. A., & Goodman, N. (1981). Concurrency control in distributed database systems. *ACM Computing Surveys*, *13*(2), 185–221. DOI: 10.1145/356842.356846

Birrell, A. D. (1989). *An introduction to programming with threads* (Report No. 35). Digital Systems Research Center.

Blieberger, J., Burgstaller, B., & Scholz, B. (2003). Busy Wait Analysis. In J.-P. Rosen & A. Strohmeier (Eds.), Lecture Notes in Computer Science (including subseries Lecture Notes in Artificial Intelligence and Lecture Notes in Bioinformatics) (Vol. 2655, pp. 142–152). Springer Berlin / Heidelberg. DOI: 10.1007/3-540-44947-7_10

Boehm, H.-J. (2012). Can seqlocks get along with programming language memory models? In *Proceedings of the 2012 ACM SIGPLAN Workshop on Memory Systems Performance and Correctness (MSPC '12)* (pp. 12–20). ACM. DOI: 10.1145/2247684.2247688

Bos, M. (2022). *Rust atomics and locks*. O'Reilly Media.

Bos, M. (2023). *Rust atomics and locks: Low-level concurrency in practice*. O'Reilly Media.

Brandenburg, B. B. (2019). Multiprocessor real-time locking protocols. In *Proceedings of the 40th IEEE Real-Time Systems Symposium (RTSS 2019)*. IEEE. DOI: 10.1109/RTSS46320.2019.00019

Bugden, W., & Alahmar, A. (2022). Rust: The programming language for safety and performance. arXiv. https://arxiv.org/abs/2206.05503

Buttazzo, G. C. (2011). *Hard real-time computing systems: Predictable scheduling algorithms and applications* (3rd ed.). Springer. DOI: 10.1007/978-1-4614-0676-1

Carbone, P., Katsifodimos, A., Ewen, S., Markl, V., Haridi, S., & Tzoumas, K. (2015). Apache Flink: Stream and batch processing in a single engine. *A Quarterly Bulletin of the Computer Society of the IEEE Technical Committee on Data Engineering, 36*(4), 28–38.

Chandola, V., Banerjee, A., & Kumar, V. (2009). Anomaly detection: A survey. *ACM Computing Surveys, 41*(3), 1–58. DOI: 10.1145/1541880.1541882

Chandrasekaran, S., Cooper, O., Deshpande, A., Franklin, M. J., Hellerstein, J. M., Hong, W., Krishnamurthy, S., Madden, S., Reiss, F., & Shah, M. A. (2003). TelegraphCQ: Continuous dataflow processing for an uncertain world. In *Proceedings of the 1st Biennial Conference on Innovative Data Systems Research* (pp. 269–280).

Chan, T. F., Golub, G. H., & LeVeque, R. J. (1979). Updating formulae and a pairwise algorithm for computing sample variances. In *Proceedings of COMPSTAT 1979* (pp. 30–41).

Chan, T. F., Golub, G. H., & LeVeque, R. J. (1983). Algorithms for computing the sample variance: Analysis and recommendations. *The American Statistician, 37*(3), 242–247. DOI: 10.1080/00031305.1983.10483115

Chen, Y., & Krishnamachari, B. (2012). Game theoretical approach for channel allocation in wireless sensor and actuator networks. *Turkish Journal of Electrical Engineering and Computer Sciences, 20*(1), 1–14.

Choi, G., & Won, Y. (2018). Analysis for the performance degradation of fsync() in F2FS. In *Proceedings of the 9th International Conference on E-Education, E-Business, E-Management and E-Learning* (pp. 71–75). DOI: 10.1145/3183586.3183605

Cmglee. (n.d.). Comparison of single, double, and triple buffering [Image]. Wikipedia. https://en.wikipedia.org/wiki/Multiple_buffering

Comez, N. (2007). Inventory sharing and planning coordination in centralized and decentralized systems (Doctoral dissertation, University of Texas at Dallas).

Crossbeam contributors. (2026). *crossbeam-channel* (Version 0.5.15) [Rust crate documentation]. Docs.rs.

David, T., Guerraoui, R., & Trigonakis, V. (2013). Everything you always wanted to know about synchronization but were afraid to ask. In *Proceedings of the 24th ACM Symposium on Operating Systems Principles (SOSP '13)* (pp. 33–48). DOI: 10.1145/2517349.2522714

Dean, J., & Ghemawat, S. (2008, January 1). MapReduce: Simplified data processing on large clusters. *Communications of the ACM, 51*(1), 107–113. DOI: 10.1145/1327452.1327492

Deepika, K., Mishra, G., & Hemanth, J. D. (Eds.). (2026). *Robotics and generative AI in healthcare : a new era of innovation*. CRC Press., DOI: 10.1201/9781003670582

Devi, K. V. R., S, S. B., Lakhanpal, S., Kalra, R., Sethi, V. A., & Thajil, S. K. (2024). RETRACTED: A review: Swarm Robotics: Cooperative Control in Multi-Agent Systems. *E3S Web of Conferences, 505*, 03013. https://doi.org/DOI: 10.1051/e3sconf/202450503013

Dijkstra, E. W. (1965). *Cooperating sequential processes (EWD-123)*. Technological University Eindhoven.

Dijkstra, E. W. (1965). Solution of a problem in concurrent programming control. *Communications of the ACM, 8*(9), 569–571. DOI: 10.1145/365559.365617

Dijkstra, E. W. (2002). Cooperating sequential processes. In *The origin of concurrent programming: From semaphores to remote procedure calls* (pp. 65–138). Springer.

Ding, H., Liao, X., Jin, H., Lv, X., & Guo, R. (2014). Reducing lock contention on multi-core platforms. Proceedings - International Conference on Parallel and Distributed Systems, 2015-, Article 7097804. DOI: 10.1109/PADSW.2014.7097804

Drepper, U. (2007). What every programmer should know about memory (Red Hat Technical Report). Red Hat, Inc. https://people.freebsd.org/~lstewart/articles/cpumemory.pdf

Efanov, A. A., Ivliev, S. A., & Shagraev, A. G. (2021). Welford's algorithm for weighted statistics. In *2021 3rd International Youth Conference on Radio Electronics, Electrical and Power Engineering (REEPE)* (pp. 1–5). IEEE. DOI: 10.1109/REEPE51337.2021.9387973

Effelsberg, W., & Haerder, T. (1984). Principles of database buffer management. *ACM Transactions on Database Systems, 9*(4), 560–595. DOI: 10.1145/1994.2022

Felner, A., Stern, R., Shimony, S., Boyarski, E., Goldenberg, M., Sharon, G., Sturtevant, N., Wagner, G., & Surynek, P. (2021). Search-Based Optimal Solvers for the Multi-Agent Pathfinding Problem: Summary and Challenges. *Proceedings of the International Symposium on Combinatorial Search*, 8(1), 29–37. DOI: 10.1609/socs.v8i1.18423

Fielding, R., Gettys, J., Mogul, J., Frystyk, H., Masinter, L., Leach, P., & Berners-Lee, T. (1999). Hypertext Transfer Protocol — HTTP/1.1. *RFC, 2616.* Advance online publication. DOI: 10.17487/rfc2616

Fiestas, J. I., & Bustamante, R. E. (2019, August). A survey on the performance of different mutex and barrier algorithms. In *2019 IEEE XXVI International Conference on Electronics, Electrical Engineering and Computing (INTERCON)* (pp. 1-4). IEEE. DOI: 10.1109/INTERCON.2019.8853605

Fisher, G., & Yeh, C. (2017). Comparing Producer-Consumer Implementations in Go, Rust, and C. https://stanford-cs242.github.io/f17/assets/chapters/2017/gsfisher-chrisyeh.pdf

Flitton, M. (2021). *Rust Web Programming: A hands-on guide to developing fast and secure web apps with the Rust programming language* (1st ed.). Packt Publishing., DOI: 10.0000/9781800566095

Gaud, F., Genevès, S., Lachaize, R., Lepers, B., Mottet, F., Muller, G., Quéma, V., & Morel, M. (2010). Efficient workstealing for multicore event-driven systems. DOI: 10.1109/ICDCS.2010.55

GeeksforGeeks. (2015, December 29). Concurrency control in DBMS. https://www.geeksforgeeks.org/dbms/concurrency-control-in-dbms/

GeeksforGeeks. (2020, March 20). Mutual exclusion in synchronization. https://www.geeksforgeeks.org/operating-systems/mutual-exclusion-in-synchronization/

GeeksforGeeks. (2024, April 30). Synchronization examples. https://www.geeksforgeeks.org/operating-systems/synchronization-examples/

Georges, A., Buytaert, D., & Eeckhout, L. (2007). Statistically rigorous Java performance evaluation. In *Proceedings of the 22nd annual ACM SIGPLAN conference on Object-oriented programming systems and applications (OOPSLA '07)* (pp. 57–76). ACM. DOI: 10.1145/1297027.1297033

Goetz, B., Peierls, T., Bloch, J., Bowbeer, J., Holmes, D., & Lea, D. (2006). *Java concurrency in practice.* Addison-Wesley.

Gong, X., Zhang, F., Lu, T., & You, W. (2022). Comparative analysis of three outlier detection methods in univariate data sets. In *3rd International Conference on Electronic Communication and Artificial Intelligence (IWECAI)* (pp. 209-213). https://doi.org/DOI: 10.1109/IWECAI55315.2022.00048

Gubbi, J., Buyya, R., Marusic, S., & Palaniswami, M. (2013). Internet of Things (IoT): A vision, architectural elements, and future directions. *Future Generation Computer Systems, 29*(7), 1645–1660. DOI: 10.1016/j.future.2013.01.010

Gülmez, B.. (2023). A survey of security in Rust: From memory safety to concurrency. *ACM Computing Surveys, 55*(13s), 1–36. DOI: 10.1145/3579853

Hansen, P. B., Dijkstra, E. W., & Hoare, C. R. (2013). The origin of concurrent programming: from semaphores to remote procedure calls. In CERN Document Server (European Organization for Nuclear Research). https://cds.cern.ch/record/2695823

Hellerstein, J. L., Diao, Y., Parekh, S., & Tilbury, D. M. (2004). *Feedback control of computing systems*. John Wiley & Sons. DOI: 10.1002/047166880X

Henning, S., & Hasselbring, W. (2019). Scalable and reliable multi-dimensional aggregation of sensor data streams. In 2019 IEEE International Conference on Big Data (Big Data) (pp. 4605–4614). IEEE. DOI: 10.1109/BigData47090.2019.9006452

Henning, S., & Hasselbring, W. (2020). Scalable and reliable multi-dimensional sensor data aggregation in data streaming architectures. *Data-Enabled Discovery and Applications, 4*(1), Article 5.

Herlihy, M., Shavit, N., Luchangco, V., & Spear, M. (2021). Spin locks and contention. The Art of Multiprocessor Programming, 147-182.

Herlihy, M., & Shavit, N. (2012). *The art of multiprocessor programming* (Rev. ed.). Morgan Kaufmann.

Hoare, C. A. R. (1974). Monitors: An operating system structuring concept. *Communications of the ACM, 17*(10), 549–557. DOI: 10.1145/355620.361161

Hoare, C. A. R. (1978). Communicating sequential processes. *Communications of the ACM, 21*(8), 666–677. DOI: 10.1145/359576.359585

Hodge, V. J., & Austin, J. (2004). A survey of outlier detection methodologies. *Artificial Intelligence Review, 22*(2), 85–126. DOI: 10.1023/B:AIRE.0000045502.10941.a9

Hua, B., Ouyang, W., Jiang, C., Fan, Q., & Pan, Z. (2021). Rupair: Towards Automatic Buffer Overflow Detection and Rectification for Rust. *Annual Computer Security Applications Conference*. https://doi.org/DOI: 10.1145/3485832.3485841

Huang, J., Du, D., Duan, Q., Zhang, Y., Zhao, Y., Luo, H., & Liu, Q. (2014). Modeling and analysis on congestion control for data transmission in sensor clouds. *International Journal of Distributed Sensor Networks, 10*(3), 453983. Advance online publication. DOI: 10.1155/2014/453983

Huber, P. J., & Ronchetti, E. (2009). *Robust statistics* (2nd ed.). J. Wiley & Sons. DOI: 10.1002/9780470434697

Hunt, P., Konar, M., Junqueira, F. P., & Reed, B. (2010). ZooKeeper: Wait-free coordination for internet-scale systems. In *Proceedings of the 2010 USENIX Annual Technical Conference*. USENIX Association.

Hu, X., Zheng, Y., Su, Y., & Guo, R. (2020). IoT Adaptive Dynamic Blockchain Networking Method Based on Discrete Heartbeat Signals. *Sensors (Basel), 20*(22), 6503. Advance online publication. DOI: 10.3390/s20226503 PMID: 33202552

Isloor, S. S., & Marsland, T. A. (n.d.). The deadlock problem: An overview. University of Alberta. https://webdocs.cs.ualberta.ca/~tony/OldPapers/dead.pdf

Jacobs, S., & Fasse, J. (2025). Verifying atomics under memory consistency models. *Formal Methods in System Design, 62*(1), 45–73. DOI: 10.1007/s10703-024-00412-9

Ji, Z., Gong, J., & Feng, J. (2021). A Novel Deep Learning Approach for Anomaly Detection of Time Series Data. *Scientific Programming, 2021*, 6636270. Advance online publication. DOI: 10.1155/2021/6636270

Johnson, P., & Kaihlavirta, V. (2018). *Learning Rust : a comprehensive guide to writing Rust applications* (1st ed.). Packt.

Jones, E. F. (2009, October 13). Scheduling 101 – the basics of best practices. Project Management Institute. https://www.pmi.org/learning/library/schedule-101-basic-best-practices-6701

Jung, R., Jourdan, J. H., Krebbers, R., & Dreyer, D. (2018). RustBelt: Securing the foundations of the Rust programming language. Proceedings of the ACM on Programming Languages, 2(POPL), Article 66. https://doi.org/DOI: 10.1145/3158154

Jung, R., Jourdan, J.-H., Krebbers, R., & Dreyer, D. (2017). RustBelt: Securing the foundations of the Rust programming language. *Proceedings of the ACM on Programming Languages, 2*(POPL), 1–34.

Kanrar, S., Chaki, N., & Chattopadhyay, S. (2018). *Concurrency control in distributed systems using mutual exclusion*. Springer. DOI: 10.1007/978-981-10-5559-1

Karger, D., Lehman, E., Leighton, T., Panigrahy, R., Levine, M., & Lewin, D. (1997). Consistent hashing and random trees: Distributed caching protocols for relieving hot spots on the World Wide Web. In *Proceedings of the Twenty-Ninth Annual ACM Symposium on Theory of Computing (STOC '97)* (pp. 654–663). DOI: 10.1145/258533.258660

Kegel, D. (1999). The C10K problem. http://www.kegel.com/c10k.html

Kevin Andrian Santoso, O., Kwee, C., Chua, W., Nabiilah, G. Z., & Rojali, . (2023). Rust's Memory Safety Model: An Evaluation of Its Effectiveness in Preventing Common Vulnerabilities. *Procedia Computer Science, 227*, 119–127. DOI: 10.1016/j.procs.2023.10.509

Klabnik, S., & Nichols, C. (2023). *The Rust programming language* (2nd ed.). No Starch Press.

Kolodin, D. (2019). Background Tasks and Thread Pools in Microservices. In *Hands-On Microservices with Rust*. Packt Publishing, Limited.

Kwak, J., Uam, J., & Sung, Y. (2015). Integration Method of Proxy and Producer-Consumer Patterns. In J. Juang (Ed.), Proceedings of the 3rd International Conference on Intelligent Technologies and Engineering Systems (ICITES2014 (Vol. 345, pp. 239–244). Springer International Publishing AG. DOI: 10.1007/978-3-319-17314-6_31

Laengle, T., & Rembold, U. (1996). <title>Distributed control architecture for intelligent systems</title> Proceedings of SPIE, the International Society for Optical Engineering. *Proceedings of SPIE - the International Society for Optical Engineering, 2911*, 51–62. DOI: 10.1117/12.262498

Lamport, L. (1977). Concurrent reading and writing. *Communications of the ACM, 20*(11), 806–811. DOI: 10.1145/359863.359878

Lamport, L. (1978). Time, clocks, and the ordering of events in a distributed system. *Communications of the ACM, 21*(7), 558–565. DOI: 10.1145/359545.359563

Lampson, B. W., & Redell, D. D. (1980). Experience with processes and monitors in Mesa. *Communications of the ACM, 23*(2), 105–117. DOI: 10.1145/358818.358824

Lavender, R. G., & Schmidt, D. C. (1996). Active object: An object behavioral pattern for concurrent programming. In *Pattern languages of program design 2* (pp. 483–499).

Lee, P. P., Bu, T., & Chandranmenon, G. (2009, October). A lock-free, cache-efficient shared ring buffer for multi-core architectures. In *Proceedings of the 5th ACM/IEEE Symposium on Architectures for Networking and Communications Systems* (pp. 78-79). DOI: 10.1145/1882486.1882508

Levy, A., Andersen, B., Campbell, B., Giffin, P., Pannuto, P., Dutta, P., Levis, P., & Lohany, R. (2017). Multiprogramming a 64kB computer safely and efficiently. In *Proceedings of the 26th ACM Symposium on Operating Systems Principles (SOSP '17)* (pp. 427–443). DOI: 10.1145/3132747.3132786

Li, J., Maier, D., Tufte, K., Papadimos, V., & Tucker, P. A. (2005). Semantics and evaluation techniques for window aggregates in data streams. In *Proceedings of the 2005 ACM SIGMOD International Conference on Management of Data* (pp. 311–322). ACM. https://doi.org/DOI: 10.1145/1066157.1066193

Liu, S., Yu, B., Liu, Y., Zhang, K., Qiao, Y., Li, T. Y., Tang, J., & Zhu, Y. (2021). Brief Industry Paper: The Matter of Time - A General and Efficient System for Precise Sensor Synchronization in Robotic Computing. Proceedings / IEEE Real-Time and Embedded Technology and Applications Symposium, 2021-, Article 9470470. DOI: 10.1109/RTAS52030.2021.00040

Liu, C. L., & Layland, J. W. (1973). Scheduling algorithms for multiprogramming in a hard-real-time environment. *Journal of the Association for Computing Machinery*, *20*(1), 46–61. DOI: 10.1145/321738.321743

Love, R. (2010). *Linux kernel development* (3rd ed.). Addison-Wesley.

Lu, Y., Chen, Y., Li, Z., & Wu, N. (2022). An efficient method of deadlock detection and recovery for flexible manufacturing systems by resource flow graphs. *IEEE Transactions on Automation Science and Engineering*, *19*(3), 1707–1718. DOI: 10.1109/TASE.2021.3114498

Lyu, S. (2021). *Practical Rust web chapters : building cloud and web-based applications*. Apress., DOI: 10.1007/978-1-4842-6589-5

Maglione, G. L., Berretta, L., Godfrey, S. B., Apostolidis, S., Kapoutsis, A., Kosmatopoulos, E., & Tremori, A. (2021). M&S-based testbed to support V&V of autonomous resources task coordinator. In Turi, M., Mazal, J., Fagiolini, A., & Vašík, P. (Eds.), *Modelling and Simulation for Autonomous Systems (MESAS 2020)* (Vol. 12619, pp. 123–138). Springer., DOI: 10.1007/978-3-030-70740-8_8

Managing the chapter schedule. (2016). Bookboon.

Mao, H., Alizadeh, M., Menache, I., & Kandula, S. (2016). Resource management with deep reinforcement learning. In *Proceedings of the 15th ACM Workshop on Hot Topics in Networks (HotNets '16)* (pp. 50–56). DOI: 10.1145/3005745.3005750

Mao, H., Schwarzkopf, M., Venkatakrishnan, S. B., Meng, Z., & Alizadeh, M. (2019). Learning scheduling algorithms for data processing clusters. In *Proceedings of the ACM Special Interest Group on Data Communication (SIGCOMM '19)* (pp. 270–288). DOI: 10.1145/3341302.3342080

Matplotlib Development Team. (2026). Matplotlib documentation. Matplotlib. https://matplotlib.org/stable/

Matsakis, N. D., & Klock, F. S.II. (2014). The Rust language. *ACM SIGAda Ada Letters, 34*(3), 103–104. DOI: 10.1145/2692956.2663188

McKenney, P. E. (2024). Is parallel programming hard, and, if so, what can you do about it? https://kernel.org/pub/linux/kernel/people/paulmck/perfbook/perfbook.html

McNamara, T. S. (2021). *Rust in action : systems programming concepts and techniques*. Manning.

Michael, M. M., & Scott, M. L. (1996). Simple, fast, and practical non-blocking and blocking concurrent queue algorithms. In *Proceedings of the fifteenth annual ACM symposium on Principles of distributed computing (PODC '96)* (pp. 267–275). ACM. DOI: 10.1145/248052.248106

Microsoft. (2026). PowerShell documentation. Microsoft Learn. https://learn.microsoft.com/powershell/

Microsoft. (2026). Start-Process (Microsoft.PowerShell.Management). Microsoft Learn. https://learn.microsoft.com/powershell/module/microsoft.powershell.management/start-process

Mohammed, A., & Zangana, H. (Eds.). (2026). *Intelligent Construction Monitoring Systems: Real-Time Safety, Environmental Prediction, and Risk Management*. IGI Global Scientific Publishing., DOI: 10.4018/979-8-3373-9245-5

MuhannadAjjan. (n.d.). A 24-byte keyboard circular buffer [Image]. Wikipedia. https://en.wikipedia.org/wiki/Circular_buffer

Muhr, D., & Affenzeller, M. (2022). Outlier/Anomaly Detection of Univariate Time Series: A Dataset Collection and Benchmark. In Kotsis, G., Gamper, J., Khalil, I., Wrembel, R., & Tjoa, A. M. (Eds.), *BIG DATA ANALYTICS AND KNOWLEDGE DISCOVERY, DAWAK 2022* (Vol. 13428, pp. 163–169). Springer International Publishing AG., DOI: 10.1007/978-3-031-12670-3_14

Mutex in std:sync - Rust. (2026). Rust-Lang.org. https://doc.rust-lang.org/std/sync/struct.Mutex.html

Natarajan, N. (2006). Communication and synchronization primitives for distributed programs. *IEEE Transactions on Software Engineering, SE-11*(4), 396–416. DOI: 10.1109/TSE.1985.232229

Nikolaev, R. (2019). A scalable, portable, and memory-efficient lock-free FIFO queue. In S. Gilbert (Ed.), Proceedings of the 33rd International Symposium on Distributed Computing (DISC 2019) (LIPIcs, Vol. 146, Article 28). Schloss Dagstuhl. https://doi.org/DOI: 10.4230/LIPIcs.DISC.2019.28

Nikolaev, R., & Ravindran, B. (2022). wCQ: A fast wait-free multi-producer/multi-consumer concurrent queue. In *Proceedings of the 27th ACM SIGPLAN Symposium on Principles and Practice of Parallel Programming (PPoPP '22)* (pp. 53–67). ACM. DOI: 10.1145/3503221.3508420

Ongaro, D., & Ousterhout, J. (2014). In search of an understandable consensus algorithm. In 2014 USENIX Annual Technical Conference (USENIX ATC '14). USENIX Association.

Ousterhout, K., Wendell, P., Zaharia, M., & Stoica, I. (2013). Sparrow: Distributed, low latency scheduling. In *Proceedings of the 24th ACM Symposium on Operating Systems Principles (SOSP '13)* (pp. 69–84). DOI: 10.1145/2517349.2522716

Pai, V. S., Druschel, P., & Zwaenepoel, W. (1999). Flash: An efficient and portable web server. In Proceedings of the USENIX Annual Technical Conference (USENIX ATC '99) (pp. 199–212). USENIX Association.

Pan, X., & University of California, Berkeley, Electrical Engineering & Computer Sciences. (2017). Parallel machine learning using concurrency control (Master's thesis, University of California, Berkeley).

Pan, Y., Liu, G., & Xiao, L. (2023). Deadlock prevention through resource ordering. *Journal of Systems and Software, 196*, 111569. DOI: 10.1016/j.jss.2023.111569

Pemasinghe, S., & Rajapaksha, S. (2022). Comparison of CPU scheduling algorithms: FCFS, SJF, SRTF, Round Robin, priority-based, and multilevel queuing. DOI: 10.1109/R10-HTC54060.2022.9929533

Pieper, R., Löff, J., Hoffmann, R. B., Griebler, D., & Fernandes, L. G. (2021). High-level and efficient structured stream parallelism for rust on multi-cores. *Journal of Computer Languages, 65*, 101054. DOI: 10.1016/j.cola.2021.101054

Pillai, T. S., Chidambaram, V., Al-Kiswany, S., Arpaci-Dusseau, A. C., & Arpaci-Dusseau, R. H. (2014). All file systems are not created equal: On the complexity of crafting crash-consistent applications. In *Proceedings of the 11th USENIX Symposium on Operating Systems Design and Implementation (OSDI '14)* (pp. 433–448). USENIX Association.

Prakash, S., Lee, Y. H., & Johnson, T. (1994). A nonblocking algorithm for shared queues using compare-and-swap. *IEEE Transactions on Computers*, *43*(5), 548–559. DOI: 10.1109/12.280802

Python Software Foundation. (2026). csv: CSV file reading and writing. In Python 3 documentation. https://docs.python.org/3/library/csv.html

Ramírez-Gallego, S., Krawczyk, B., García, S., Woźniak, M., & Herrera, F. (2017). A survey on data preprocessing for data stream mining. *Neurocomputing*, *239*, 39–57. DOI: 10.1016/j.neucom.2017.01.078

Rand Chapter Developers. (2026). Rand crate documentation (v0.8). Docs.rs. https://docs.rs/rand/0.8

Rao, X., Yang, J., & Xu, L. (2024). Understanding the usage of unsafe in real-world Rust systems. *IEEE Transactions on Software Engineering*, *50*(2), 307–325. DOI: 10.1109/TSE.2023.3267891

Ricart, G., & Agrawala, A. K. (1981). An optimal algorithm for mutual exclusion in computer networks. *Communications of the ACM*, *24*(1), 9–17. DOI: 10.1145/358527.358537

Riekenbrauck, N., Weisgut, M., Lindner, D., & Rabl, T. (2024, May). A three-tier buffer manager integrating CXL device memory for database systems. In 2024 IEEE 40th International Conference on Data Engineering Workshops (ICDEW) (pp. 395-401). IEEE. DOI: 10.1109/ICDEW61823.2024.00063

Rousseeuw, P. J., & Hubert, M. (2011). Robust statistics for outlier detection. *Wiley Interdisciplinary Reviews. Data Mining and Knowledge Discovery*, *1*(1), 73–79. DOI: 10.1002/widm.2

Rust Chapter Developers. (2026). *std:sync:mpsc*. In *The Rust standard library* (Version 1.94.0).

Rust Chapter Developers. (n.d.). Rust by Example. Rust Documentation. https://doc.rust-lang.org/rust-by-example/

Rust Chapter. (2024a). Module std:fs. The Rust Standard Library Documentation. https://doc.rust-lang.org/std/fs/

Rust Chapter. (2024b). Module std:sync. The Rust Standard Library Documentation. https://doc.rust-lang.org/std/sync/

Rust Chapter. (2024c). Module std:sync:mpsc. The Rust Standard Library Documentation. https://doc.rust-lang.org/std/sync/mpsc/

Rust Chapter. (2024d). Module std:thread. The Rust Standard Library Documentation. https://doc.rust-lang.org/std/thread/

Rust Team. (2024). std:sync:mpsc — Rust standard library. Rust Documentation. https://doc.rust-lang.org/std/sync/mpsc/

Rust Team. (2024). std:sync:Mutex — Rust standard library. Rust Documentation. https://doc.rust-lang.org/std/sync/struct.Mutex.html

Sai, A., Reddy, D., & Perumalla Raghavendra, G. Yashwanth Kiran, & R, R. V. (2022). Producer-Consumer problem using Thread pool. 2022 3rd International Conference for Emerging Technology (INCET). https://doi.org/DOI: 10.1109/INCET54531.2022.9824031

Samson, C. F. (2024). Understanding OS-Backed Event Queues, System Calls, and Cross-Platform Abstractions. In *Asynchronous Programming in Rust*. Packt Publishing, Limited.

Schmid, P., Besta, M., & Hoefler, T. (2020). High-performance concurrent data structures. *ACM Computing Surveys*, *53*(3), 1–38. DOI: 10.1145/3377526

Schwarzkopf, M., Konwinski, A., Abd-El-Malek, M., & Wilkes, J. (2013). Omega: Flexible, scalable schedulers for large compute clusters. In *Proceedings of the 8th ACM European Conference on Computer Systems (EuroSys '13)* (pp. 351–364). DOI: 10.1145/2465351.2465386

Serde Developers. (2024). Serde: Serialization framework for Rust. Retrieved from https://serde.rs/

serde_json contributors. (2024). serde_json: A JSON serialization library for Rust. https://docs.rs/serde_json/

Sethi, P., & Sarangi, S. R. (2017). Internet of Things: Architectures, protocols, and applications. *Journal of Electrical and Computer Engineering*, *2017*, 9324035. Advance online publication. DOI: 10.1155/2017/9324035

Shi, Z., Møldrup, L., Mathur, U., & Pavlogiannis, A. (2026). The complexity of testing message-passing concurrency. Proceedings of the ACM on Programming Languages, 10(POPL), Article 103. DOI: 10.1145/3776643

Shwe, H. Y., Gacanin, H., & Adachi, F. (2010). Multi-layer WSN with power efficient buffer management policy. In *2010 IEEE International Conference on Communication Systems* (pp. 36–40). DOI: 10.1109/ICCS.2010.5686107

Silberschatz, A., Galvin, P. B., & Gagne, G. (2018). *Operating system concepts* (10th ed.). Wiley.

Sirejiding, S., Bayramli, B., Lu, Y., Huang, S., Lu, H., & Ding, Y. (2024). Adaptive task-wise message passing for multi-task learning: A spatial interaction perspective. *IEEE Transactions on Circuits and Systems for Video Technology*, *34*(10), 9499–9514. DOI: 10.1109/TCSVT.2024.3399613

Solanki, A., Parekh, A., Chawda, G., & S, M. G. (2021). Lightron : A GUI Integrated, Rust Based Web Server. *International Journal of Scientific Research in Science and Technology*, 554–560. DOI: 10.32628/CSEIT2174127

Song, N. Y., Son, Y., Han, H., & Yeom, H. Y. (2016). Efficient memory-mapped I/O on fast storage device. [TOS]. *ACM Transactions on Storage*, *12*(4), 1–27. DOI: 10.1145/2846100

Steere, D. C., Goel, A., Gruenberg, J., McNamee, D., Pu, C., & Walpole, J. (1999). A feedback-driven proportion allocator for real-rate scheduling. In *Proceedings of the 3rd USENIX Symposium on Operating Systems Design and Implementation (OSDI '99)* (pp. 145–158).

Stevens, W. R., & Rago, S. A. (2013). Advanced programming in the UNIX environment (3rd ed.). Addison-Wesley. Blandy, J., & Orendorff, J. (2021). Programming Rust (2nd ed.). O'Reilly Media.

Stevens, W. R., & Rago, S. A. (2013). *Advanced programming in the UNIX environment* (3rd ed.). Addison-Wesley.

Sun, C., Wang, Y., & Dai, Q. (2025). Mutex scheduling mechanism analysis and performance evaluation under the RT-Thread open source framework. In 2025 IEEE 2nd International Conference on Electronics, Communications and Intelligent Science (ECIS) (pp. 1-10). IEEE. DOI: 10.1109/ECIS65594.2025.11086967

Sutter, H. (2007, December 11). Effective concurrency: Use lock hierarchies to avoid deadlock.

Tanenbaum, A. S., & Bos, H. (2015). Modern operating systems. (4th edition / Andrew S. Tanenbaum, Herbert Bos.). Prentice Hall.

Tanenbaum, A. S., & Bos, H. (2014). *Modern operating systems* (4th ed.). Pearson.

Terzian, G. (2020, May 11). Rust concurrency: the single-writer principle, applied. Medium. https://medium.com/@polyglot_factotum/rust-concurrency-the-single-writer-principle-applied-aada2cdc6fb0

The Rust Chapter Developers. (2026). rand (Version 0.10.0) [Software documentation]. Docs.rs. https://docs.rs/rand/latest/rand/trait.RngExt.html#method.random:~:text=Return%20a%20random,distribution

The Rust Chapter Developers. (2026a). Rust standard library documentation. Rust Chapter. https://doc.rust-lang.org/std/

The Rust Chapter Developers. (2026b). std:collections:HashSet. Rust Chapter. https://doc.rust-lang.org/std/collections/struct.HashSet.html

The Rust Chapter Developers. (2026c). std:collections:VecDeque. Rust Chapter. https://doc.rust-lang.org/std/collections/struct.VecDeque.html

The Rust Chapter Developers. (2026d). std:process:Command. Rust Chapter. https://doc.rust-lang.org/std/process/struct.Command.html

The Rust Chapter Developers. (2026e). std:sync:Arc. Rust Chapter. https://doc.rust-lang.org/std/sync/struct.Arc.html

The Rust Chapter Developers. (2026f). std:sync:atomic:AtomicBool. Rust Chapter. https://doc.rust-lang.org/std/sync/atomic/struct.AtomicBool.html

The Rust Chapter Developers. (2026g). std:sync:Condvar. Rust Chapter. https://doc.rust-lang.org/std/sync/struct.Condvar.html

The Rust Chapter Developers. (2026h). std:sync:mpsc. Rust Chapter. https://doc.rust-lang.org/std/sync/mpsc/

The Rust Chapter Developers. (2026i). std:sync:Mutex. Rust Chapter. https://doc.rust-lang.org/std/sync/struct.Mutex.html

The Rust Chapter Developers. (2026j). std:thread. Rust Chapter. https://doc.rust-lang.org/std/thread/

The Rust Chapter Developers. (2026k). std:time:Instant. Rust Chapter. https://doc.rust-lang.org/std/time/struct.Instant.html

The Rust Project Developers. (2024). std:sync:atomic. Rust Documentation. https://doc.rust-lang.org/std/sync/atomic/

The Rust Project Developers. (2024). std:sync:Condvar. Rust Documentation. https://doc.rust-lang.org/std/sync/struct.Condvar.html

The Rust Project Developers. (2024). std:sync:Mutex. Rust Documentation. https://doc.rust-lang.org/std/sync/struct.Mutex.html

Tsoukalos, M. (2022, June 22). How to develop multi-threaded code: Mihalis Tsoukalos explains all you need to know about Rust and concurrency so you can assemble efficient multi-threaded tools. *Linux Format*, *292*, 90.

Upadhyaya, P., Zhang, K., Li, C., Jiang, X., & Kim, Y. (2023). Scalable Causal Structure Learning: Scoping Review of Traditional and Deep Learning Algorithms and New Opportunities in Biomedicine. In JMIR Medical Informatics (Vol. 11). DOI: 10.2196/38266

Urrea, C. (2025). Hybrid deep learning–reinforcement learning for adaptive human–robot task allocation in Industry 5.0. *Systems*, *13*(8), 631. DOI: 10.3390/systems13080631

Vo, V. A., Le, D. H., & Choo, K.-K. R. (2023). Optimized aggregation tree protocols for IoT sensor networks. *IEEE Internet of Things Journal*, *10*(9), 7921–7935. DOI: 10.1109/JIOT.2022.3201234

Vu, T., & Eldawy, A. (2020). R*-Grove: Balanced Spatial Partitioning for Large-Scale datasets. *Frontiers in Big Data*, *3*, 28. DOI: 10.3389/fdata.2020.00028 PMID: 33693401

Wang, C., Tu, T., Qin, S., Wu, G., Gao, F., & Wan, M. (2024, October). Understanding Atomics and Memory Ordering Issues in Real-World Rust Software. In 2024 IEEE 35th International Symposium on Software Reliability Engineering (ISSRE) (pp. 582-592). IEEE. DOI: 10.1109/ISSRE62328.2024.00061

Welford, B. P. (1962). Note on a method for calculating corrected sums of squares and products. *Technometrics*, *4*(3), 419–420. DOI: 10.1080/00401706.1962.10490022

Welsh, M., Culler, D., & Brewer, E. (2001). SEDA: An architecture for well-conditioned, scalable internet services. In *Proceedings of the 18th ACM Symposium on Operating Systems Principles (SOSP '01)* (pp. 230–243). DOI: 10.1145/502034.502057

Wu, T., & Wang, W. (2020). An improvement mechanism for heartbeat-driven MAC synchronization in wireless body sensor networks. *International Journal of Communication Systems*, *33*(4), e4224. DOI: 10.1002/dac.4224

Xia, Y.. (2022). XPC: Architectural support for secure and efficient cross-process call. In *Proceedings of the 49th Annual International Symposium on Computer Architecture (ISCA '22)*. ACM/IEEE. DOI: 10.1145/3470496.3527400

Xu, G., Shi, Y., Guan, X., & Chen, H. (2014). Energy-efficient data aggregation in wireless sensor networks. *Wireless Communications and Mobile Computing*, *14*(6), 643–654. DOI: 10.1002/wcm.2225

Xu, H.. (2020). Memory-safety challenge considered solved? An empirical study with all Rust CVEs. In *Proceedings of the 35th IEEE/ACM International Conference on Automated Software Engineering (ASE '20)*. IEEE/ACM. DOI: 10.1145/3324884.3416587

Xu, H., Pang, G., Wang, Y., & Wang, Y. (2023). Deep isolation forest for anomaly detection. *IEEE Transactions on Knowledge and Data Engineering*, *35*(12), 12591–12604. DOI: 10.1109/TKDE.2023.3270293

Yu, L., Song, X., & Zhang, Y. (2019). An empirical study of Rust-for-Linux: The success, dissatisfaction, and compromise. In *Proceedings of the 2019 ACM Joint European Software Engineering Conference and Symposium on the Foundations of Software Engineering (ESEC/FSE '19)*. ACM. DOI: 10.1145/3338906.3340448

Zhou, X., Arulraj, J., Pavlo, A., & Cohen, D. (2021, June). Spitfire: A three-tier buffer manager for volatile and non-volatile memory. In *Proceedings of the 2021 International Conference on Management of Data* (pp. 2195-2207). DOI: 10.1145/3448016.3452819

Zhou, Y., Guo, Y., Zhang, W., & Zhao, P. (2026). Synchronization, circuit implementation, and a three-layer image encryption scheme for a novel 5D hyperchaotic system. *Nonlinear Dynamics*, *114*(6), 416. Advance online publication. DOI: 10.1007/s11071-026-12288-z

About the Contributors

Mohammed Aquil Mirza is a senior member of IEEE, a researcher and educator specializing in AI, AIoT, robotics, sensor networks, and medical imaging. He is a Dean and Professor at Liaoning Technical University, China and a Senior Lecturer at The Hong Kong Polytechnic University, Hong Kong. His research has been funded by major grants, and his work has earned several national and international awards such as the IEEE Best Student Paper awards, Young Investigator Award, etc. With a strong academic background and innovative research, he continues to inspire future engineers.

***

Mohammad Al-Khaldy is an Assistant professor of Business Intelligence and Data Analytics at University of Petra, Jordan. He completed his PhD in Atrificial Intelligence and Data Science for the University of Hull -United Kingdom - 2017. Alkhaldy's research interests include data analytics, machine learning, predictive analytics, NLP, and decision support systems. His research has been published extensively in academic journals and conferences. Alkhaldy taught a variety of courses in Artificial intelligence and business intelligence, including Data Mining, Business Analytics, Computer Programming, Intelligent business systems, Data Visualization, and machine learning. Alkhaldy is a member of Jordan Computers Society, and Arab Robotics & AI Association. He also a reviewer for several other academic conferences and journals.

Mingyuan Tang is an undergraduate student majoring in Information Security at The Hong Kong Polytechnic University, Hong Kong.

Hewa Majeed Zangana is an Assistant Professor currently affiliated with Duhok Polytechnic University (DPU) in Iraq. He holds a Doctorate in Philosophy (PhD) degree in ITM, which he is currently pursuing at DPU. He has also held administrative positions such as Curriculum Division Director - Presidency of DPU, Head of the Computer Science Department at Nawroz University, and Acting Dean of the College of Computer and IT at Nawroz University. Hewa Majeed Zangana's research interests span a wide range of topics in computer science, including network systems, information security, mobile communication, data communication, and intelligent systems. He has published extensively in peer-reviewed journals and has published many academic books with IGI Global, several of which are indexed in Scopus and Web of Science (Clarivate). In addition to his research contributions, Hewa Majeed Zangana has been actively involved in editorial roles. He has also been a member of various scientific committees and administrative bodies at Nawroz University.

JingyanZeng is an information security student at Hong Kong Polytechnic University. This recent work establishes a robust benchmarking framework to optimize pipeline efficiency and system stability using Rust.

Index

N

O

P

R

S

T

W

Z

www.ingramcontent.com/pod-product-compliance
Lightning Source LLC
LaVergne TN
LVHW081532300926
849161LV00006B/1391